PROGRESS IN EVOLUTION EQUATIONS

PROGRESS IN EVOLUTION EQUATIONS

GASTON M. N'GUÉRÉKATA
EDITOR

Nova Science Publishers, Inc.
New York

Library of Congress Cataloging-in-Publication Data

Progress in evolution equations / Gaston M. N'Guerekata (editor).
p. cm.
ISBN 978-1-60456-328-3 (hardcover)
1. Evolution equations. I. N'Guerekata, Gaston M., 1953-
QA371.P77 2008
515'.353--dc22 2008013800

Published by Nova Science Publishers, Inc. ✢ New York

CONTENTS

PREFACE

This book presents new research from around the world on the theory and methods of linear and nonlinear evolution equations as well as their further applications. It includes the asymptotic behavior of solutions to evolution equations. Other nonlinear differential equations and applications to natural sciences are also discussed.

Chapter 1 –The authors consider a class of elliptic-parabolic-hyperbolic degenerate equations of the form: $b(u)_t$ - $a(u,\ \varphi\ (u)_x)_x = f$ with homogeneous Dirichlet conditions and initial conditions. Existence of entropy solutions is proved for nondecreasing continuous functions b and φ vanishing at zero; and for a continuous function a vanishing at zero and nondecreasing with respect to the second variable, under Alt and Luckhaus Structure condition (1983)

Chapter 2 – In this note, the authors obtain a new and general comparison principle for a class of coupled systems of fully nonlinear parabolic equations under nonlocal and nonlinear boundary conditions, which extends some existing results.

Chapter 3 – In the bi-level's information model of evolutionary dynamics [1] (with the random processes at microlevel and the dynamic processes at macrolevel), the evolutionary changes arise when the microlevel's randomness affects the macrolevel, creating a new information at macrolevel and renovating the operator of the macrodynamic equations. The dynamics [1] also bring a sequential cooperation of the renovated states into a hierarchical network. This paper studies a *complexity* of the evolutionary changes, caused by both the state's renovation and their cooperation, establishing the information complexity measures (MC). It is shown that a common indicator of the complexity, is the *specific* entropy's speed (related to the increment of the model's volume), rather than the entropy, as it was accepted before. The complexity mechanism applies the minimax variation principle (VP) of information macrodynamics (IMD), which describes a consequent evolutionary transition from a local, unstable process's movement to a local, stable process, associated with the current influx of information and its accumulation. This transition enables the production of the cooperative phenomena and, in particular, the contributions from different superimposing processes, measured by the MC's cooperative complexity. The MC, arising as an indicator of these phenomena at the unification (or decomposition) of the system's processes, is defined by the *invariant information* measure, allowing for both analytical formulation and computer evaluation. An optimal multi-dimensional consolidation process, satisfying the VP, forms the information hierarchical network (IN) consisting of the model eigenvalues' sequential cooperation in triples. The MC of such an optimal cooperative triplets' structure is measured by the IN's triplet *code* (as an algorithm of the minimal program, which evaluates the IN's hierarchical structure by the triplet's information contributions in bits of information). The cooperative IN allows for the automatic arrangement and measurement of the MC-local complexities for a multi-dimensional process, taking into account their time-space locations and the mutual dependencies, providing the MC *hierarchical invariant* information measure by quantity and quality in the triplet's code. MC covers Kolmogorov's complexity, which measures a deterministic order over a stochastic disorder by a minimal program, as well as the statistical complexity. MC provides a precise complexity measure of a *dynamic irreversible process*, evaluating the aforementioned forms of complexities in bits of information. The considered geometrical space curvature conceals information of the cooperative structures in the cells' form, whose code, in particular, measures the cooperative complexity. The process' trajectory, located in this geometrical space, acquires a sequence of the code cells.

Chapter 4 – A state of equilibrium of a plasma flow in a nuclear fusion reactor is achieved as a minimizer of the magnetic flux energy over a class of rearrangements of a prescribed profile magnetic flux function. It is known that a Lipschitz continuous minimizer exists when the cross section of the reactor is convex. The classical Lagrange Multiplier Rule with a functional constraint is modified and is applied to the Kruskal-Kulsrud Principle to derive a non-degenerate Euler-Lagrange equation satisfied by a minimizer of the energy functional that arises in magnetohydrodynamics (MHD). A sufficient condition for the existence of a smooth minimizer is presented in this work. Under this sufficient condition, an explicit Euler-Lagrange equation that is compatible with numerical algorithms is obtained within the framework of the Lagrange Multiplier Rule.

Chapter 5 – The authors consider a BBM-Burgers system with a homogeneous nonlinearity. For this system the authors obtain decay estimates exploiting properties of the semigroup generated by the linear part and estimating the associated integral equation.

Chapter 6 – The paper deals with the investigation of second order weakly nonlinear variational hyperbolic inequality in domain bounded in time variable and unbounded in spatial variables. The authors have obtained the conditions of existence and uniqueness of the solution of the inequality with initial data. These conditions do not include any restrictions as to a behavior at infinity of the solution, the nonhomogeneous term and the initial data. The classes of the existence and the uniqueness are spaces of locally integrable functions.

Chapter 7 – This paper deals with the initial problem $u_{it} = \Delta u_i + f_i(u_1,\dots,u_m) + g_i(u_1,\dots,u_m)|\nabla u_i|^2$ with $u_i|_{\partial\Omega} = 0$ and $u_i(x,0) = \phi_i(x)$, $i = 1,\dots,m$, in a bounded domain $\Omega \subset R^n$. Under suitable assumptions on f_i, g_i, the authors prove that, if $\phi_I \geq (1+\varepsilon_0)\psi_i$ in $D_i \subset \Omega$ for some small $\varepsilon_0 > 0$, then the solutions blow up in finite time, where ψ_i is a positive solution of $\Delta\psi_i + f_i(\psi) + g_i(\psi)|\nabla\psi_i|^2 = 0$ in D_i with $\psi_i|_{\partial D_i} = 0$ for $i = 1,\dots,m$. If $0 \leq \phi_i(x) \leq \lambda\psi_i(x)$ in Ω with $\lambda < 1$, then the solution exists for all $t > 0$ and decay exponentially in t, where ψ_i is a positive solution of $\Delta\psi_i + f_i(\psi) + g_i(\psi)|\nabla\psi_i|^2 = 0$ in Ω with $\psi_i|_{\partial\Omega} = 0$ for $i = 1,\dots,m$.

Chapter 8 – In this paper, the compound Burgers-Korteweg-de Vries (cBKdV) equation with initial-boundary conditions is investigated in a reproducing kernel space. Its exact solution is represented in the form of series. In the mean time, the approximate solution $u_n(t,x)$ is obtained by the n-term intercept of series and is proved to converge to the exact solution. Moreover, the error between the approximate solution $u_n(t,x)$and the exact solution $u(t,x)$is monotone decreasing. Some numerical examples are studied to demonstrate the accuracy of the present method. Results obtained by the method are compared with the exact solution of each example and are found to be in good agreement with each other.

Chapter 9 – The authors establish the critical global existence curve and critical Fujita curve for a degenerate parabolic system with nonlinear boundary conditions in multi-dimension.

Chapter 10 – Let **T** be a time scale. The authors study the existence of positive solutions for the nonlinear four-point singular boundary value problem with p-Laplacian dynamic delay differential equations on time scales, subject to some boundary conditions. By using the fixed-point index theory, the existence of positive solution and many positive solutions for nonlinear four-point singular boundary value problem with p-Laplacian operator are obtained.

Chapter 11 – The authors prove new asymptotical stability and instability theorems for non autonomous 2 x 2 system of first-order differential equations by using a new version of the classical Levinson asymptotic theorem for 2 x 2 systems. The proof of this version is based on the construction of approximate fundamental solution of the original system in the special form with unknown phase function and the error estimates formulated in the terms of generalized characteristic functional. In the case of constant matrix A generalized characteristic functional turns to the usual characteristic polynomial and by choosing phase functions as eigenvalues of the matrix A the error could be eliminated. As an-other application the authors derive a transition probability formula for the two level atom in the external electromagnetic field described by Schrödinger system.

Chapter 12 – Travelling wave solution for Ibragimov-Shabat equation, is obtained by using an improved sine-cosine method and the Wu's elimination method. An infinite number of conserved quantities for the above equation are also obtained by solving a set of coupled Riccati equations.

In: Progress in Evolution Equations
Editor: Gaston M. N'Guerekata, pp. 1-18
ISBN: 978-1-60456-328-3

Chapter 1

Entropy Solutions of Nonlinear Elliptic-Parabolic-Hyperbolic Degenerate Problems in One Dimension

Stanislas Ouaro*
Laboratoire d'Analyse Mathématique des Equations (LAME), UFR.
Sciences Exactes et Appliquées, Université de Ouagadougou,
03 BP 7021 Ouaga 03, Ouagadougou, Burkina Faso

Abstract

We consider a class of elliptic-parabolic-hyperbolic degenerate equations of the form: $b(u)_t - a(u, \varphi(u)_x)_x = f$ with homogeneous Dirichlet conditions and initial conditions. Existence of entropy solutions is proved for nondecreasing continuous functions b and φ vanishing at zero; and for a continuous function a vanishing at zero and nondecreasing with respect to the second variable, under Alt and Luckhaus Structure condition (1983)

2000 Mathematics Subject Classification: 35K65, 35L65.

Keys words: Elliptic, Parabolic, Hyperbolic, Degenerate, Weak solution, Entropy solution, Mild solution, Semi-group. [1]

1. Introduction

Let I be an open bounded interval of $\mathbb{R}$. We consider the initial-boundary-value problem:

$$(EP)\quad \begin{cases} b(u)_t - a(u, \varphi(u)_x)_x = f \text{ in } \mathbb{Q} =]0,T[\times I \\ b(u) = v_0 \qquad \text{on} \quad \{0\}\times I \\ u = 0 \qquad \text{on } \Gamma =]0,T[\times \partial I, \end{cases}$$

*E-mail address: souaro@univ-ouaga.bf
[1]Suported by ICTP under SIDA.

with $T > 0$, where

$$\begin{cases} a : (z,\xi) \in \mathbb{R}\times\mathbb{R}\to\mathbb{R} \text{ is continuous, nondecreasing in } \xi\in\mathbb{R} \text{ with } a(0,0)=0; \\ b : \mathbb{R}\to\mathbb{R} \text{ is continuous, nondecreasing and } b \text{ surjective with } b(0)=0; \\ \varphi : \mathbb{R}\to\mathbb{R} \text{ is continuous, nondecreasing with } \varphi(0)=0. \end{cases}$$

Whenever u is such that $b(u)$ is constant, (EP) degenerates into an elliptic problem of the form:

$$\begin{cases} -a(u,\varphi(u)_x)_x = f & \text{in } \mathbb{Q} =]0,T[\times I \\ b(u) = v_0 & \text{on } \quad \{0\}\times I \\ u = 0 & \text{on } \Gamma =]0,T[\times\partial I. \end{cases} \tag{1}$$

If the function $b = Id$, on each part where u is such that $\varphi(u)$ is constant, (EP) degenerates to a scalar conservation law of the form:

$$\begin{cases} u_t - a(u,0)_x = f & \text{in } \mathbb{Q} =]0,T[\times I \\ u = u_0 & \text{on } \quad \{0\}\times I \\ u = 0 & \text{on } \Gamma =]0,T[\times\partial I. \end{cases} \tag{2}$$

It is then clear that we include in (EP), some first order hyperbolic problems, for which (even under assumptions of regularity on data) there is no hope of getting classical global solutions.

It is well known that, for such equations, the above problems are ill-posed in the sense that there is no uniqueness. It is necessary to introduce Kruzhkov solutions in order to obtain existence and uniqueness results (see [13]).
So in the general case we do not expect to have well-posed problems.
Since b and φ are not strictly increasing, the above formulations include stefan problems, filtration problems, etc, in the one dimensional case. Such formulations involve a large class of problems and an important literature has been developed. The problems considered in this paper has been solved when $b = id$ by Benilan and Touré [6]. The case $a(k,\nabla\varphi(\xi)) = \phi(k) + \nabla\varphi(\xi)$ where $\phi \in C(\mathbb{R},\mathbb{R}^N)$, $\varphi \in C(\mathbb{R})$ and $\varphi(0) = 0$, $\phi_j(0) = 0$, $1 \le j \le N$ has been studied by Carrillo [8] in a bounded domain of $\mathbb{R}^N$. Note also that the case $\varphi = id$ has been studied by several authors in a bounded domain of $\mathbb{R}^N$, $(N \ge 1)$ (see [2, 7, 8, 11, etc.]). See also [5, 9, 10] and the corresponding references for the semigroup approach.

In order to study existence and uniqueness of solutions for the time-dependent problem we study the steady-state problem called "stationary problem in sense of Benilan":

$$(SP) \quad \begin{cases} b(u) - a(u,\varphi(u)_x)_x = f & \text{in} \quad I \\ u = 0 & \text{on } \Gamma = \partial I, \end{cases}$$

which is also denoted by $(SP(f)(b,a,\varphi))$ or $(SP)(b,a,\varphi,f)$. In [15], the author studied (SP). We will quickly recall the results in section 2 where we will develop the concept of existence and uniqueness of mild solution for the time-dependent problem (EP). In section 3, we deal with the existence of entropy solutions of (EP) under Alt and Luckhaus structure condition.

2. Mild Solution

Under the following mild conditions on a, b and φ, we prove existence of entropy solutions of (EP):

$$(H_1) \qquad \lim_{|\xi|\to\infty} \inf_{|k|<R} |\, a(k,\xi)\,| = +\infty \quad \forall R > 0.$$

$$(H_2) \begin{cases} (a(r,\xi)-a(s,\eta)).(\xi-\eta)+M(r,s)(1+|\xi|^2+|\eta|^2)\,|\varphi(r)-\varphi(s)| \geq \\ \Gamma(\varphi(r),\varphi(s)).\xi+\widehat{\Gamma}(\varphi(r),\varphi(s)).\eta \end{cases}$$

for all r, s, ξ, $\eta \in \mathbb{R}$, where $M : \mathbb{R}\times\mathbb{R} \to \mathbb{R}^+$, $\Gamma, \widehat{\Gamma} : \mathbb{R}\times\mathbb{R}\to\mathbb{R}$ are continuous.

$$(H_3) \begin{cases} \text{There exists } \widehat{a} : \mathbb{R}\times\mathbb{R}\to\mathbb{R} \text{ continuous, nondecreasing with respect} \\ \text{to the second variable and such that } \widehat{a}(b(u),\varphi(u)_x) = a(u,\varphi(u)_x). \end{cases}$$

Note that assumption (H_3) is called Alt and Luckhaus structure condition.

$$(H_4) \begin{cases} (a(z,\xi)-a(z,0)).(\xi) \geq \lambda\,|\xi|^2 \\ |a(r,\xi)|^2 \leq C\left(1+|\xi|^2\right), \end{cases}$$

for r, z, $\xi \in \mathbb{R}$; where $\lambda > 0$ and $C > 0$.
Let γ be a maximal monotone operator defined on $\mathbb{R}$. We denote by γ_0 the main section of γ, i.e.

$$\gamma_0(s) = \begin{cases} \text{the element of minimal absolute value of } \gamma(s) \text{ if } \gamma(s) \neq \emptyset, \\ +\infty \text{ if } [s,+\infty)\cap D(\gamma) = \emptyset, \\ -\infty \text{ if } (-\infty,s]\cap D(\gamma) = \emptyset. \end{cases}$$

Throughout this paper, we use the operator H^+ (also denoted by $sign^+$ in the literature) defined by

$$H^+(s) = \begin{cases} 1 & \text{if } \; s > 0 \\ [0,1] & \text{if } \; s = 0 \\ 0 & \text{if } \; s < 0. \end{cases}$$

Then

$$H_\varepsilon(s) = \min(\frac{s^+}{\varepsilon}, 1),$$

$$H_0(s) = \begin{cases} 1 & \text{if } s > 0 \\ 0 & \text{otherwise.} \end{cases}$$

Furthermore we use

$$H_{Max}(s) = \begin{cases} 1 & \text{if } s \geq 0 \\ 0 & \text{otherwise.} \end{cases}$$

Furthermore, we define

$$H(k) = a(k,0) \text{ for } k \in \mathbb{R} \text{ and } h = a(u, \varphi(u)_x). \tag{3}$$

Let us, for completeness, recall the stationary problem results given in [15].

Definition 1 *Let $f \in L^\infty(I)$; a weak solution of (SP) is a measurable function u such that $b(u) \in L^1(I)$, $\varphi(u) \in W^{1,\infty}(I)$ and $\varphi(u) = 0$ on ∂I, $h \in L^2(I)$ and*

$$b(u) - a(u, \varphi(u)_x)_x = f \text{ in } \mathcal{D}'(I),$$

or equivalent to

$$\int_I \{b(u)\xi + h\xi_x\}\, dx = \int_I f\xi dx \quad \text{for any } \xi \in H_0^1(I) \cap L^\infty(I).$$

Remark 2 *We easily check that if u is a weak solution of $(SP)(b, a, \varphi, f)$ then $-u$ is a weak solution of $(SP)(\widetilde{b}, \widetilde{a}, \widetilde{\varphi}, -f)$ where $\widetilde{b}(s) = -b(-s)$, $\widetilde{\varphi}(s) = -\varphi(-s)$, and $\widetilde{a}(s,k) = -a(-s,-k)$.*

Definition 3 *Let $f \in L^\infty(I)$; an entropy solution of (SP) is a weak solution u satisfying:*
(i) there exists $h \in C(I)$ such that $h = a(u, \varphi(u)_x)$ a.e on I,
(ii) the following entropy inequalities

$$(a) \int_I H_0(u-k)\{(H(k)-h)\xi_x + \xi(f - b(u))\}\, dx \geq 0$$

for any $(k,\xi) \in \mathbb{R} \times (H_0^1(I) \cap L^\infty(I))$ such that $\xi \geq 0$, andfor any $(k,\xi) \in \mathbb{R} \times (H^1(I) \cap L^\infty(I))$ such that $\xi \geq 0$ and such that $k \geq 0$;

$$(b) \int_I H_0(k-u)\{(H(k)-h)\xi_x + \xi(f - b(u))\}\, dx \leq 0$$

for any $(k,\xi) \in \mathbb{R} \times (H_0^1(I) \cap L^\infty(I))$ such that $\xi \geq 0$, andfor any $(k,\xi) \in \mathbb{R} \times (H^1(I) \cap L^\infty(I))$ such that $\xi \geq 0$ and such that $k \leq 0$.

Remark 4 *It is easy to see that if u is an entropy solution of $(SP)(b, a, \varphi, f)$ then $(-u)$ is an entropy solution of $(SP)(\widetilde{b}, \widetilde{a}, \widetilde{\varphi}, \widetilde{f})$ where $\widetilde{b}(r) = -b(-r)$, $\widetilde{a}(r,k) = -a(-r,-k)$, $\widetilde{\varphi}(r) = -\varphi(-r)$ and $\widetilde{f} = -f$.*

In [15], we proved under general assumptions on the data, without Alt and Luckhaus structure condition (see [1]) that (SP) admit a unique entropy solution which permit us to define the L^1 operator associated with the evolution problem (EP) by $A_b b(u) = -a(u, \varphi(u)_x)_x$ satisfiying:

$$\begin{cases} v \in A_b b(u) \text{ if and only if } b(u) \in L^1(I), v \in L^\infty(I) \text{ and } u \text{ is entropy} \\ \\ \text{solution of the stationary problem } (SP) \text{ with } f = v + b(u). \end{cases}$$

We proved for this operator, the following result that we present as a lemma.

Lemma 5 *Suppose that (H_1) and (H_2) are satisfied. Then the operator A_b defined above satisfies the following:*

1. *A_b is $T-$accretive in $L^1(I)$ i.e $\|(x-\widetilde{x})^+\|_{L^1} \leq \|(x-\widetilde{x}+\lambda(A_b x - A_b \widetilde{x}))^+\|_{L^1}$ $\forall \lambda \geq 0$ and $x, \widetilde{x} \in D(A_b)$.*
2. *For any $\lambda > 0$, the range $R(I+\lambda A_b)$ of $I+\lambda A_b$ is dense in $L^1(I)$.*
3. *The domain $D(A_b)$ of A_b is dense in $L^1(I)$.*

As usual, in the theory of evolution equations governed by accretive operators, we consider an approximation of (EP) by an implicit time discretization (also used by Alt and Luckhaus [1]),

$$(PDE) \quad \begin{cases} \dfrac{b(u_I) - b(u_{i-1})}{t_I - t_{i-1}} = a(u_I, \varphi(u_I)_x)_x + f_I \\ \\ u_I \in L^\infty(I), b(u_I) \in L^1(I), \varphi(u_I) \in W^{1,\infty}(I) \\ \\ \varphi(u_I) = 0 \text{ on } \partial I \text{ for } i = 1, ..., n, \end{cases}$$

where

$$(DE) \quad \begin{cases} t_0 = 0 < t_1 < ... < t_n \leq T, t_I - t_{i-1} \leq \varepsilon, T - t_n \leq \varepsilon \\ \\ f_1, ..., f_n \in L^\infty(I), \sum_I \displaystyle\int_{t_{i-1}}^{t_I} \|f(t) - f_I\|_1 \, dt \leq \varepsilon \\ \\ u_0 \in L^\infty(I), \|v_0 - b(u_0)\|_1 \leq \varepsilon, \text{ for } \varepsilon > 0. \end{cases}$$

This method is actually the method of nonlinear semi-group theory. Naturally, we are led to give the following definition according to [7] (see also [3, 8]).

Definition 6 *A mild solution of (EP) is a measurable function $u : \mathbb{Q} \to \mathbb{R}$ satisfying $v = b(u) \in C([0,T]; L^1(I))$, $v(0) = v_0$ and, $\forall \varepsilon > 0$, there exists $(t_0, ..., t_n, f_1, ..., f_n, u_0, ..., u_n)$ satisfying (DE) and (PDE), such that $\|v(t) - b(u_I)\|_1 \leq \varepsilon$ $\forall t \in]t_{i-1}, t_I]$, $i = 1, ..., n$.*

Using nonlinear semigroup theory (cf. [3, 5]) by interpreting problem (EP) in the form of evolution equation in $L^1(I)$

$$(CP)\quad \begin{cases} \dfrac{dv}{dt}+A_b v \ni f \; in \; [0,T] \\ \\ v(0)=v_0 \end{cases}$$

with $v=b(u)$;
one deduces immediately from the preceding lemma, the following theorem:

Theorem 7 *Suppose that* (H_1) *and* (H_2) *are satisfied. Then, for any* $f \in L^1(\mathbb{Q})$, $v_0 \in \overline{D(A_b)} = L^1(I)$, *there exists a unique mild solution u of*

$$(EP(f,v_0)(A_b))\begin{cases} \dfrac{dv}{dt}+A_b v \ni f \; in \; [0,T] \\ \\ v(0)=v_0, \end{cases}$$

and $v=b(u)\in C([0,T];L^1(I))$.
Moreover, let u_I *be the mild solution of* $(EP(f_I,v_{0_I})(A_b))$ *for* $i=1,2$.
Then

$$\int_Q \alpha\{(v_1-v_2)\xi'+(f_1-f_2)\xi\}\, dxdt \geq 0$$

for any nonnegative $\xi \in D(0,T)$, *and for some* $\alpha \in H^+(v_1-v_2)$ *almost everywhere in* Q. *In particular,* $v_1 \leq v_2$ *for* $v_{0_1} \leq v_{0_2}$ *and* $f_1 \leq f_2$, *where* $v_I = b(u_I)$.

Proof. See, for instance, [5] Propositions 1.28 and 2.2.

Remark 8 *The concept of uniqueness considered here is the uniqueness of* $b\,(u)$; *on the other hand, if* b *is one to one, the uniqueness of* $b\,(u)$ *is equivalent to that of* u.

If v_0 and f verify

$$v_0 \in L^\infty(I),\ f \in L^1(\mathbb{Q}) \text{ and } \int_0^T \|f(t,.)\|_{L^\infty(I)} < \infty, \tag{4}$$

then, the mild solution is bounded; more precisely, one has the following estimation:

Proposition 9 *Let us suppose (4) is satisfied and u being the mild solution of* (EP), *then* $u \in L^\infty(Q)$ *and*

$$\|b(u)\|_{L^\infty(\mathbb{Q})} \leq \|v_0\|_{L^\infty(I)} + \int_0^T \|f(t,.)\|_{L^\infty(I)}\, dt. \tag{5}$$

Proof. For the proof of this proposition, we interprete the problem (EP) in the form of the evolution equation (CP) in $L^1(I)$ with $v=b(u)$; and, in this case, the result follow immediately by using proposition 1-4 of [6].

3. Entropy Solutions

For any continuous and nondecreasing or nonincreasing function ψ, we define the proper lower semi-continuous and convex or upper semi-continuous and concave function

$$B_\psi(s) = \begin{cases} \int_0^s \psi(\varphi \circ (b^{-1})_0(r))dr & \text{for } s \in \overline{(\psi \circ \varphi) \circ b^{-1}} \\ +\infty & \text{otherwise,} \end{cases}$$

and we have $(\psi \circ \varphi) \circ b^{-1} \subset \partial B_\psi$.
We now define according to [1, 7], a weak solution of (EP) by

Definition 10 *Let $f \in L^2(0,T;H^{-1}(I))$ and $v_0 \in L^1(I)$. A weak solution of problem (EP) is a measurable function u which satisfies the following:*

$$b(u) \in L^1(\mathbb{Q}),\ b(u)_t \in L^2(0,T;H^{-1}(I)), \tag{6}$$

$$\varphi(u) \in L^2(0,T;H_0^1(I)),\ h = a(u,\varphi(u)_x) \in L^2(\mathbb{Q}), \tag{7}$$

$$b(u)_t - h_x = f \text{ in } D'(\mathbb{Q}) \tag{8}$$

$$b(u(0,x)) = v_0(x) \text{ a.e on } I. \tag{9}$$

The condition (9) should be understood in the following sense:

$$\int_0^T \langle b(u)_t, \xi\rangle\, dt = -\int_{\mathbb{Q}} b(u)\xi_t dxdt - \int_I v_0 \xi(0)dx$$

for any $\xi \in L^2(0,T;H_0^1(I)) \cap W^{1,1}(0,T;L^\infty(I))$, such that $\xi(T) = 0$, where $\langle , \rangle$ represent the duality pairing between $H^{-1}(I)$ and $H_0^1(I)$.
In [1] (see also [7]), the existence of a weak solution is related to an energy estimate on the initial data. In our case, such an energy estimate is

$$B_{id}(v_0) \in L^1(I).$$

We now have to define entropy solution:

Definition 11 *Let $f \in L^2(0,T;H^{-1}(I)) \cap L^1(\mathbb{Q})$ and $v_0 \in L^1(I)$. An entropy solution of problem (EP) is a weak solution u which satisfies the following:*

$$\begin{cases} \int_{\mathbb{Q}} H_0(u-k)\{\xi_x(h - H(k)) - (b(u) - b(k))\xi_t - f\xi\}\, dxdt - \\ -\int_I (v_0 - b(k))^+ \xi(0)dx \leq 0 \end{cases} \tag{10}$$

for any $(k,\xi) \in \mathbb{R} \times (L^2(0,T;H^1(I)) \cap W^{1,1}(0,T;L^\infty(I)))$ *such that* $k \geq 0$, $\xi \geq 0$, *and* $\xi(T) = 0$ *and for any* $(k,\xi) \in \mathbb{R} \times (L^2(0,T;H_0^1(I)) \cap W^{1,1}(0,T;L^\infty(I)))$ *such that* $\xi \geq 0$, *and* $\xi(T) = 0$;

$$\begin{cases} \int_{\mathbb{Q}} H_0(k-u)\{\xi_x(h-H(k)) - (b(u)-b(k))\xi_t - f\xi\}\,dxdt + \\ + \int_{\mathbb{R}} (v_0 - b(k))^- \xi(0)dx \geq 0 \end{cases} \tag{11}$$

for any $(k,\xi) \in \mathbb{R} \times (L^2(0,T;H^1(I)) \cap W^{1,1}(0,T;L^\infty(I)))$ *such that* $k \leq 0$, $\xi \geq 0$, *and* $\xi(T) = 0$ *and for any* $(k,\xi) \in \mathbb{R} \times (L^2(0,T;H_0^1(I)) \cap W^{1,1}(0,T;L^\infty(I)))$ *such that* $\xi \geq 0$, *and* $\xi(T) = 0$.

Remark 12 *It is easy to see that if* u *is an entropy solution of* $(EP)(b,a,\varphi,f)$ *then* $(-u)$ *is an entropy solution of* $(EP)(\widetilde{b},\widetilde{a},\widetilde{\varphi},\widetilde{f})$ *where* $\widetilde{b}(r) = -b(-r)$, $\widetilde{a}(r,k) = -a(-r,-k)$, $\widetilde{\varphi}(r) = -\varphi(-r)$ *and* $\widetilde{f} = -f$.

In order to prove existence of weak (entropy) solutions, we need an energy estimate similar to the one given in Lemma 1.5 of [1].

Lemma 13 *Let* $\psi \in C^{0,1}(\mathbb{R})$, *let* ψ *be monotone, let* $v_0 \in L^1(I)$ *such that* $B_\psi(v_0) \in L^1(I)$, *let* u *be a measurable function such that* $b(u)$ *satisfies (6) and (9) and such that* $\varphi(u)$ *satisfies (7). Then*

$$B_\psi(b(u)) \in L^\infty(0,T;L^1(I))$$

and, for almost every $t \in [0,T]$,

$$\begin{cases} \int_I B_\psi(b(u(t)))\xi(t)dx - \int_I B_\psi(v_0)\xi(0)dx = \int_0^t \langle b(u)_t, \psi(\varphi(u))\xi\rangle\,dt + \\ \int_0^t \int_I B_\psi(b(u))\xi_t dxdt \end{cases} \tag{12}$$

for any $\xi \in C^{0,1}(\overline{\mathbb{Q}})$ *such that* $\psi(\varphi(u))\xi \in L^2(0,T;H_0^1(I))$.

Proof. For the proof of lemma 13, see the proof of lemma 4 of [8].

To complete this section we prove the following theorem:

Theorem 14 *Let* (H_1), (H_2), (H_3) *and* (H_4) *hold. Let* $v_0 \in L^\infty(I)$, *let* $f \in L^\infty(\mathbb{Q})$ *and let* u *be the unique mild solution of* (EP). *Then* u *is an entropy solution of* (EP).

Proof. For any $M \in \mathbb{N}^*$, we define $\tau = \frac{T}{M}$. For $i = 0,1,...,M$ we define $t_I = i \times \tau$. Let f_1, f_2, ..., $f_M \in L^\infty(I)$ be such that

$$\|f_I\|_{L^\infty(I)} \leq \|f\|_{L^\infty(Q)},$$

$$\sum_{i=1}^{M} \int_{t_{i-1}}^{t_I} \|f(t) - f_I\|_{L^1(I)}\,dt \overset{M\to+\infty}{\longrightarrow} 0.$$

For $i = 1,...,M$, let v_I be the unique solution of

$$\tau f_I + v_{i-1} \in (I + \tau A_b) v_I,$$

i.e, there exists a measurable function u_I such that $v_I = b(u_I)$ and u_I is an entropy solution of $(SP(\tau f_I + v_{i-1})(b, \tau a, \varphi))$ for $1 \leq i \leq M$.

Define

$$f^\tau(t) = f_I \text{ for } t_{i-1} < t \leq t_I \text{ and } 1 \leq i \leq M.$$

Similarly we define $v^\tau = b(u^\tau)$ where

$$u^\tau = u_I \text{ for } t_{i-1} < t \leq t_I \text{ and } 1 \leq i \leq M.$$

Since u is a mild solution, we know (see [5]) that

$$\|v(t) - v^\tau\|_{L^\infty(0,T;L^1(I))} \overset{M \to +\infty}{\longrightarrow} 0,$$

we deduce the existence of a subsequence of τ, still denoted by τ such that $b(u^\tau) \overset{\tau \to 0}{\longrightarrow} v$ a.e in $\mathbb{Q}$; and, since $v^\tau = b(u^\tau)$ is uniformly bounded in $L^\infty(\mathbb{Q})$ (see Proposition 9), from the Lebesgue Theorem we deduce that

$$\|v(t) - v^\tau\|_{L^p(\mathbb{Q})} \overset{M \to +\infty}{\longrightarrow} 0 \text{ for } 1 \leq p < +\infty.$$

Now, since u_I is an entropy solution (and therefore a weak) solution of problem $(SP(\tau f_I + b(u_{i-1}))(b, \tau a, \varphi))$ for $1 \leq i \leq M$, we have

$$\int_I \{(b(u_I) - b(u_{i-1}) - \tau f_I)\xi + \tau h_I \xi_x\}\, dx = 0$$

for any $\xi \in H_0^1(I)$, where $h_I = a(u_I, \varphi(u_I)_x)$. In particular, for $\xi = \varphi(u_I)$ we have

$$\int_I \{(b(u_I) - b(u_{i-1}))\varphi(u_I) + \tau a(u_I, \varphi(u_I)_x)\varphi(u_I)_x\}\, dx = \tau \int_I f_I \varphi(u_I) dx$$

which gives

$$\begin{cases} \int_I (b(u_I) - b(u_{i-1}))\varphi(u_I) dx + \tau \int_I [a(u_I, \varphi(u_I)_x) - a(u_I, 0)]\, \varphi(u_I)_x dx + \\ \tau \int_I a(u_I, 0)\varphi(u_I)_x dx = \tau \int_I f_I \varphi(u_I) dx. \end{cases}$$

Using (H_4), we obtain

$$\begin{cases} \int_I (b(u_I) - b(u_{i-1}))\varphi(u_I) dx + \lambda\tau \int_I |\varphi(u_I)_x|^2\, dx + \\ \tau \int_I a(u_I, 0)\varphi(u_I)_x dx \leq \tau \int_I f_I \varphi(u_I) dx. \end{cases}$$

Since B_{id} is convex we deduce that

$$\begin{cases} \int_I [B_{id}(b(u_I)) - B_{id}(b(u_{i-1}))]dx + \lambda\tau \int_I |\varphi(u_I)_x|^2 dx + \\ \tau \int_I a(u_I,0)\varphi(u_I)_x dx \leq \tau \int_I f_I \varphi(u_I)dx, \end{cases}$$

where $B_{id}(b(u_I)) \in L^\infty(I)$. Then we deduce that

$$\begin{cases} \int_I [B_{id}(b(u_M)) - B_{id}(v_0)]dx + \lambda\tau \sum_{i=1}^{M} \int_I |\varphi(u_I)_x|^2 dx + \\ \tau \sum_{i=1}^{M} \int_I a(u_I,0)\varphi(u_I)_x dx \leq \tau \sum_{i=1}^{M} \int_I f_I \varphi(u_I)dx. \end{cases}$$

Hence

$$\|\varphi(u^\tau)\|_{L^2(0,T;H_0^1(I))} \leq C$$

since

$$\int_I a(u,0)\varphi(u)_x dx = \int_I (G(\varphi(u)))_x dx = 0$$

where

$$G(s) = \int_0^s a((\varphi^{-1})_0(r),0)dr;$$

and therefore there exists a subsequence of τ, still denoted by τ, such that

$$\varphi(u^\tau) \overset{\tau\to 0}{\rightharpoonup} w \text{ weakly in } L^2(0,T;H_0^1(I)).$$

Since $b(u^\tau)$ converges in $L^2(\mathbb{Q})$ and $\varphi(u^\tau)$ converges weakly in $L^2(\mathbb{Q})$ and since $\varphi \circ b^{-1}$ is a maximal monotone operator (in $L^2(\mathbb{Q})$), we deduce that

$$w \in \varphi \circ b^{-1}(v),$$

whence there exists $\widetilde{u} \in b^{-1}(v)$ such that $w = \varphi(\widetilde{u})$. Then we get

$$u = \left((b+\varphi)^{-1}\right)_0 (v+w) = \left((b+\varphi)^{-1}\right)_0 (b(\widetilde{u}) + \varphi(\widetilde{u})).$$

Obviously, u is a measurable function and we have

$$v = b(u) \text{ and } w = \varphi(u).$$

Now let $\xi \in D(\overline{\mathbb{Q}})$ be such that $\xi = 0$ on $([0,T] \times \partial I) \cap (\{T\} \times I)$. Then

$$\begin{cases} \int_{\mathbb{Q}} f^\tau(t)\xi(t)dxdt = \int_{\mathbb{Q}} \left[\frac{b(u^\tau(t)) - b(u^\tau(t-\tau))}{\tau}\xi(t) + a(u^\tau(t),\varphi(u^\tau(t))_x)\xi_x(t)\right] dxdt \\ = \int_0^{T-\tau} \int_I b(u^\tau(t)) \frac{\xi(t) - \xi(t+\tau)}{\tau} dxdt + \int_{\mathbb{Q}} a(u^\tau(t),\varphi(u^\tau(t))_x)\xi_x(t)dxdt \\ + \frac{1}{\tau} \int_{T-\tau}^{T} \int_I b(u^\tau(t))\xi(t)dxdt - \frac{1}{\tau} \int_0^\tau \int_I v_0 \xi(t)dxdt, \end{cases} \tag{13}$$

where $b(u^\tau(t)) = v_0$ for $t \le 0$.
For the passage to the limit as $\tau \to 0$ in the equality above, we use a classical pseudomonotonicity argument. To this end, let $\widetilde{u}^\tau$ be the piecewise linear function defined by

$$\widetilde{u}^\tau(t) = b(u_{i-1}) + \frac{t - t_{i-1}}{t_I - t_{i-1}}(b(u_I) - b(u_{i-1})) \text{ for } t \in [t_{i-1}, t_I],\ i = 1, ..., M.$$

Since $\widetilde{u}^\tau \to b(u)$ in $C([0,T];L^1(I))$ as $\tau \to 0$, then

$$g^\tau = -a(u^\tau, \varphi(u^\tau)_x)_x = f^\tau - \widetilde{u}^\tau_t \to f - b(u)_t = g \text{ weakly in } L^2(0,T;H^{-1}(I)),$$

according to (H_4), since $\|\varphi(u^\tau)\|_{L^2(0,T;H_0^1(I))} \le C$ and u_i is an entropy solution of $(SP(\tau f_I + b(u_{i-1}))(b, \tau a, \varphi))$ for $i = 1, \ldots, M$.
Next, note that

$$\begin{cases} \langle g^\tau, \varphi(u^\tau)\rangle = \langle f^\tau, \varphi(u^\tau)\rangle - \int_0^T \int_I \varphi(u^\tau)\widetilde{u}^\tau_t dxdt \le \langle f^\tau, \varphi(u^\tau)\rangle - \\ \int_I [B_{id}(b(u^\tau_M)) - B_{id}(b(u^\tau)(0))]\,dx = \langle f^\tau, \varphi(u^\tau)\rangle - \int_I [B_{id}(b(u^\tau_M)) - B_{id}(v_0)]\,dx. \end{cases}$$

Due to the lower semi-continuity of B_I (since id is continuous and nondecreasing), by using Fatou's Lemma, it follows according to Lemma 13 that

$$\begin{cases} \limsup_\tau \langle g^\tau, \varphi(u^\tau)\rangle \le \langle f, \varphi(u)\rangle - \int_I [B_{id}(b(u)(T)) - B_{id}(v_0)]\,dx \le \langle f, \varphi(u)\rangle \\ -\int_0^T \langle b(u)_t, \varphi(u)\rangle. \end{cases}$$

Therefore,

$$\limsup_\tau \langle g^\tau, \varphi(u^\tau)\rangle \le \langle g, \varphi(u)\rangle.$$

Moreover, due to the monotonicity of a, for any $\xi \in L^2(0,T;H_0^1(I))$,

$$\int_Q [a(u^\tau, \varphi(u^\tau)_x) - a(u^\tau, \xi_x)].(\varphi(u^\tau) - \xi)_x dxdt \ge 0,$$

which is equivalent to

$$\begin{cases} \int_Q a(u^\tau, \varphi(u^\tau)_x).(\varphi(u^\tau) - \xi)_x dxdt \ge \int_Q a(u^\tau, \xi_x).(\varphi(u^\tau) - \xi)_x dxdt = \\ \int_Q \widehat{a}(b(u^\tau), \xi_x).(\varphi(u^\tau) - \xi)_x dxdt. \end{cases}$$

Note also that $\widehat{a}$ is continuous from $\mathbb{R} \times \mathbb{R} \to \mathbb{R}$, hence $\widehat{a}(b(u^\tau), \xi_x)$ converges to $\widehat{a}(b(u), \xi_x)$ a.e. According to (H_4), we may apply Lebesgue's dominated convergence theorem to obtain

$$\begin{cases} \liminf_{\tau} \langle g^\tau, \varphi(u^\tau) - \xi \rangle \geq \int_Q \widehat{a}(b(u), \xi_x).(\varphi(u) - \xi)_x dxdt \\ = \int_Q a(u, \xi_x).(\varphi(u) - \xi)_x dxdt. \end{cases}$$

Combining the preceding estimates yields

$$\langle g, \varphi(u) - \xi \rangle \geq \int_Q a(u, \xi_x).(\varphi(u) - \xi)_x dxdt,$$

for all $\xi \in L^2(0,T;H_0^1(I))$. Choosing $\xi = \varphi(u) + \sigma\phi$, $\sigma \in \mathbb{R}$, $\phi \in L^2(0,T;H_0^1(I))$, we obtain

$$\langle g, \sigma\phi \rangle \leq \sigma \int_Q a(u, \sigma\phi_x + \varphi(u)_x).\phi_x dxdt.$$

Dividing the preceding inequality by $\sigma > 0$, resp. $\sigma < 0$ and passing to the limit with $\sigma \downarrow 0$, resp. $\sigma \uparrow 0$, we obtain

$$\langle g, \phi \rangle = \int_Q a(u, \varphi(u)_x).\phi_x dxdt,$$

for any $\phi \in L^2(0,T;H_0^1(I))$.
Consequently, as $\tau \to 0$ in (13), we get

$$\int_Q f\xi dxdt = \int_Q \{-b(u)\xi_t + a(u, \varphi(u)_x)\xi_x\} dxdt - \int_I v_0\xi(0)dx.$$

Hence u is a weak solution of (EP).
Now, since u_I is an entropy solution of $(SP(\tau f_I + b(u_{i-1}))(b, \tau a, \varphi))$ for $1 \leq i \leq M$, we have

$$\int_I H_0(u_I - k)\{\tau(H(k) - h_I)\xi_x + \xi(\tau f_I + b(u_{i-1}) - b(u_I))\} dx \geq 0$$

for any $(k,\xi) \in \mathbb{R} \times (H_0^1(I) \cap L^\infty(I))$ such that $\xi \geq 0$, and for any $(k,\xi) \in \mathbb{R} \times (H^1(I) \cap L^\infty(I))$ such that $\xi \geq 0$ and such that $k \geq 0$.
Therefore, we have

$$\begin{cases} \int_Q H_0(u^\tau(t) - k)\left[\frac{b(u^\tau(t)) - b(u^\tau(t-\tau))}{\tau} - f^\tau(t)\right]\xi(t)dxdt + \\ \int_Q H_0(u^\tau(t) - k)[h^\tau(t) - H(k)]\xi_x dxdt \leq 0 \end{cases} \quad (14)$$

for any $(k,\xi) \in \mathbb{R} \times D(\overline{\mathbb{Q}})$ such that $\xi \geq 0$, $\xi(T) = 0$ and $\xi = 0$ on $[0,T] \times \partial I$; and for any $(k,\xi) \in \mathbb{R} \times D(\overline{\mathbb{Q}})$ such that $\xi \geq 0$, $\xi(T) = 0$ and $k \geq 0$, where $h^\tau(t) = a(u^\tau(t), \varphi(u^\tau(t))_x)$.
Now let us observe that for such functions ξ, we have

$$\begin{cases} \int_{\mathbb{Q}} H_0(u^\tau(t)-k)\frac{b(u^\tau(t))-b(u^\tau(t-\tau))}{\tau}\xi(t)dxdt = \\ \int_{\mathbb{Q}} \frac{H_0(u^\tau(t)-k)(b(u^\tau(t))-b(k))}{\tau}\xi(t)dxdt - \\ \int_{\mathbb{Q}} \frac{H_0(u^\tau(t)-k)(b(u^\tau(t-\tau))-b(k))}{\tau}\xi(t)dxdt \geq \\ \int_{\mathbb{Q}} \frac{H_0(u^\tau(t)-k)(b(u^\tau(t))-b(k))}{\tau}\xi(t)dxdt - \\ \int_{\mathbb{Q}} \frac{H_0(u^\tau(t-\tau)-k)(b(u^\tau(t-\tau))-b(k))}{\tau}\xi(t)dxdt, \end{cases} \tag{15}$$

since $H_0(u^\tau(t)-k)(b(u^\tau(t-\tau))-b(k)) \leq H_0(u^\tau(t-\tau)-k)(b(u^\tau(t-\tau))-b(k))$.
Also,

$$\begin{cases} \int_{\mathbb{Q}} \frac{H_0(u^\tau(t)-k)(b(u^\tau(t))-b(k))}{\tau}\xi(t)dxdt - \\ \int_{\mathbb{Q}} \frac{H_0(u^\tau(t-\tau)-k)(b(u^\tau(t-\tau))-b(k))}{\tau}\xi(t)dxdt = \\ \int_{\mathbb{Q}} H_0(u^\tau(t)-k)(b(u^\tau(t))-b(k))\frac{\xi(t)-\xi(t+\tau)}{\tau}dxdt + \\ \int_{\mathbb{Q}} \frac{H_0(u^\tau(t)-k)(b(u^\tau(t))-b(k))}{\tau}\xi(t+\tau)dxdt - \\ \int_{\mathbb{Q}} \frac{H_0(u^\tau(t-\tau)-k)(b(u^\tau(t-\tau))-b(k))}{\tau}\xi(t)dxdt = \\ \int_{\mathbb{Q}} H_0(u^\tau(t)-k)(b(u^\tau(t))-b(k))\frac{\xi(t)-\xi(t+\tau)}{\tau}dxdt + \\ \frac{1}{\tau}\int_0^{T-\tau}\int_I H_0(u^\tau(t)-k)(b(u^\tau(t))-b(k))\xi(t+\tau)dxdt - \\ \frac{1}{\tau}\int_0^{\tau}\int_I H_0(u^\tau(t-\tau)-k)(b(u^\tau(t-\tau))-b(k))\xi(t)dxdt - \\ \frac{1}{\tau}\int_\tau^{T}\int_I H_0(u^\tau(t-\tau)-k)(b(u^\tau(t-\tau))-b(k))\xi i(t)dxdt = \\ \int_{\mathbb{Q}} H_0(u^\tau(t)-k)(b(u^\tau(t))-b(k))\frac{\xi(t)-\xi(t+\tau)}{\tau}dxdt - \frac{1}{\tau}\int_0^{\tau}\int_I (v_0-b(k))^+\xi(t)dxdt. \end{cases}$$

Then we deduce from (14) and (15) that

$$\begin{cases} \int_{\mathbb{Q}} H_0(u^\tau(t)-k)\left[(b(u^\tau(t))-b(k))\frac{\xi(t)-\xi(t+\tau)}{\tau}-f^\tau(t)\xi(t)\right]dxdt+ \\ \int_{\mathbb{Q}} H_0(u^\tau(t)-k)\left[h^\tau(t)-H(k)\right]\xi_x dxdt-\frac{1}{\tau}\int_0^\tau\int_I (v_0-b(k))^+\xi(t)dxdt\leq 0, \end{cases} \tag{16}$$

for any $(k,\xi)\in\mathbb{R}\times D(\overline{\mathbb{Q}})$ such that $\xi\geq 0$, $\xi(T)=0$ and $\xi=0$ on $[0,T]\times\partial I$; and for any $(k,\xi)\in\mathbb{R}\times D(\overline{\mathbb{Q}})$ such that $\xi\geq 0$, $\xi(T)=0$ and $k\geq 0$.
Since H_0 is bounded, there exists a subsequence of τ still denoted by τ such that

$$H_0(u^\tau-k)\overset{\tau\to 0}{\rightharpoonup}\chi_{u,k} \text{ in } L^\infty(\mathbb{Q}) \text{ weak} * .$$

It is easy to see that

$$H_0(u^\tau-k)\in H^+(b(u^\tau)+\varphi(u^\tau)-b(k)-\varphi(k)).$$

Then, since $b(u^\tau)+\varphi(u^\tau)$ converges to $b(u)+\varphi(u)$ in $L^2(Q)$, we deduce that

$$\chi_{u,k}\in H^+(b(u)+\varphi(u)-b(k)-\varphi(k)).$$

By using (H_3), (H_4) and taking limit as $\tau\to 0$ in (16), we get by the same arguments as in the proof of existence of weak solutions:

$$\begin{cases} \int_{\mathbb{Q}} \chi_{u,k}\{(b(k)-b(u))\xi_t-f\xi+[h-H(k)]\xi_x\}dxdt- \\ \int_I (v_0-b(k))^+\xi(0)dx\leq 0, \end{cases} \tag{17}$$

for any $(k,\xi)\in\mathbb{R}\times D(\overline{\mathbb{Q}})$ such that $\xi\geq 0$, $\xi(T)=0$ and $\xi=0$ on $[0,T]\times\partial I$; and for any $(k,\xi)\in\mathbb{R}\times D(\overline{\mathbb{Q}})$ such that $\xi\geq 0$, $\xi(T)=0$ and $k\geq 0$.
Define

$$I=\int_{\mathbb{Q}}\chi_{u,k}\{(b(k)-b(u))\xi_t+[a(u,\varphi(u)_x)-a(k,0)]\xi_x\}dxdt$$

and

$$J=\int_{\mathbb{Q}}H_0(u-k)\{(b(k)-b(u))\xi_t+[a(u,\varphi(u)_x)-a(k,0)]\xi_x\}dxdt.$$

Since $\chi_{u,k}\in H^+(b(u)+\varphi(u)-b(k)-\varphi(k))$, we prove easily by using (H_3) that $I=J$ and then from (17) follows

$$\begin{cases} \int_{\mathbb{Q}}\{(-\chi_{u,k})f\xi+H_0(u-k)\{(b(k)-b(u))\xi_t+[a(u,\varphi(u)_x)-a(k,0)]\xi_x\}\}dxdt- \\ \int_I (v_0-b(k))^+\xi(0)dx\leq 0, \end{cases} \tag{18}$$

for any $(k,\xi) \in \mathbb{R} \times D(\overline{\mathbb{Q}})$ such that $\xi \geq 0$, $\xi(T) = 0$ and $\xi = 0$ on $[0,T] \times \partial I$; and for any $(k,\xi) \in \mathbb{R} \times D(\overline{\mathbb{Q}})$ such that $\xi \geq 0$, $\xi(T) = 0$ and $k \geq 0$.
Now, since $b+\varphi$ is continuous, let $[k^{(m)}, k^{(M)}] = (b+\varphi)^{-1}(b(k)+\varphi(k))$ and let $k_n > k^{(M)}$, $k_n \searrow k^{(M)}$. Then $\chi_{u,k_n} \nearrow H_0(u-k^{(M)})$ almost everywhere in $\mathbb{Q}$. From inequality (17) and from the Lebesgue Theorem we deduce that

$$\begin{cases} \int_{\mathbb{Q}} H_0(u-k^{(M)}) \left\{ (b(k^{(M)}) - b(u))\xi_t - f\xi + \left[h - H(k^{(M)}) \right] \xi_x \right\} dxdt - \\ \int_I (v_0 - b(k^{(M)}))^+ \xi(0) dx \leq 0, \end{cases} \tag{19}$$

for any nonnegative $\xi \in D(\overline{\mathbb{Q}})$ such that $\xi(T) = 0$ and such that $\xi = 0$ on $[0,T] \times \partial I$.
Since $b(k) = b(k^{(m)}) = b(k^{(M)})$ and $\varphi(k) = \varphi(k^{(m)}) = \varphi(k^{(M)})$ for any $k \in (b+\varphi)^{-1}(b(k)+\varphi(k))$, from assumption (H_3) we have $H(k) = H(k^{(m)}) = H(k^{(M)})$. Hence we deduce from (19) that

$$\begin{cases} \int_{\mathbb{Q}} \left\{ -H_0(u-k^{(M)}) f\xi + H_0(u-k) \left\{ (b(k) - b(u))\xi_t + [a(u,\varphi(u)_x) - a(k,0)]\xi_x \right\} \right\} dxdt \\ - \int_I (v_0 - b(k))^+ \xi(0) dx \leq 0, \end{cases} \tag{20}$$

for any $k \in (b+\varphi)^{-1}(b(k)+\varphi(k))$ and for any nonnegative $\xi \in D(\overline{\mathbb{Q}})$ such that $\xi(T) = 0$ and such that $\xi = 0$ on $[0,T] \times \partial I$.
Now, let $k_n < k^{(m)}$, $k_n \nearrow k^{(m)}$. Then $\chi_{u,k_n} \searrow H_{Max}(u-k^{(m)})$ almost everywhere in $\mathbb{Q}$. Replacing k by k_n in (18) and letting $n \to +\infty$, we deduce that

$$\begin{cases} \int_{\mathbb{Q}} H_{Max}(u-k^{(m)}) \left\{ (b(k^{(m)}) - b(u))\xi_t - f\xi + \left[h - H(k^{(m)}) \right] \xi_x \right\} dxdt - \\ \int_I (v_0 - b(k^{(m)}))^+ \xi(0) dx \leq 0, \end{cases}$$

whence we get according again to (H_3)

$$\begin{cases} \int_{\mathbb{Q}} \left\{ -H_{Max}(u-k^{(m)}) f\xi + H_0(u-k) \left\{ (b(k) - b(u))\xi_t + [a(u,\varphi(u)_x) - a(k,0)]\xi_x \right\} \right\} dxdt \\ - \int_I (v_0 - b(k))^+ \xi(0) dx \leq 0, \end{cases} \tag{21}$$

for any $k \in (b+\varphi)^{-1}(b(k)+\varphi(k))$ and for any nonnegative $\xi \in D(\overline{\mathbb{Q}})$ such that $\xi(T) = 0$ and such that $\xi = 0$ on $[0,T] \times \partial I$.
Now, let η be smooth enough, $0 \leq \eta \leq 1$ and let $\xi \in D(\overline{\mathbb{Q}})$ such that $\xi \geq 0$, $\xi(T) = 0$ and $\xi = 0$ on $[0,T] \times \partial I$. We substitute $\xi\eta$ in (21) and we substitute $\xi(1-\eta)$ in (20), whence we get

$$\begin{cases} \displaystyle\int_{\mathbb{Q}} \left\{ (H_0(u-k^{(M)}) - H_{Max}(u-k^{(m)})) f\xi\eta - H_0(u-k^{(M)}) f\xi \right\} dxdt + \\ \displaystyle\int_{\mathbb{Q}} H_0(u-k) \left[(b(k)-b(u))\xi_t + (h-H(k))\,\xi_x \right] dxdt - \\ \displaystyle\int_I (v_0 - b(k))^+ \xi(0) dx \leq 0, \end{cases}$$

for any $k \in (b+\varphi)^{-1}(b(k)+\varphi(k))$, for any nonnegative $\xi \in D(\overline{\mathbb{Q}})$ such that $\xi(T) = 0$ and $\xi = 0$ on $[0,T] \times \partial I$, and for any $0 \leq \eta \leq 1$, with η smooth enough. In particular, by density, it is enough to have $\eta \in L^\infty(\mathbb{Q})$; then we choose

$$\eta = \frac{H_0(u-k^{(M)}) - H_0(u-k)}{H_0(u-k^{(M)}) - H_{Max}(u-k^{(m)})}$$

and the preceding inequality becomes

$$\begin{cases} \displaystyle\int_{\mathbb{Q}} H_0(u-k) \left\{ -f\xi + (b(k)-b(u))\xi_t + (h-H(k))\,\xi_x \right\} dxdt - \\ \displaystyle\int_I (v_0 - b(k))^+ \xi(0) dx \leq 0, \end{cases} \tag{22}$$

for any $k \in (b+\varphi)^{-1}(b(k)+\varphi(k))$, and hence for any $k \in \mathbb{R}$, and for any nonnegative $\xi \in D(\overline{\mathbb{Q}})$ such that $\xi(T) = 0$ and $\xi = 0$ on $[0,T] \times \partial I$.
Finally, let $\xi \in D(\overline{\mathbb{Q}})$ with $\xi \geq 0$, $\xi(T) = 0$, let $k \geq 0$ and let $\xi_n \in D(I)$, with $0 \leq \xi_n \leq 1$, $\xi_n(x) \to 1$ for any $x \in I$. We substitute $\xi\xi_n$ in (22) and $\xi(1-\xi_n)$ in (18) to get

$$\begin{cases} \displaystyle\int_{\mathbb{Q}} -(H_0(u-k)\xi_n + \chi_{u,k}(1-\xi_n)) f\xi dxdt + \\ \displaystyle\int_{\mathbb{Q}} H_0(u-k) \left\{ (b(k)-b(u))\xi_t + (h-H(k))\,\xi_x \right\} dxdt - \\ \displaystyle\int_I (v_0 - b(k))^+ \xi(0) dx \leq 0. \end{cases}$$

By letting $n \to +\infty$, we get

$$\begin{cases} \displaystyle\int_{\mathbb{Q}} H_0(u-k) \left\{ -f\xi + (b(k)-b(u))\xi_t + (h-H(k))\,\xi_x \right\} dxdt - \\ \displaystyle\int_I (v_0 - b(k))^+ \xi(0) dx \leq 0, \end{cases}$$

for any nonnegative k in $\mathbb{R}$ and for any nonnegative $\xi \in D(\overline{\mathbb{Q}})$ such that $\xi(T) = 0$.
Arguing as above we prove

$$\begin{cases} \int_{\mathbb{Q}} H_0(k-u)\{-f\xi+(b(k)-b(u))\xi_t+(h-H(k))\xi_x\}\,dxdt+ \\ \\ \int_I (b(k)-v_0)^+\xi(0)dx \geq 0, \end{cases}$$

for any $(k,\xi) \in \mathbb{R} \times D(\overline{\mathbb{Q}})$ such that $\xi \geq 0$, $\xi(T) = 0$ and $\xi = 0$ on $[0,T] \times \partial I$ and for any $(k,\xi) \in \mathbb{R} \times D(\overline{\mathbb{Q}})$ such that $\xi \geq 0$, $\xi(T) = 0$ and $k \leq 0$.

Remark 15 *As one can see in the proof of existence theorem of entropy solutions, assumption (H_3) seems to be essential. We do not know actually if the mild soltution of (EP) can be an entropy solution of (EP) without assumption (H_3). But, when $\varphi = id$ or $\varphi^{-1} \in C(\mathbb{R})$, Ammar and Wittbold proved existence of renormalized solution in a bounded domain of $\mathbb{R}^N$ without assumption (H_3) (see [2]) for a problem like (EP). Also, Carrillo has proved existence and uniqueness of entropy solutions in a bounded domain of $\mathbb{R}^N$ of a problem like (EP) where the vector field a is such that $a(k,\xi) = \phi(k) + \xi$ without assumption (H_3) (see [8]), but the mathematics arguments for their results need the continous dependence of mild solution with respect to the data which is an open problem for (EP) even in one dimension due to the strongly nonlinearity of the problem. In our case, assumption (H_3) is not needed in order to get existence and uniqueness of entropy solution for the stationary problem (see [15]) and for uniqueness of entropy solution of (EP) ([16]). A other interesting open problem is the high dimension case. The reasonning for the one dimension case is not necessarily true for several space dimension due to the fact that assumption (H_1) which is the main assumption of the stationary problem (see [15]) is only true in one dimension.*

Acknowledgment

The author wants express his deepest thanks to the editor and anonymous referee's for comments and suggestions on the paper.

References

[1] H.W. Alt & S. Luckhaus; Quasi-linear elliptic-parabolic differential equations, *Math. Z.* **183** (1983) 311-341.

[2] K. Ammar & P. Wittbold; Existence of renormalized solutions of degenerate elliptic-parabolic problems. *Proc. R. Soc. Edinb., Sect. A, Math* **133**, No. 3, 477-496 (2003).

[3] Ph. Bénilan, M.G. Crandall & A. Pazy; *Evolution equation governed by accretive operators* (book to appear).

[4] Ph. Bénilan & SN. Kruskhov; Quasi linear first order equation with continuous Nonlinearities; *Russian Acad. Sci. Dokl. Math.* Vol 50 , No.3 1995, 391-396.

[5] Ph. Bénilan; Equations d'évolution dans un espace de Banach quelconque et applications, Thèse d'état, Orsay (1972).

[6] Ph. Bénilan & H. Touré; Sur l'équation générale $u_t = a(.,u,\varphi(u,.)_x)_x + v$ dans L^1. II-Etude du problème d'évolution. *Ann. Inst. Henri Poincaré*, Vol 12, Analyse non linéaire. No.6, 1995, 727-761.

[7] Ph. Bénilan & P. Wittbold; On mild and weak solutions of elliptic-parabolic problems. *Adv. Diff. Equ.* (1996), 1053-1073.

[8] J. Carrillo; Entropy solutions for nonlinear degenerate problems. *Arch. Ration. Mech. Anal.* **147**, No.4, 269-361 (1999).

[9] M.G. Crandall; The semigroup approach to first order quasilinear equations in several spaces variables, *Israel. J. Math* **12**, 108-122 (1972).

[10] M.G. Crandall & T. Liggett; Generation of semi-groups of nonlinear transformations in general Banach spaces, *Amer. J. Math* **93**, 265-298 (1971).

[11] J. Carrillo & P. Wittbold; Uniqueness of renormalized solutions of degenerate elliptic-parabolic problems. *J. Diff. Eqns* **156**, 93-121 (1999).

[12] J.I. Diaz & R. Kershner; On a nonlinear degenerate parabolic equation on filtration or evaporation through a porous medium, *J. Diff. Eqns* **69** (1987),368-403.

[13] S.N. Kruzhkov; First-order quasilinear equations with several space variables, *Math. USSR-Sb.* **10** (1970), 217-242.

[14] J.L. Lions; *Quelques méthodes de résolutions de problèmes aux limites non linéaires* . Dunod. Paris. 1969.

[15] S. Ouaro; Entropy solutions of a stationary problem associated to a nonlinear parabolic strongly degenerate problem in one space dimension. *An. Univ. Craiova. Math. Inform.* **33** (2006), 108-131.

[16] S. Ouaro; *Uniqueness of entropy solutions of nonlinear elliptic-parabolic-hyperbolic problems in one dimension. In preparation.*

In: Progress in Evolution Equations
Editor: Gaston M. N'Guerekata, pp. 19-26
ISBN: 978-1-60456-328-3

Chapter 2

COMPARISON PRINCIPLE FOR A CLASS OF COUPLED SYSTEMS OF FULLY NONLINEAR PARABOLIC EQUATIONS UNDER NONLOCAL BOUNDARY CONDITIONS*

***Huan-Yu Wang*† and Fang Li‡**
Department of Mathematics, University of Science and Technology of China, Hefei 230026, People's Republic of China

Abstract

In this note, we obtain a new and general comparison principle for a class of coupled systems of fully nonlinear parabolic equations under nonlocal and nonlinear boundary conditions, which extends some existing results.

Key words: Parabolic equations, Coupled system; Nonlocal and nonlinear boundary conditions; Comparison principle.

1. Introduction

The qualitative properties of the solutions to various linear or nonlinear parabolic equations or parabolic systems, including various coupled systems of parabolic equations, have received much attention over the past two decades (cf. e.g., [1-12]). Some of these properties of parabolic equations can hardly be obatined without corresponding comparison principles.

Based on these works, especially [4, 9, 10], we investigate further the comparison principle for a class of coupled systems of fully nonlinear parabolic equations subject to nonlocal boundary conditions, and give a new and more general comparison principle for the coupled systems of parabolic equations.

*The work was supported partly by the National Natural Science Foundation of China 10571165, the NCET-04-0572 and Research Fund for the Key Program of the Chinese Academy of Sciences.

†E-mail address: huanyuw@mail.ustc.edu.cn

‡E-mail address: fangli@ustc.edu.cn

Let Ω be a bounded domain with the boundary $\partial\Omega$ in R^n, $n \geq 1$. Denote

$$D_T = \Omega \times (0,T),\ S_T = \partial\Omega \times (0,T),$$
$$\mathbf{u} = \mathbf{u}(x,t) = (u_1(x,t),\cdots,u_N(x,t)),$$

where T is a constant.

We consider the following coupled systems of fully nonlinear parabolic equations with nonlocal boundary conditions

$$\begin{aligned}
\frac{\partial u_i}{\partial t} &= f_i(x,t,\mathbf{u},\nabla u_i,\nabla^2 u_i), \quad x \in \Omega,\ t > 0,\\
B_i u_i &= \int_\Omega K_i(x,y,t,\mathbf{u}(y,t))\,dy, \quad x \in \partial\Omega,\ t > 0, \qquad (1.1)\\
u_i(x,t) &= u_{i_0}(x,t), \quad x \in \Omega,\ i = 1,2,\cdots,N,
\end{aligned}$$

where

$$B_i = \alpha_i(x,t)\frac{\partial}{\partial \nu} + 1\ , \quad \alpha_i \geq 0,\ i = 1,2,\cdots,N,$$

$\frac{\partial}{\partial \nu}$, ∇u_i and $\nabla^2 u_i$ is the outward normal derivative on $\partial\Omega$, the gradient of u_i and the Hessian matrix of u_i in space variables respectively.

2. Results and Proofs

As in [7], we set

$$\mathbf{f}(x,t,\mathbf{u}) = (f_1(x,t,\mathbf{u},\nabla u_i,\nabla^2 u_i),\cdots,f_N(x,t,\mathbf{u},\nabla u_N,\nabla^2 u_N)),$$

$$S_{n\times n} = \begin{pmatrix} S_{11} & \cdots & S_{1n} \\ \vdots & \ddots & \vdots \\ S_{n1} & \cdots & S_{nn} \end{pmatrix},\ Z_{n\times n} = \begin{pmatrix} Z_{11} & \cdots & Z_{1n} \\ \vdots & \ddots & \vdots \\ Z_{n1} & \cdots & Z_{nn} \end{pmatrix}.$$

Definition 2.1. ([7]) A vector function $\mathbf{f}(x,t,\mathbf{u})$ is said to be quasi-monotone nondecreasing in some subset $\mathcal{J}$ of R^N if for each $i = 1,2,\cdots,N$, on writing $\mathbf{u}$ in the split form $\mathbf{u} = (u_i,[\mathbf{u}]_{N-1}), f_i(x,t,u_i,[\mathbf{u}]_{N-1},\nabla u_i,\nabla^2 u_i)$ is obtained as nondecreasing with respect to the components of $[\mathbf{u}]_{N-1}$ for all $\mathbf{u} \in \mathcal{J}$.

(H1) $\mathbf{f}(x,t,\mathbf{u})$ is quasi-monotone nondecreasing in a given subset of R^N, and for every $(x_0,t_0) \in D_T$, $\mathbf{u},P,S_{jk},Z_{jk}$ $(j,k = 1,2,\cdots,n)$, the quadratic

$$\sum_{j,k=1}^{n}(S_{jk}-Z_{jk})\lambda_j\lambda_k \geq 0,\ \forall\lambda = (\lambda_1,\cdots,\lambda_n) \in R^n,$$

implies

$$f_i(x_0,t_0,\mathbf{u},P,S) \geq f_i(x_0,t_0,\mathbf{u},P,Z), \qquad i = 1,2,\cdots,N.$$

(H2) There exists a constant $C > 0$ such that $\mathbf{u} \geq \mathbf{v}$ implies

$$f_i(x,t,\mathbf{u},P,S) - f_i(x,t,\mathbf{v},P,S) \leq C\sum_{j=1}^{N}(u_j - v_j),\ i = 1,2,\cdots,N.$$

(H3) $K_i(x,y,t,\mathbf{u})$ is continuously differentiable in $\mathbf{u}$ and suitably smooth in x,y,t.

$$K_{i(j)} := \frac{\partial K_i}{\partial u_j},\ i,j = 1,2,\cdots,N,$$

satisfying

$$K_{i(j)}(x,y,t,\mathbf{u}(y,t)) \geq 0,$$
$$\int_\Omega \sum_{j=1}^{N} K_{i(j)}(x,y,t,\mathbf{u}(y,t))\,dy < 1,\ x \in \partial\Omega,\ y \in \Omega,\ t > 0.$$

(H4) $$0 \leq K_{i(j)}(x,y,t,\mathbf{u}(y,t)) \leq M, \quad x \in \partial\Omega,\ y \in \Omega,\ t > 0,$$

where $M > 0$ is a constant.

Let $\mathbf{C}^{1,2}(D_T)$ denote the set of functions being once continuously differentiable in t and twice continuously differentiable in x for $(x,t) \in D_T$ and $\mathbf{C}^{0,1}(\overline{D}_T)$ denote the set of functions being continuous in t and once continuously differentiable in x for $(x,t) \in \overline{D}_T$.

Theorem 2.2. *Let the hypotheses* (H1), (H2) *and* (H3) *hold, and let* $\mathbf{u} = (u_1,\cdots,u_N)$, $\mathbf{v} = (v_1,\cdots,v_N) \in (\mathbf{C}^{1,2}(D_T) \cap \mathbf{C}(\overline{D}_T))^N$ *satisfy*

$$\begin{aligned}
\frac{\partial u_i}{\partial t} &\geq f_i(x,t,\mathbf{u},\nabla u_i,\nabla^2 u_i),\ (x,t) \in D_T,\\
\frac{\partial v_i}{\partial t} &\leq f_i(x,t,\mathbf{v},\nabla v_i,\nabla^2 v_i),\ (x,t) \in D_T,\\
B_i u_i &\geq \int_\Omega K_i(x,y,t,\mathbf{u}(y,t))\,dy,\ (x,t) \in S_T,\\
B_i v_i &\leq \int_\Omega K_i(x,y,t,\mathbf{v}(y,t))\,dy,\ (x,t) \in S_T,\\
u_i(x,0) &\geq v_i(x,0), x \in \overline{\Omega}.\ i = 1,2,\cdots,N,
\end{aligned} \tag{2.1}$$

Then $\mathbf{u}(x,t) \geq \mathbf{v}(x,t)$ *in* $\overline{D}_T$.

Proof. We first suppose that

$$\begin{aligned}
\frac{\partial u_i}{\partial t} &\geq f_i(x,t,\mathbf{u},\nabla u_i,\nabla^2 u_i), \quad (x,t) \in D_T,\\
\frac{\partial v_i}{\partial t} &< f_i(x,t,\mathbf{v},\nabla v_i,\nabla^2 v_i), \quad (x,t) \in D_T,\\
B_i u_i &\geq \int_\Omega K_i(x,y,t,\mathbf{u}(y,t))\,dy, \quad (x,t) \in S_T,\\
B_i v_i &< \int_\Omega K_i(x,y,t,\mathbf{v}(y,t))\,dy, \quad (x,t) \in S_T,\\
u_i(x,0) &> v_i(x,0), \quad x \in \overline{\Omega},\ i = 1,2,\cdots,N,
\end{aligned} \tag{2.2}$$

We claim that

$$\mathbf{u}(x,t) > \mathbf{v}(x,t) \quad \text{in } \overline{D}_T. \tag{2.3}$$

The proof of claim (2.3) is given as follows.

Set

$$\mathbf{w} = (w_1(x,t), \cdots, w_N(x,t)) = (u_1(x,t) - v_1(x,t), \cdots, u_N(x,t) - v_N(x,t)),$$

and

$$A = \{t;\, t \le T,\ \mathbf{w}(x,s) > 0 \text{ for } x \in \overline{\Omega} \text{ and } 0 \le s \le t\}.$$

Then, by $\mathbf{w}(x,0) = \mathbf{u}(x,0) - \mathbf{v}(x,0) > 0$ $(x \in \overline{\Omega})$, we see that there exists $\bar{t} = \sup A$ and $0 < \bar{t} \le T$. Therefore,

$$\mathbf{w}(x,0) \ge 0,\ (x,t) \in \overline{D}_{\bar{t}}.$$

We assume, contrary to (2.3), that $\bar{t} < T$, so there exists a point $\bar{x} \in \overline{\Omega}$ and $w_i(x,t)$ such that $w_i(x,t) > 0,\ (x,t) \in \overline{\Omega} \times [0,\bar{t})$ and $w_i(\bar{x},\bar{t}) = 0$, i.e.,

$$w_i(\overline{x,t}) = \min_{\overline{D}_{\bar{t}}} w_i(x,t).$$

If $(\overline{x,t}) \in S_T$, it follows from

$$w_i(\overline{x,t}) = \min_{\overline{D}_{\bar{t}}} w_i(x,t),\ w_i(x,t) > 0 = w_i(\overline{x,t}),\ (x,t) \in D_{\bar{t}},$$

that $\dfrac{\partial w_i(\overline{x,t})}{\partial \nu} \le 0$. In view of the boundary condition in (2.2), it is easy to verify that

$$\begin{aligned} 0 &\ge \alpha_i \frac{\partial w_i(\bar{x},\bar{t})}{\partial \nu} \\ &= \alpha_i \frac{\partial w_i(\bar{x},\bar{t})}{\partial \nu} + w(\bar{x},\bar{t}) \\ &> \int_\Omega \sum_{j=1}^{N} K_{i(j)}(x,y,t,\xi(y,t))\,dy \\ &\ge 0, \text{ on } S_{\bar{t}}, \end{aligned}$$

where $\xi(y,t)$ is an intermediate value between $\mathbf{u}$ and $\mathbf{v}$. This is a contradiction.

If $(\bar{x},\bar{t}) \in D_T$, it follows from

$$w_i(\overline{x,t}) = \min_{\overline{D}_{\bar{t}}} w_i(x,t),\ w_i(x,t) > 0 = w_i(\overline{x,t}),\ (x,t) \in D_{\bar{t}},$$

that $\dfrac{\partial w_i}{\partial t}(\bar{x},\bar{t}) \le 0$, $\dfrac{\partial w_i}{\partial x_j}(\bar{x},\bar{t}) = 0$, $j = 1,2,\cdots,N$, and

$$\sum_{j,k=1}^{n} \frac{\partial^2 w_i}{\partial x_j \partial x_k}(\bar{x},\bar{t})\lambda_j\lambda_k \ge 0,\ \forall \lambda = (\lambda_1,\cdots,\lambda_n) \in R^n.$$

By (H1) and (2.2), we obtain

$$\begin{aligned}
0 &\geq \frac{\partial w_i}{\partial t}(\bar{x},\bar{t}) \\
&> f_i(\bar{x},\bar{t},u_i(\bar{x},\bar{t}),[\mathbf{u}]_{N-1}(\bar{x},\bar{t}),\nabla u_i(\bar{x},\bar{t}),\nabla^2 u_i(\bar{x},\bar{t})) \\
&\quad -f_i(\bar{x},\bar{t},v_i(\bar{x},\bar{t}),[\mathbf{v}]_{N-1}(\bar{x},\bar{t}),\nabla v_i(\bar{x},\bar{t}),\nabla^2 v_i(\bar{x},\bar{t})) \\
&\geq f_i(\bar{x},\bar{t},v_i(\bar{x},\bar{t}),[\mathbf{v}]_{N-1}(\bar{x},\bar{t}),\nabla v_i(\bar{x},\bar{t}),\nabla^2 u_i(\bar{x},\bar{t})) \\
&\quad -f_i(\bar{x},\bar{t},v_i(\bar{x},\bar{t}),[\mathbf{v}]_{N-1}(\bar{x},\bar{t}),\nabla v_i(\bar{x},\bar{t}),\nabla^2 v_i(\bar{x},\bar{t})) \\
&\geq 0,
\end{aligned}$$

yielding a contradiction. Thus, $\mathbf{w}(x,t)>0,\ (x,t)\in D_T$. This proves $\mathbf{w}(x,t)>0,\ (x,t)\in \overline{D}_T$.

For the general case where (2.1) holds, we let

$$\mathbf{q}(x,t) := (q_1(x,t),\cdots,q_N(x,t)),\ q_i = v_i - \varepsilon e^{\mu t},\ i=1,2,\cdots,N,$$

where $\mu > NN_0$ is a constant.

Applying the second inequality and (H2), we obtain

$$\begin{aligned}
\frac{\partial q_i}{\partial t} &= \frac{\partial v_i}{\partial t} - \varepsilon e^{\mu t} \\
&\leq f_i(x,t,\mathbf{v},\nabla v_i,\nabla^2 v_i) - \varepsilon e^{\mu t} \\
&\leq f_i(x,t,\mathbf{q},\nabla v_i,\nabla^2 v_i) + \varepsilon NN_0 e^{\mu t} - \varepsilon \mu e^{\mu t} \\
&< f_i(x,t,\mathbf{q},\nabla v_i,\nabla^2 v_i) \\
&= f_i(x,t,\mathbf{q},\nabla q_i,\nabla^2 q_i),\ i=1,2,\cdots,N.
\end{aligned}$$

Moreover, (H3) and the boundary condition implies that

$$\begin{aligned}
B_i q_i &= B_i v_i - \varepsilon e^{\mu t} \\
&\leq \int_\Omega K_i(x,y,t,\mathbf{v}(y,t))\,dy - \varepsilon e^{\mu t} \\
&= \int_\Omega K_i(x,y,t,\mathbf{q}(y,t))\,dy + \varepsilon e^{\mu t}\int_\Omega \sum_{j=1}^{N} K_{i(j)}(x,y,t,\zeta(y,t))\,dy - \varepsilon e^{\mu t} \\
&< \int_\Omega K_i(x,y,t,\mathbf{q}(y,t))\,dy,\ \text{on } S_T,\ i=1,2,\cdots,N,
\end{aligned}$$

where $\zeta(y,t)$ is a value relying on $\mathbf{v}$. The initial conditions in (2.1) yields that

$$u_i(x,0) \geq v_i(x,0) > q_i(x,0),\ x\in\overline{\Omega},\ i=1,2,\cdots,N.$$

Therefore, with $\mathbf{v}$ replaced by $\mathbf{q}$, (2.2) holds. This means that

$$u_i(x,t) > q_i(x,t) - \varepsilon e^{\mu t},\ (x,t)\in\overline{D}_T,\ i=1,2,\cdots,N.$$

Taking $\varepsilon \to 0^+$, we have $\mathbf{u}\geq\mathbf{v}$ in $\overline{D}_T$, which leads to the desired result. □

More generally, we have the following comparison principle.

Theorem 2.3. *Let the hypothesis* (H1), (H2) and (H4) hold, and let $\mathbf{u} = (u_1, \cdots, u_N)$, $\mathbf{v} = (v_1, \cdots, v_N) \in (\mathbf{C}^{1,2}(D_T) \cap \mathbf{C}^{0,1}(\overline{D}_T))^N$ satisfy

$$\begin{aligned}
\frac{\partial u_i}{\partial t} &\geq f_i(x,t,\mathbf{u},\nabla u_i,\nabla^2 u_i), \quad (x,t) \in D_T,\\
\frac{\partial v_i}{\partial t} &\leq f_i(x,t,\mathbf{v},\nabla v_i,\nabla^2 v_i), \quad (x,t) \in D_T,\\
B_i u_i &\geq \int_\Omega K_i(x,y,t,\mathbf{u}(y,t))\,dy, \quad (x,t) \in S_T,\\
B_i v_i &\leq \int_\Omega K_i(x,y,t,\mathbf{v}(y,t))\,dy, \quad (x,t) \in S_T,\\
u_i(x,0) &\geq v_i(x,0), \quad x \in \overline{\Omega}, \ i = 1, 2, \cdots, N.
\end{aligned} \tag{2.4}$$

Then $\mathbf{u}(x,t) \geq \mathbf{v}(x,t)$ in $\overline{D}_T$.

Proof. Let

$$u_i(x,t) = f(x)p_i(x,t), \quad v_i(x,t) = f(x)q_i(x,t), \ i = 1,2,\cdots,N,$$

where $f(x) > 0$, $x \in \overline{\Omega}$ and $f|_{\Omega_0} = 1$ (Ω_0 is a domain containing $\partial\Omega$ in R^n), satisfying

$$\int_\Omega f(y)dy < \frac{1}{MN}.$$

Direct calculation gives

$$\begin{aligned}
\frac{\partial p_i}{\partial t} &\geq \widetilde{f}_i(x,t,\mathbf{p},\nabla p_i,\nabla^2 p_i), \quad (x,t) \in D_T,\\
\frac{\partial q_i}{\partial t} &\leq \widetilde{f}_i(x,t,\mathbf{q},\nabla q_i,\nabla^2 q_i), \quad (x,t) \in D_T,\\
B_i p_i &\geq \int_\Omega K_i(x,y,t,\mathbf{p}(y,t))\,dy, \quad (x,t) \in S_T,\\
B_i q_i &\leq \int_\Omega K_i(x,y,t,\mathbf{q}(y,t))\,dy, \quad (x,t) \in S_T,\\
p_i(x,0) &\geq q_i(x,0), \quad x \in \overline{\Omega}, \ i = 1,2,\cdots,N,
\end{aligned} \tag{2.5}$$

where

$$\begin{aligned}
\widetilde{f}_i(x,t,\mathbf{u},P,S) &= \frac{1}{f(x)} f_i(x,t,fu,u_i\nabla f + fP, u_i + \nabla^2 f + (\nabla f)'P + P'\nabla f + fS)\\
&\in \mathbf{C}[\overline{D}_T \times R^N \times R^n \times R^{n^2}, R],\\
\widetilde{K}_i(x,y,t,\mathbf{u}) &= K_i(x,y,t,f(y)\mathbf{u}).
\end{aligned}$$

With f_i, K_i, u_i and v_i replaced by $\widetilde{f}_i$, $\widetilde{K}_i$, $f\widetilde{u}_i$ and q_i respectively, $i = 1,2,\cdots,N$, the hypotheses in Theorem 2.3 also hold. Therefore (2.5) implies that

$$p_i(x,t) \geq q_i(x,t), \ (x,t) \in \overline{D}_T, \ i = 1,2,\cdots,N.$$

This means that $\mathbf{u} \geq \mathbf{v}$ in D_T, which completes the proof. □

Corollary 2.4. *Let* (H1), (H2) *and* (H4) *hold, and system* (1.1) *has a solution. Then the solution for system* (1.1) *is unique.*

Example 2.5. Consider the following coupled system of reaction-diffusion equations:

$$\begin{aligned} &\frac{\partial u_i}{\partial t} - L_i u_i = g_i(x,t,\mathbf{u},\nabla u_i),\ x \in \Omega, t > 0,\\ &B_i u_i = \int_\Omega (x^2+y^2) \sum_{j=1}^{N} \frac{2u_j(y,t) - \sin 2u_j(y,t)}{4} dy,\ x \in \partial\Omega,\ t > 0, \\ &\widetilde{u}_i(x,0) = u_{i_0}(x),\ x \in \overline{\Omega},\ i = 1,2,\cdots,N, \end{aligned} \tag{2.6}$$

where the elliptic operator L_i is given by

$$L_i = \sum_{j,k=1}^{n} a_{jk}^{(i)}(x,t)\frac{\partial^2}{\partial x_j \partial x_k} + \sum_{j=1}^{n} b_j^{(i)}(x,t)\frac{\partial}{\partial x_j}, \quad i = 1,2,\cdots,N,$$

and for each $i = 1,2,\cdots,N$, $g_i(\cdot,u_1,\cdots,u_N,\cdot)$ is nondecreasing with respect to u_j $(j \neq i)$ and there exists $d > 0$ such that

$$g_i(x,t,\mathbf{u},P) - g_i(x,t,\mathbf{v},P) \le d \sum_{j=1}^{N} (u_j - v_j), \quad \mathbf{u} \ge \mathbf{v},\ \ \mathbf{u},\ \mathbf{v} \in R^N,\ P \in R^N.$$

Clearly, (H4) holds. Therefore, if system (2.7) has a solution, then the solution is unique.

References

[1] C. I. Byrnes, D. S. Gilliam, A.Isidori and V. I. Shubov, Zero dynamics modeling and boundary feedback design for parabolic systems, M*ath. Comput. Modelling* **44** (2006), no. 9-10, 857–869.

[2] S. Carl, S. Heikkila and J. W. Jerome, Trapping regions for discontinuously coupled systems of evolution variational inequalities and application, *J. Math. Anal. Appl.* **282**(2003),421-435.

[3] C. -P. Danet, Some applications of parabolic comparison principles to the study of decay estimates, *Acta Math. Univ. Comenianae* Vol. LXXV, 2(2006), 227-232.

[4] J. Liang, H. Y. Wang and T. J. Xiao, *A comparison principle for nonlocal delay coupled systems,* preprint.

[5] C. V. Pao, Numerical analysis of coupled systems of nonlinear parabolic equations, *SIAM J. Numer. Anal.* **36** (1999), no. 2, 393–416.

[6] T. Vejchodsky, Comparison principle for a nonlinear parabolic problem of a nonmonotone type, *Appl. Math. (Warsaw)* **29** (2002), no. 1, 65–73.

[7] N. Umeda, Existence and nonexistence of global solutions of a weakly coupled system of reaction-diffusion equations, *Commun. Appl. Anal.* **10** (2006), no. 1, 57–78.

[8] R. N. Wang, T. J. Xiao and J. Liang, Asymptotic behavior of solutions for systems of periodic reaction-diffusion equations in unbounded domains, *Int. J. Evol. Equations*, **1** (2005), no.3, 281–198.

[9] R. N. Wang, T. J. Xiao and J. Liang, A comparison principle for nonlocal coupled systems of fully nonlinear parabolic equations. *Appl. Math. Lett.* **19**(2006), 1272-1277.

[10] R. N. Wang, T. J. Xiao and J. Liang, *Coupled nonlocal periodic parabolic systems with time delays,* preprint.

[11] T. J. Xiao and J. Liang, The Cauchy Problem for Higher Order Abstract Differential Equations, *Lecture Notes in Math.* **1701**, Springer, Berlin, 1998.

[12] T. J. Xiao and J. Liang, Second order parabolic equations in Banach spaces with dynamic boundary conditions, *Trans. Amer. Math. Soc.* **356** (2004), 4787-4809.

In: Progress in Evolution Equations
Editor: Gaston M. N'Guerekata, pp. 27-63

ISBN: 978-1-60456-328-3

Chapter 3

INFORMATION COMPLEXITY OF EVOLUTIONARY DYNAMICS

Vladimir S. Lerner*

13603 Marina Pointe Drive, Suite C-608, Marina Del Rey, CA 90292, USA

Abstract

In the bi-level's information model of evolutionary dynamics [1] (with the random processes at microlevel and the dynamic processes at macrolevel), the evolutionary changes arise when the microlevel's randomness affects the macrolevel, creating a new information at macrolevel and renovating the operator of the macrodynamic equations. The dynamics [1] also bring a sequential cooperation of the renovated states into a hierarchical network. This paper studies a *complexity* of the evolutionary changes, caused by both the state's renovation and their cooperation, establishing the information complexity measures (MC). It is shown that a common indicator of the complexity, is the *specific* entropy's speed (related to the increment of the model's volume), rather than the entropy, as it was accepted before. The complexity mechanism applies the minimax variation principle (VP) of information macrodynamics (IMD), which describes a consequent evolutionary transition from a local, unstable process's movement to a local, stable process, associated with the current influx of information and its accumulation. This transition enables the production of the cooperative phenomena and, in particular, the contributions from different superimposing processes, measured by the MC's cooperative complexity. The MC, arising as an indicator of these phenomena at the unification (or decomposition) of the system's processes, is defined by the *invariant information* measure, allowing for both analytical formulation and computer evaluation. An optimal multi-dimensional consolidation process, satisfying the VP, forms the information hierarchical network (IN) consisting of the model eigenvalues' sequential cooperation in triples. The MC of such an optimal cooperative triplets' structure is measured by the IN's triplet *code* (as an algorithm of the minimal program, which evaluates the IN's hierarchical structure by the triplet's information contributions in bits of information). The cooperative IN allows for the automatic arrangement and measurement of the MC-local complexities for a multi-dimensional process, taking into account their time-space locations and the mutual dependencies, providing the MC *hierarchical invariant* information measure by quantity and quality in the triplet's code. MC covers Kolmogorov's complexity, which measures a deterministic order over a stochastic disorder by a minimal program, as well as the

* E-mail address: vslerner@yahoo.com

statistical complexity. MC provides a precise complexity measure of a *dynamic irreversible process*, evaluating the aforementioned forms of complexities in bits of information. The considered geometrical space curvature conceals information of the cooperative structures in the cells' form, whose code, in particular, measures the cooperative complexity. The process' trajectory, located in this geometrical space, acquires a sequence of the code cells.

Keywords: Cooperative macrodynamics; Variation principle; Information measure; Hierarchical complexity

Introduction

Basic complexity measures have been developed in algorithmic computation theory [2-5], important indicators of complexity have been proposed in physics [6-9]; numerous other publications [10-20] are connected with these basics. These complexity's measures focus on the evaluating complexity for an already formed complex system. We intend to analyze an *origin* of complexity in the evolution process [1].

Evolution arises in the interactive dynamic process, accompanied by creation of new phenomena, which are potential sources of a complexity. The universality of information language allows the *generalization of* the description of various interactions in terms of the *information* interactions, considered independent of their specific forms and nature. Focus on interactive *informational* dynamics leads to a study of a *dynamic* complexity resulting from the interactions of information flows, measured by the specific information speeds. That's why the dynamic information complexity should be connected with the information speeds rather than just with a quantity of information in the above publications.

An intuitive notion of a system's complexity, which distinguishes complexity from simplicity, is associated with the assembling of the system's elements into a joint system during a cooperative process. This means that the system's complexity is naturally connected to its ability to cooperate, which depends on the phenomena and parameters of *cooperative* dynamics. It has been pointed out repeatedly that algorithmic complexity [2-5] does not fit the intuitive notion of complexity [6-7]. A system's complexity, resulting from the cooperative dynamics of a *multiple* set of interacting processes (elements), having an adequate information measure, has not been studied yet [10-20]. The main questions are: What is a general mechanism of cooperation and the condition of its origination? Does there exist a general measure of a *dynamic* complexity independent of a particular physical-chemical nature of the cooperative dynamics with a variety of their phenomena and parameters? How can the dynamic *multi-dimensional* cooperative complexity be defined and measured?

The answers for these questions require a new approach leading us to a *unified notion of dynamic information complexity*, measured in terms of quantities and qualities of information by a corresponding information code.

Compared to the known publications, we analyze the complexity as an *attribute* of the process's *cooperative dynamics,* considering both the phenomenological concept and formal measure of the complexity.

Studying the *regularities* of collective dynamics, accompanied by a formation of cooperative structures, can be formalized using a variation principle (VP), applied to the informational path *functional* [21] and equations of informational macrodynamics (IMD)[22].

The paper's *objective* consists of a definition and formulation of the complexity's information measure, the analysis of the complexity's origination in cooperative dynamics, and both the analytical and computational measure's connections to the informational dynamic's parameters. This objective is implemented by introducing a formal notion of MC complexity (sec.1), considering the MC-emergence in an elementary cooperative process (sec.2), generalizing the MC complexity measure in view of the formal multi-cooperative mechanism in the information hierarchical network (sec. 3), and applying the information geometry's space equations to determine an intensity of information attraction and the complexity (sec. 4).

1. The Notion of Information Macrocomplexity and Its Relevance

Definition 1.1. Let's have two sources of information S_i, S_k concentrated in volumes V_i, V_k accordingly, which are able to interact by some portions of their information (entropy) ΔS_{ik}, accompanied by a change of a shared volume ΔV_{ik}, and characterized by the corresponding information speeds $-\frac{\partial \Delta S_{ik}}{\partial t} = H_{ik}$.

Then the *information measure of complexity* for the interacting sources: MC_{ik} is defined by an information speed, concentrated in shared volume $\Delta V_{ik} : MC_{ik} = \frac{H_{ik}}{\Delta V_{ik}}$, or an instant entropy's concentration in this volume: $\frac{\partial \Delta S_{ik}}{\Delta V_{ik} \partial t}$ (the entropy production), which evaluates the specific information contribution, transferred during the source's interaction in *dynamics.*

Definition 1.2. Let's consider an increment of complexity: $\delta(MC_{ik})$, generated by an increment of the specific information (entropy) speed:

$$\delta(\frac{\partial \Delta S_{ik}}{\Delta V_{ik} \partial t}) = \delta(\frac{\partial \Delta S_{ik}}{\partial t}) / \Delta V_{ik} - \frac{\partial \Delta S_{ik}}{\partial t} \frac{(\delta \Delta V_{ik})}{(\Delta V_{ik})^2} = \frac{(\delta \Delta V_{ik})}{(\Delta V_{ik})} [\delta(\frac{\partial \Delta S_{ik}}{\partial t}) / (\delta \Delta V_{ik}) - \frac{\partial \Delta S_{ik}}{\Delta V_{ik} \partial t}],$$

which at a small ΔV_{ik}^{δ} satisfies $\delta \Delta V_{ik} \cong \delta V_{ik}^{\delta}$, and we get $MC_{ik}^{\delta} = \delta MC_{ik} + MC_{ik}$, where $\delta MC_{ik} = -\delta(\frac{\partial \Delta S_{ik}}{\Delta V_{ik} \partial t}), \delta H_{ik} / \delta V_{ik}^{\delta} = MC_{ik}^{\delta}$ at $-\frac{\partial \Delta S_{ik}}{\partial t} = H_{ik}$.

Then MC_{ik}^{δ} is *the information measure of a differential complexity, defined by* the increment of the information flow $-\frac{\partial \Delta S_{ik}}{\partial t}$ per a small volume increment δV_{ik}^{δ} (within the shared volume ΔV_{ik}), or $MC_{ik}^{\delta} = \frac{\partial H_{ik}}{\partial t} / \frac{\partial \Delta V_{ik}}{\partial t}$ is defined by the ratio of the above speeds.

The MC_{ik}^{δ} automatically includes both the MC_{ik} and its increment δMC_{ik}. The information transition includes transferring both the information flow and the volume. At $\delta^* V_{ik}^{\delta} = \frac{\delta V_{ik}^{\delta}}{\delta \Delta V_{ik}} \to +1$, the shared volume ΔV_{ik} increases, and at $\delta^* V_{ik}^{\delta} = \frac{\delta V_{ik}^{\delta}}{\delta \Delta V_{ik}} \to -1$, the shared volume decreases, shrinks. The shrinkage is associated with the S_i, S_k volumes assembly (cooperation) while the volume enlargement is associated with the volume's source's disassembly. According to the second thermodynamic law, the internal (per volume) entropy production can only be increased: $\delta(\frac{\partial \Delta S_{ik}}{\partial t}) / \Delta V_{ik} > 0$, at a fixed ΔV_{ik} and any changes of the S_i, S_k external entropies. This leads to $\delta H_{ik} < 0$ at a fixed ΔV_{ik}. Therefore at δV_{ik}^{δ}<0, $\delta H_{ik} < 0$, the corresponding differential information complexity MC_{ik}^{δ} >0, and at δV_{ik}^{δ}>0 we have MC_{ik}^{δ}<0. In both cases, an existence of the information flows $H_{ik} \neq 0$ and its increment $\delta H_{ik} \neq 0$, which determine the involvement of both S_i, S_k in the information transition, are a *necessary* condition that connects S_i, S_k and can assemble they into a cooperative. The corresponding MC_{ik}^{δ} >0 we call the *cooperative complexity measure*.

Comments 1.0. At $H_{ik} < 0$ and $\delta \Delta V_{ik} / V_{ik} = -1$, we get also $MC_{ik} > 0$ at the cooperation. A single source S_i, possessing the volume's concentration of entropy $s_i = \frac{S_i}{V_i}$, can generate outside a total potential increment $\delta s_i = \frac{\delta V_i}{V_i}(\frac{\delta S_i}{\delta V_i} - s_i)$. Using the corresponding *information* measures: $S_i = -H_i, s_i = -h_i$ at $|\frac{\delta V_i}{V_i}| = 1$ we get the increment in the form $MC_i^{\delta} = \delta h_i + h_i$, where $MC_i^{\delta} = \frac{\delta H_i}{\delta V_i}$ is the source's *potential* differential complexity, which can also be written through the ratio of the source's information speed to volume speed: $MC_i^{\delta} = \frac{\dot{H}_i}{\dot{V}_i}$. In particular, at $\frac{\delta V_i}{V_i} = -1$ and $\delta H_i < 0$ the complexity $MC_i^{\delta} > 0$. ●

Let's have *a set of elements (objects)* $\Delta v_j \in \Delta V$, $j = 1, ..., k, .., n$ with the internal concentrations of information $\frac{\Delta h_j}{\Delta v_j} = h_j^v$ (the object's information capacity) for each *j* and the element's possibility of a free movement within a volume ΔV. Suppose within the set exists an element k with the *internal* $\frac{\Delta h_k}{\Delta v_k} = h_k^v$, whose increment δh_k^v is a *source* of information transferred to other elements in a communication process. This means that an *external* increment of the element's k information δh_k^e (related to the element's surface) (at $\delta h_k^v = \delta h_k^e$) is transmitted to some other element j, satisfying relation $\delta h_k^e \geq \delta h_j^e$, where δh_j^e is the j-external increment (related to the element's surface) that affects its internal h_j^v according to the corresponding equality $\delta h_j^e = \delta h_j^v$.

Proposition 1.0.1. The information indicator of the element's k involvement in the cooperative association with other elements of the above set serves the ratio

$$\delta_h(\Delta h_k) = \frac{dh_k}{dt} / \frac{\Delta v_k}{\Delta t_k} > 0, \tag{1.0.1}$$

where $-\frac{d(\Delta s_k)}{dt} = \frac{dh_k}{dt}$ is the element's information flow and $\frac{\Delta v_k}{\Delta t_k}$ is the volume's fixed speed (both of them assume to be controllable); *then* the above ratio (1.0.1) is applicable

(1)-at a fixed volume Δv_k and at a fixed $\frac{d(\Delta h_k)}{dt}$ within the time interval Δt_k; and

(2)-if the transmission of the information flow is accompanied by transferring of the element's information volume, *then* the ratio (1.0.1) acquires the form

$$\delta_{h,v} h_k^v = \frac{\delta h_k}{\Delta v_k} - \frac{h_k \delta v_k}{(\Delta v_k)^2} = \frac{\delta v_k}{\Delta v_k}\left(\frac{\delta h_k}{\delta v_k} - h_k^v\right) = \frac{\delta v_k}{\Delta v_k}\left(\frac{\dot{h}_k}{\dot{v}_k} - h_k^v\right), \tag{1.0.2}$$

Indeed. Considering the increment δh_k^v at a fixed volume Δv_k, we come to relations $\delta_h\left(\frac{\Delta h_k}{\Delta v_k}\right) = \frac{\delta(\Delta h_k)}{\Delta v_k}, \delta(\Delta h_k) = \frac{d(\Delta h_k)}{dt}\Delta t_k$, where Δt_k is a time interval at a fixed $\frac{d(\Delta h_k)}{dt}$ and at a fixed Δv_k. We get (1.0.1). When the transmission includes both the information flow and the element's information volume, then (1.0.1) leads to (1.0.2). ●

Corollary 1.0.1. At a small Δv_k relation (1.0.2) admits $\frac{\delta v_k}{\Delta v_k} \rightarrow 1$ and it leads to $\frac{\dot{h}_k}{\dot{v}_k} = \delta_{h,v} h_k^v + h_k^v$. In this case, the information transition includes a relative increment of information volume. At $\frac{\delta v_k}{\Delta v_k} \rightarrow +1$, the element's initial volume increases, and at

$\frac{\delta v_k}{\Delta v_k} \to -1$, the element's volume could be transferred to others during the transitions. In both (1.0.1, 1.0.2), function $\delta h_k^v > 0$ characterizes the element source's *informational potential* as its ability to involve other elements in the information connection. ●

An ability of the element's for the information connections by adjoining other elements into a cooperative we *call* the element's *potential complexity* MC_{pk}^{δ}, which is measured in both formulas by the ratio of the information speed to the volume speed. Specifically, at transferring the information flow: $MC_{pk}^{\delta} = \frac{dh_k}{dt} / \frac{\Delta v_k}{\Delta t_k} > 0$, this function coincides with the element information potential. With changing both the element's information flow and the volume, the potential complexity also includes the element's volume internal concentration of information h_k^v : $MC_{phv}^{\delta} = \frac{\dot{h}_k}{\dot{v}_k} = (\delta_{h,v} h_k^v + h_k^v) > 0$. The positivity of both above functions can be reached at the same signs of $\frac{dh_k}{dt}$ and $\frac{\Delta v_k}{\Delta t_k}$ (or $\dot{v}_k$). More generally, at $\frac{\delta v_k}{\Delta v_k} \neq |1|$, both functions acquire the form

$$MC_{phv}^{\delta v} = \frac{\dot{h}_k}{\dot{v}_k} = \frac{\delta_{h,v} h_k^v}{\delta v_k} \Delta v_k + h_k^v . \tag{1.0.3}$$

Comments 1.0.2. At $MC_{pk}^{\delta} \to 0$, the information flow, transferred by the element, decreases, reaching zero in a limit, when the element is unable to make any connections to others and becomes independent, even though it could be located close to its neighbors. At $MC_{phv}^{\delta} \to 0$, we get the same result, even though $\delta_{p,v} h_k^v > 0$, because from $MC_{phv}^{\delta}=0$ follows equality $\delta_{h,v} h_k^v = -h_k^v$, that means this increment is consumed by the element itself without transferring its information outside. Therefore, a positivity of above complexity measures is an *indicator* of the element's a *cooperative activity,* which characterizes the element's ability to transfer its information to other elements. ●

*Proposition 1.0.2.*The information increment *transferred between* elements (k, j) follows from (1.0.3) in the form: $h_{k,j}^v = \frac{dh_{k,j} / dt}{\Delta v_{k,j} / \Delta t_{k,j}} \geq 0$, with

$$\delta_h h_{k,j}^v = \frac{dh_k / dt}{\Delta v_k / \Delta t_k} - \frac{dh_j / dt}{\Delta v_j / \Delta t_j} = MC_k^{\delta} - MC_j^{\delta} = \Delta MC_k^{\delta} \geq 0 , \tag{1.0.4}$$

where the elements are coordinated in such a way that one of them (k) is a source of information for others ($j,m,...,n$), which serve as the information consumers, and $MC_j^{\delta}=\frac{dh_j/dt}{\Delta v_j/\Delta t_j}>0$ becomes an *indicator* of the *element* j *cooperative activity*, which for this element characterizes its ability to be involved in communication with a source (and possibly with other elements). (Generally, we assume that an object possesses an ability for a *self-controllable* cooperative activity *or* an opposite ability for inactivity). ●

Corollary 1.0.2. With changing both the element's information flow and the volume, the element's *j* cooperative activity acquires the form analogous to (1.0.3):

$$MC_{jhv}^{\delta v}=\frac{\dot{h}_j}{\dot{v}_j}=\frac{\delta_{h,v}h_j^v}{\delta v_j}\Delta v_j+h_j^v. \qquad (1.04a)$$

Condition (1.0.4), applied to transferring between the elements both the information flows and volumes, leads to

$$\delta_{h,v}h_{k,j}^v=\delta_{h,v}(h_k^v-h_j^v)=\frac{\delta v_k}{\Delta v_k}\left(\frac{\dot{h}_k}{\dot{v}_k}-h_k^v\right)-\frac{\delta v_j}{\Delta v_j}\left(\frac{\dot{h}_j}{\dot{v}_j}-h_j^v\right) \qquad (1.0.5)$$

At $\frac{\delta v_k}{\Delta v_k}=-\frac{\delta v_j}{\Delta v_j}$ the element j's volume relative increment decreases by transferring it to the increasing element k's relative volume increment. In this case, $\dot{v}_k>0$ and $\dot{v}_j<0$, that it's why $\frac{\dot{h}_k}{\dot{v}_k}>0$ and $\frac{\dot{h}_j}{\dot{v}_j}<0$. For $h_j^v=\frac{\Delta h_j}{\Delta v_j}$, we have $\Delta v_j=-\Delta v_k\frac{\delta v_j}{\delta v_k}$ and $h_j^v<0$ at $\Delta h_j>0$. At $\frac{\delta v_k}{\Delta v_k}=-\frac{\delta v_j}{\Delta v_j}=1$, a total volume of the element j passes to the element's volume k, merging with that element's volume. This merge corresponds to a binding of both element's information flows and volumes. At this case, relation (1.0.5) acquires form

$$\frac{\dot{h}_k}{\dot{v}_k}-\frac{\dot{h}_j}{\dot{v}_j}=(\delta_{h,v}h_k^v+h_k^v)-(\delta_{h,v}h_j^v+h_k^v), \qquad (1.0.6)$$

where $\Delta v_j = -\delta v_j$. This means that each element's connection is measured by the same form of indicator $MC_j^{\delta} = \frac{\dot{h}_j}{\dot{v}_j} > 0$, which is subtracting from the source's potential complexity measure $MC_{phv}^{\delta} = MC_k^{\delta} = \frac{\dot{h}_k}{\dot{v}_k}$.

Comments 1.0.3. Decreasing of a current difference ($MC_{phv}^{\delta} - MC_{jhv}^{\delta}$) means *arising* of the information connections between the elements' $\Delta v_j \in \Delta V$, or growing the linking information flows and the volumes that bind them. A total number of the elements, satisfying condition (1.0.4), or (1.0.6) (in the MC_{phv}^{δ} environment) measures a current cooperative complexity of the set. Each partial transition of the initial relative flow $\frac{\dot{h}_k}{\dot{v}_k}$ to the following elements' $\frac{\dot{h}_j}{\dot{v}_j}, j, m, ..., n$ decreases the potential complexity and increases the *number* of elements merged into the cooperative. A current difference: $(\frac{\dot{h}_k}{\dot{v}_k} - \frac{\dot{h}_j}{\dot{v}_j}) = \Delta MC_{kj}^{\delta}$, represented by (1.0.5) or (1.0.6), measures an *increase* of cooperative complexity in the *dynamics*, accompany by a growing number of the involved elements.

We call ΔMC_{kj}^{δ} the *dynamic* cooperative complexity. A total decline of a current potential complexity $\Delta MC_{phv}^{\delta} = MC_{phv}^{\delta} - \sum_{j}^{N} MC_{jhv}^{\delta}$ is measured by the sum of cooperative activities of the elements N involved in the information transitions. At $\Delta MC_{piv}^{\delta}=0$, the potential complexity covers the cooperative activities of all involved elements, even though their individual activities are not equal. If all involved elements are joined into a complex, then MC_{pkv}^{δ} measures the complexity of this cooperative. If the individual measure of activity $MC_j^{\delta} = \frac{\dot{h}_j}{\dot{v}_j}$ is equal for all connected n-elements $j, m, ..., n$, and the potential complexity's measure covers the sum of the element's individual measures, satisfying relation $MC_{pkv}^{\delta} - nMC_j^{\delta}=0$, $n = MC_{pkv}^{\delta} / MC_j^{\delta}$, $N=n$, then all n elements, joined into the cooperative, get the equal divided source's information contributions, as the equal information consumers. Even if the individual's MC_j^{δ} are not equal, their total number, deducted from MC_{pkv}^{δ}, measures the number of elements.Thus, both the potential complexity $MC_{pkv}^{\delta o}$ and the individual element's cooperative activity MC_j^{δ}, as well as the dynamic complexity, are measured by the *specific* information speed related to the speed of volume at the information transitions, which join the elements. A maximal cooperative complexity $\max_{h,v} MC^{\delta}_{pkv} = \max_{h,v} \frac{\dot{h}_k}{\dot{v}_k} = \max_{h,v}(\delta h^v_k + h^v_k) > 0$ limits the complexity of the cooperative.

The IMD Basics and Both Complexities Measure

The IMD studies the *dynamics of information*, generated by multiple interactions, using their macroscopic dynamic model, whose extremals model the informational macrodynamic process. Conventional information science, considering generally an information *process*, traditionally uses the probability measure for the random *states* and corresponding Shannon's entropy measure as the uncertainty's function of the states.

The *process*' integral information measure [21,22,1] evaluates a process' trajectory by an integral probability, defined on the random trajectories analogous to Feynman's path functional [23]. The corresponding uncertainty functional

$$\Delta S(\tilde{x}_t, \tilde{x}^1_t) \overset{def}{=} M_{s,\tilde{x}}[\ln q^t_s(\tilde{x}_t \to \tilde{x}^1_t)], q^t_s = \frac{P_{s,\tilde{x}}[d\tilde{x}_t]}{P^1_{s,\tilde{x}}[d\tilde{x}^1_t]} \tag{1.1}$$

represents an integral *entropy* measure of the *information distance* between the given process' trajectories $\tilde{x}_t$, $\tilde{x}^1_t$, where $q^t_s(\tilde{x}_t \to \tilde{x}^1_t)$ is the Radon-Nikodim's density measure of the transformation of probability $P_{s,\tilde{x}}$ into $P^1_{s,\tilde{x}}$ with the initial conditions ($s,\tilde{x}$) evaluated by the conditional mathematical expectations $M_{s,\tilde{x}}(\bullet)$ (s is a starting moment of $\tilde{x}_t$, $\tilde{x}^1_t$). The entropy functional defines the conditional information for the compared stochastic processes $\tilde{x}_t$, $\tilde{x}^1_t$. The IMD *information model of a dynamic process* is built *as an extremal* $x_t = x_t(\tilde{x}_t(u^o))$ that minimizes the functional uncertainty between the *processes* $\tilde{x}_t$, $\tilde{x}^1_t$:

$$\min_{u\in\Omega} \Delta S(\tilde{x}_t(u), \tilde{x}^1_t) = \Delta S(\tilde{x}_t(u^o), \tilde{x}^1_t) = \Delta S(x_t, \tilde{x}^1_t), \tag{1.2}$$

where the process $\tilde{x}_t$ depends on a control u, and the extremal x_t is selected by an optimal control u^o, while both controls belong to a set of admissible controls Ω.

The equations [1, 22], which solve the variation problem (1.2), extract a maximum amount of information from a modeled process $\tilde{x}_t$ regarding some given (a programmable, or a standard) process $\tilde{x}^1_t$, minimizing a gap between them. While maximum entropy is associated with disordering and a complexity, minimum entropy means ordering and simplicity, known as Ockham's razor for selecting the essentials. The model's controls execute the variation principle (VP) (1.2) by selecting the macrotrajectory extremal's *sections* in a multi-dimensional macroprocess, divided by a "punched" microlevel's window, which is a source of new information. The additional "needle" controls, applied *between* the extremal's sections (segments), *stick them into a cooperative structure* and successively join into a multi-dimensional cooperation. The “needle” control action consists of *switching* the macroprocess from the extremal’s section, satisfying the entropy minimum, to the extremal’s piece, satisfying the entropy maximum, and back to the entropy minimal extremal’**s** section for each model’s dimension. The control’s switch closes a jump of the probability density and the corresponding entropy influx at each border’s “punched”point (DP) between the section’s discrete intervals [1]. During the jump, a local chaotic motion might arise with a *chaotic resonance*, which is able to join the current equal probable macrostates (at the section’s borders) into the *cooperative structure* (physically performing the control function). The cooperation brings a physical analogy of the states’ superposition into a *compound* state, accompanied by a nonsymmetry of the formed *ordered* macrostructure**s**. The jump is a source of the *irreversibility* at each DP and the nonlinear phenomena, accompanied by the creation of new properties. In particular, the above collective properties are also associated with forming dissipative structures [24]. Macrocomplexity MC_{ik} measures the specific quantity of information concentrated *within* and at a border of each extremal section, while the complexity for a total macrotrajectory MC^{Σ} measures a summary of specific information contributions from all of the macrotrajectory’s sections [22]. The differential complexity: MC^{δ}_{ik} characterizes the phenomena arising from the transformation from one extremal section to another by the *jump* through a microlevel’s stochastic process during the $o(\varepsilon)$-window between the *i,k* sections. These phenomena consist of transferring first the macromovement from an ordered local stable (but a nonequilibrium) macroprocess (at one extremal segment) to a disordered microlevel’s process with a local instability. The second transformation brings this microlevel process back to the macromovement at the subsequent stable extremal segment. Both transformations are analogous to Baker’s transformation [24, 25]. In the *evolutionary informational macrodynamics* [1], MC^{δ}_{ik} evaluates the contribution to complexity, generated by the *appearance* of new features during the information interaction with the environment, while MC_{ik} evaluates an *accumulation* of this complexity at each interval of an *assimilation* of these features. Later on we find the concrete forms for both complexity’s measures and constructively evaluate their values in evolutionary dynamics.

Let’s start with MC^{δ}_{ik}, whereas the increment is generated by the model’s needle control, performing the sections *cooperation* and bringing the above contributions. In this case, MC^{δ}_{ik} acquires a meaning of *cooperative complexity*. Assume we have a couple of the extremal’s sections i, j along a macroprocess with the needle control connecting them, and let’s evaluate the *information* contribution of the needle control δv_{ij} actions, implementing both of above transformations at a small time interval between the sections:

$$\delta v_{ij} = v_i(t_i, t_{i-\varepsilon}) + v_j(t_{\varepsilon}, t_j), \tag{1.3}$$

We illustrate δv_{ij} using a *finite* approximation of the needle control (Fig.1) that consists of two parts:

$$v_{ij} = v_i(t_i, t_{k-\varepsilon}) + v_j(t_k, t_j), \tag{1.3a}$$

where ε is a small time interval between the end of control $v_i(t_i, t_{k-\varepsilon})$ and the start of control $v_j(t_k, t_j)$; $v_i(t_i, t_{k-\varepsilon})$ transfers the distribution $P_{t_i,x}[dx(t_i)]$ of the macrotrajectory x_t (around $t_i : \delta P_{t_i,x}[dx(t_i)] \sim \delta t_i$) to the distribution $P_{t_k,\tilde{x}}[d\tilde{x}(t_{k-\varepsilon})]$ of the microtrajectory $\tilde{x}_t$ (around $t_k : P_{t_k,\tilde{x}}[d\tilde{x}(t_{k-\varepsilon})] \sim \delta t_k$); $v_j(t_k, t_j)$ transfers $P_{t_k,\tilde{x}}[d\tilde{x}(t_k)]$ to a renovated macrotrajectory's distribution $P_{t_j,x}[dx(t_j)]$ (around $t_j : \delta P_{t_j,x}[dx(t_j)] \sim \delta t_j$) (Fig.1).

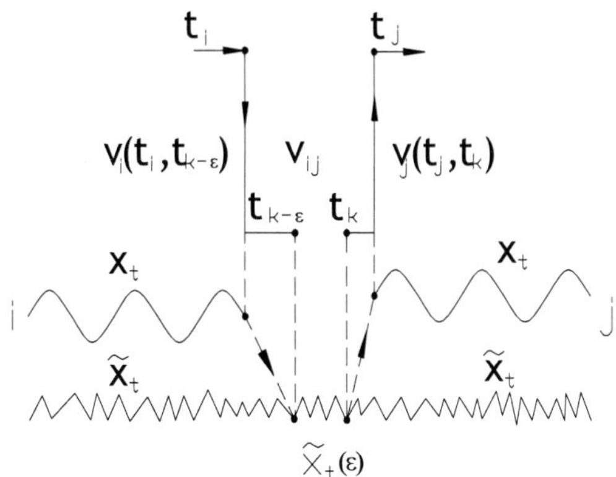

Figure 1. The illustration of transformations of the i - extremal's sections with a macroprocess x_t to a section $\tilde{x}_t(\varepsilon)$ of the microprocess $\tilde{x}_t$ and back to the j-section x_t under the controls $v(t_i, t_{k-\varepsilon}), v(t_k, t_j)$.

Proposition 1.1. From the above assumptions, we get the following MC_{ik}^{δ} complexities measures, generated by the needle control v_{ij} (1.3a):

$$MC_{ik}^{\Delta} = \Delta H_{ik} / \Delta V_{ik}^{\Delta}, \ \Delta H_{ik} = \frac{p(\tilde{x}(t_{k-\varepsilon})) - p(x(t_i))}{p(\tilde{x}(t_{k-\varepsilon}))p(x(t_i))} = \Delta p_{ik}^{*} \tag{1.4a}$$

$$MC_{kj}^{\Delta} = \Delta H_{kj} / \Delta V_{kj}^{\Delta}, \ \Delta H_{kj} = \frac{p(x(t_j)) - p(\tilde{x}(t_k))}{p(x(t_j))p(\tilde{x}(t_k))} = \Delta p_{kj}^{*} \tag{1.4b}$$

with a total information increment

$$\Delta H_{ij} = \Delta H_{ik} + \Delta H_{kj} = -\Delta(\frac{\partial S_{ij}}{\partial t}) = \Delta p^*_{ij} \tag{1.4c}$$

and the complexities

$$MC^{\delta}_{ik} = \delta H_{ik} / \delta V^{\delta}_{ik}, MC^{\delta}_{kj} = \delta H_{kj} / \delta V^{\delta}_{kj}, MC^{\delta}_{ij} = \delta H_{ij} / \delta V^{\delta}_{ij}, \tag{1.4}$$

by applying the needle control δv_{ij} (1.3) at $\Delta \to \delta$ in the relations (1.4a-c).

Proof. Considering the increments of a random entropy [26] at the (t_i, t_k) and (t_k, t_j) localities:

$$\Delta S(t_i, t_{k-\varepsilon}) = -\ln p(x(t_i)) + \ln p(\tilde{x}(t_{k-\varepsilon})) = S(t_i) - S(t_{k-\varepsilon}),$$

$$\Delta S(t_k, t_j) = -\ln p(\tilde{x}(t_k)) + \ln p(x(t_j)) = S(t_k) - S(t_j) \tag{1.5a}$$

at the indications of the elementary probabilities:

$$p(x(t_i)) = P_{t_i, x}[dx(t_i)], p(\tilde{x}(t_{k-\varepsilon})) = P_{t_k, \tilde{x}}[d\tilde{x}(t_{k-\varepsilon})],$$

$$p(\tilde{x}(t_k)) = P_{t_k, \tilde{x}}[d\tilde{x}(t_k)], p(x(t_j)) = P_{t_j, x}[dx(t_j)] \tag{1.5b}$$

we get the corresponding derivatives

$$-\frac{\partial S}{\partial t}(t_i) = p(x(t_i))^{-1}, -\frac{\partial S}{\partial t}(t_{k-\varepsilon}) = p(\tilde{x}(t_{k-\varepsilon}))^{-1},$$

$$-\frac{\partial S}{\partial t}(t_k) = p(\tilde{x}(t_k))^{-1}, -\frac{\partial S}{\partial t}(t_j) = p(x(t_j))^{-1}, \tag{1.5c}$$

where the moments $(t_i, t_{k-\varepsilon}, t_k, t_j)$ are defined by the nonrandom controls (1.3a). The entropy increments at the finite time intervals $\Delta t_{ik} = t_i - t_k$, and $\Delta t_{kj} = t_k - t_j$ are determined by the related functions (1.5c) in the forms

$$\Delta H_{ik} = -\Delta(\frac{\partial S_{ik}}{\partial t}) = -\frac{\partial S}{\partial t}(t_i) + \frac{\partial S}{\partial t}(t_{k-\varepsilon}),$$

$$\Delta H_{kj} = -\Delta(\frac{\partial S_{kj}}{\partial t}) = -\frac{\partial S}{\partial t}(t_k) + \frac{\partial S}{\partial t}(t_j) \tag{1.6a}$$

$$\Delta H_{kj} = \frac{p(x(t_j)) - p(\tilde{x}(t_k))}{p(x(t_j))p(\tilde{x}(t_k))} = \Delta p^*_{kj}. \tag{1.6b}$$

with a total increment

$$\Delta H_{ij} = \Delta H_{ik} + \Delta H_{kj} = -\Delta(\frac{\partial S_{ij}}{\partial t}) = \Delta p^*_{ij}, \ \frac{\partial S}{\partial t}(t_{k-\varepsilon}) \neq \frac{\partial S}{\partial t}(t_k), \tag{1.6c}$$

where controls $v_i(t_i, t_{k-\varepsilon})$ and $v_i(t_i, t_{k-\varepsilon})$ provide (1.6a,b) accordingly, and control v_{ij} supplies the information contribution (1.6c) which is necessary for the cooperation. ●

Comments 1.1. The relative difference Δp^*_{ik} represents the probabilistic measure of distance between stability and instability for the above states, related to the joint probability $p(\tilde{x}(t_{k-\varepsilon}), x(t_i)) = p(\tilde{x}(t_{k-\varepsilon}))p(x(t_i))$ of the independent events at the macro- and micro-trajectories $x(t_i), \tilde{x}(t_{k-\varepsilon})$. It also measures the distance between *order and disorder* in stochastic dynamics [22] and detects determinism amongst the randomness and singularities. This distance is measured directly by the increment of entropy production $\Delta(\frac{\partial S_{ik}}{\partial t})$. If the probability to determine a state: $p(\tilde{x}(t_{k-\varepsilon})) \to 0$, then Δp^*_{ik} become undefined with an indefinite entropy production; for any *finite* $0 < p(x(t_i)) \leq 1$ and $0 < p(\tilde{x}(t_k)) \leq 1$, Δp^*_{ik} takes the finite values. The Δp^*_{kj} evaluates the probabilistic transition from a random instable microstate $\tilde{x}(t_k)$ to a newly formed macrostate $x(t_j)$.

While (1.4a) evaluates the stochastic phenomena, associated with the instability, as a first step toward cooperation, the (1.4b) evaluate the information contribution at the final step for both a state, joining with a given macrostate, and a stable composite structure. A total *specific* information contribution, required for the cooperation of the sections (i, j) during a finite time interval $\Delta t_{ij} = t_i - t_j$, is evaluated by the cooperative complexity (1.4c). The related complexities possess the corresponding features.

The above distance between two different distributions is an analog of the Jensen–Shannon divergence [27]. The stochastic phenomena, associated with coupling of the sections, can be evaluated by Kolmogorov's (K) algorithmic complexity [2]. K-complexity is measured by the relative differential entropy of one object (i) with respect to other object (j), which is represented by a shortest program in bits. MC complexity measures the quantity of information (transmitted by the relative information flow), required to join the object j with the object i, which can be expressed by the algorithm of a minimal program, encoded in a common communication code (considered in sec.3). This program also measures a "difficulty" of obtaining information j from i in the transition *dynamics*. Assigning a common digital MC measure to all communicated objects allows determining the unknown constant C_h in the K complexity [2]. The above relations connect the MC^{Δ}_{ik} (for a fixed ΔV^{Δ}_{ik}, or for a concentrated system) to the K-complexity, providing also K-measure by the entropy production. A unification of a "complementary" $-\frac{\partial S}{\partial t}(t_i)$ and $\frac{\partial S}{\partial t}(t_j)$, accompanied by a decrease of the distance Δp_{ij}^*, represents an information essence of cooperation. •

Example. We illustrate the application of (1.4a) on the example from physical kinetics for a chain of connected chemical reactions (important in the evolution). Following [28], the kinetic equation between the number N_i of components for i-reaction and the number N_k of components for k-reaction is

$$\frac{\partial N_i}{\partial t} = (-\lambda_{ik} N_i + \lambda_{ki} N_k),\ \lambda_{ik} + \lambda_{ki} = 1, \tag{1.7}$$

where $-\lambda_{ik} N_i$ represents a decrease of N_i at ik-transition, while $\lambda_{ki} N_k$ represents an increase of N_k at ki-transition; $\lambda_{ik}, \lambda_{ki}$ are the probabilities of transitions for these components, being independent of time. The transition takes place during a time interval Δt_{ik} under a fixed increment of chemical potentials $\Delta\mu_{ik} = \mu_i - \mu_k$ for these reactions. The specific entropy production for this transition is

$$\frac{\partial \Delta S_v}{\partial t} = \frac{\partial N_i}{\partial t} \Delta\mu_{ik} = (-\lambda_{ik} N_i + \lambda_{ki} N_k) \Delta\mu_{ik} \tag{1.8}$$

where ΔS_v is an increment of specific entropy. Applying (1.4c) at $\Delta t_{ik} \to \partial t$ and using the derivation, we come to the cooperative complexity

$$MC_{ik}^d = -\lambda_{ik} \frac{\partial N_i}{\partial t} \Delta\mu_{ik} + \lambda_{ki} \frac{\partial N_k}{\partial t} \Delta\mu_{ik} = -\frac{\partial \Delta S_{vi}}{\partial t} + \frac{\partial \Delta S_{ki}}{\partial t}, \tag{1.9}$$

which is measured by the above increment of the entropy productions between the above reactions, generating the new components. These transitions join the above reactions in a *cooperative*. For a chain of these reactions, we get a sum, taken at the corresponding time-space's locations. The time-space distributed *information network* (sec.3) allows the automatic allocation and measurement of the MC-local complexities for a multi-dimensional process, taking into account their *mutual dependencies*.

2. The MC Complexity in a Cooperative Dynamic Process

As a rather a formal example, let us find the complexity of the model's *cooperative* dynamics, using (1.4) at

$$\lim_{\delta t\to 0} MC^{\delta} = \frac{\delta H / \delta t}{\delta V^{\delta} / \delta t} = \dot{H} / \dot{V}^{\delta} = MC_t^{\delta} , \tag{2.1}$$

Proposition 2.1. Let us consider the cooperative dynamics for the two nearest extremal's sections of the macrotrajectory, accompanied by joining their corresponding eigenvalues $(\alpha_i^t, \alpha_{i+1}^t)$ of the model's operator spectrum $A = \{\alpha_i^t\}, i = 1,..,k,.,n$ by the needle control (1.3), and following the solutions of variation problem (1.2) for the model's Hamiltonian and operator in the form

$$H = SpA = -\partial \Delta S / \partial t \tag{2.1a}$$

Then the complexity (2.1), being reduced to beginning of a segment's time interval (t_i), acquires the form

$$MC_t^{\delta}(t_o) = \dot{H}_i(t_i) / v_i(t_i) = (\mathbf{a}_o)^2 / \pi c_i^3 t_i^4 , \tag{2.2}$$

where $(\mathbf{a}_o)^2$ is the segment's information needle control's contribution, measured by the section's information invariant $\mathbf{a}_o = \alpha_{io}^t t_i$; $v_i(t_i) = \dot{V}_i / t_i = \pi \rho_i^3 t_i$ is an elementary volume increment at the time interval t_i, with a current $\dot{V}_i = \pi c_i \rho_i^2$, determined by the fixed model's space speed c_i and the volume's cross-section $F_i = \pi \rho_i^2$, $\rho_i = c_i t_i$, $\dot{V}_i = c_i F_i$.

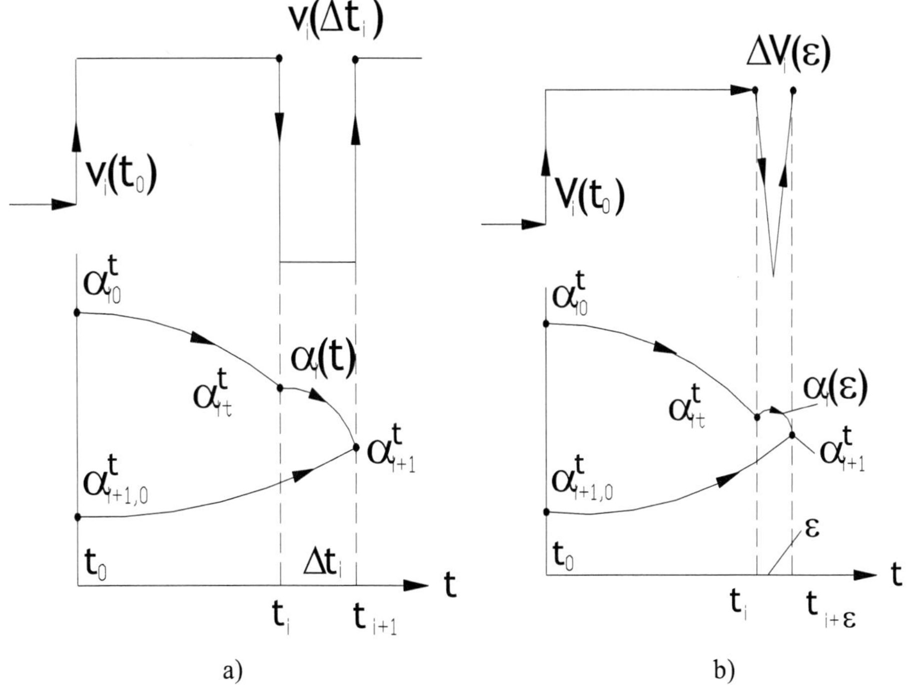

a) b)

Figure 2a,b. The illustration of cooperation of two exremal's sections under the discrete control $v_i(\Delta t_i)$ (a) and the needle control $\delta v_i(t_i)$ (b) (other indications are in the text).

Proof. According to the IMD, the process starts with applying a regular controls $v_i(t_o)$ (Fig.2a) to the initial eigenvalues α_{io}^t and $\alpha_{i+1,o}^t$, initiating the dynamics of $\alpha_i^t(t)$ and $\alpha_{i+1}^t(t)$ at both sections within the time intervals t_i and t_{i+1}, accordingly. The corresponding dynamic equations are determined by the model's invariant relations, following from the solution of the variation problem [1]:

$$\alpha_{it}^t = \alpha_{io}^t \, \mathbf{a}/\mathbf{a}_o, \; \mathbf{a} = \alpha_{i=1,t}^t t_i, \; \mathbf{a}_o(\gamma) = \alpha_{io}^t t_i,$$

$$\alpha_{i+1,t}^t = \alpha_{i+1,o}^t \, \mathbf{a}/\mathbf{a}_o, \; \mathbf{a}(\gamma) = \alpha_{i+1,t}^t t_{i+1}, \; \mathbf{a}_o = \alpha_{i+1,o}^t t_{i+1} \tag{2.2a}$$

where α_{io}^t, $\alpha_{i+1,o}^t$ are the eigenvalues at a beginning of time intervals t_i, t_{i+1} accordingly, α_{it}^t, $\alpha_{i+1,t}^t$ are the eigenvalues at the end of these time intervals. The above information invariants depend on the model parameter of uncertainty γ, evaluating a closeness of the nearest initial eigenvalues (α_{io}^t and $\alpha_{i+1,o}^t$): $\mathbf{a}_o(\gamma=0.5)=\ln 2$, $\mathbf{a}(\gamma=0.5)=0.25$ [1]. The needle control $\delta v_i(\Delta t_i)$ (Fig.2b), applied at the moment t_i^v to the eigenvalue α_{it}^t and acting during a finite time interval $\Delta t_i = t_{i+1}^v - t_i^v$, provides the consolidation of the eigenvalue α_{it}^t with $\alpha_{i+1,t+1}^t$ at the moment t_{i+1}^v. The corresponding derivation of local model's Hamiltonian $\dot{H}_i(\alpha_i^t) = \partial\alpha_i^t / \partial t$ is determined by the function $\alpha_i^t(t_{i+1}^v) = \alpha_{it}^t \exp(\alpha_{it}^t \Delta t_i)(2 - \exp(\alpha_{it}^t \Delta t_i))^{-1}$, where $\alpha_{it}^t = \alpha_i^t(t_i^v)$ is the eigenvalue at the beginning of Δt_i preceding the cooperation, $\alpha_i^t(t_{i+1}^v)$ is the eigenvalue at the moment t_{i+1}^v of cooperation. Both of them depend on the model's dimension n and γ. Using the indication

$$p = \exp(\alpha_{it}^t \Delta t_i)(2 - \exp(\alpha_{it}^t \Delta t_i))^{-1}, \tag{2.3}$$

we get the corresponding derivation

$$\dot{H}_i(\alpha_i^t) = (\alpha_{it}^t)^2 (p + p^2), \tag{2.3a}$$

where p is the parameter of the dynamic cooperation α_{it}^t with $\alpha_{i+1,t+1}^t$ at Δt_i:

$$\alpha_{it}^t \, p = \alpha_{i+1,t}^t \, . \tag{2.4}$$

By substituting relations (2.3, 2.3a) into (2.4), we represent p in (2.3-2.4) by the ratio of the initial eigenvalues: $p = \alpha_{i+1,o}^t / \alpha_{io}^t$, which relates p to the model's parameter of multiplication $\gamma_i^\alpha = \alpha_{io}^t / \alpha_{i+1,o}^t$ ($\gamma_i^\alpha = 2.21$ at γ=0.5), where $\alpha_{it}^t = \alpha_{it}^t(n, \gamma)$. This allows us to express the interval of cooperation: $\Delta t_i = p(\gamma_i^{\alpha})$ and directly obtain the increment (2.3a) during the cooperative dynamics, using the model's invariants and the initial α_{io}^t :

$$\dot{H}_i(\alpha_i^t) = \partial\alpha_i^t / \partial t(n,\gamma) = (\alpha_{io}^t)^2 (\mathbf{a}/\mathbf{a}_o)^2 \; [(\gamma_i^{\alpha})^{-1} + (\gamma_i^{\alpha})^{-2}], \tag{2.5}$$

where a maximal α_{io}^t depends on the model γ and dimension $n : \max_{\gamma\to 0} \alpha_{it}^t(n,\gamma) = \alpha_{it}^{to}(n)$ [22]. Applying the needle control at a very small time interval $\delta t_i : \delta v_i(\delta t_i) = \lim_{\delta t_i \to 0} \delta v_i(\delta t_i) = \delta v_i$, we get the corresponding derivation of local Hamiltonian $\dot{H}_i(\alpha_i^t)$ in a limit: $\dot{H}_i(\delta v) = \lim_{\delta t\to 0}(\partial\alpha_i^t / \partial t)$, which is determined by function $\alpha_i^t(\delta v) = \alpha_{it}^t \exp\alpha_{it}^t \delta t (2 - \exp\alpha_{it}^t \delta t)^{-1}$. From that we obtain the actual needle control's contribution

$$\dot{H}_i(\delta v) = \lim_{\delta t\to 0}(\partial\alpha_i^t / \partial t) = 2(\alpha_{it}^t)^2, \tag{2.6}$$

and using the information invariants (2.2), we come to

$$\dot{H}_i(\delta v) = \dot{H}_i(t_i) = 2(\mathbf{a})^2 / t_i^2 = 2(\alpha_{io}^t)^2 (\mathbf{a}/\mathbf{a}_o)^2, \tag{2.7}$$

where

$$\dot{H}_i(t_o) = 2(\alpha_{io}^t)^2 = 2(\mathbf{a}_o)^2 / t_i^2 , \tag{2.8}$$

at $\dot{H}_i(t_o) = \dot{H}_i(t_i)\,(\mathbf{a}_o/\mathbf{a})$, is the information contribution at a beginning of the section, reduced to the section's time interval t_i . The value of $\dot{H}_i(t_o)/2$ (2.8) can be applied to each consecutive starting segment.

To evaluate the elementary cooperative complexity at each extremal section's MC_t^{δ} , we use any of (2.7, 2.8) and the model's current $\dot{V}_i = \pi c_i \rho_i^2$, determined by the fixed model's space speed c_i and the volume's cross-section $F_i = \pi\rho_i^2$. Because both invariants in (2.7, 2.8) depend on a section time *interval*, we reduce the volume's increment also to the considered time interval t_i . At the radius' increment $\Delta\rho_i = c_i t_i$ and $\Delta\dot{V}_i = \pi c_i^3 t_i^2$, using the contribution (2.8), we get (2.2). ●

Comments 2.1. Let us evaluate the numerical values of $MC_t^{\delta}(t_i)$ (2.2) (at a fixed volume increment) for the model's admissible variations of the parameter $\gamma = (0.0072 - 0.8)$ and a reasonable range of the dimensions $n=(2-20)$. The MC_t^{δ} dependency on two variables (n,γ) splits the $MC_t^{\delta}(n,\gamma)$ on two functions: $MC_{t_i}^{\delta}(\gamma)$ and $MC_{t_i}^{\delta}(n)$.

Then $MC_{t_i}^{\delta}(\gamma)=(\mathbf{a}/\mathbf{a}_o)^2[(\gamma_i^{\alpha})^{-1}+(\gamma_i^{\alpha})^{-2}]$ takes the following computed values: $MC_{t_i}^{\delta}(\gamma=0.0072)\cong 0.052866353$, $MC_{t_i}^{\delta}(\gamma=0.8)\cong 0.5936826$ with the ratio's range $MC_{t_i}^{\delta}(\gamma=0.8)/MC_{t_i}^{\delta}(\gamma=0.0072)\cong 11.23$. $MC_{t_i}^{\delta}(n)=(\alpha_{n(io)}^t)^2$, at $n(io)=2-20$ (where $\alpha_{n(io)}^t$ is the initial eigenvalue of dimension $n=n(io)$) gets the following computed values: $MC_{t_i}^{\delta}(n=2)\cong 0.6058$ and $MC_{t_i}^{\delta}(n=20)\cong 58708.8$ with the ratio's span $MC_{t_i}^{\delta}(n=20)/MC_{t_i}^{\delta}(n=2)\cong 96911.2$. This means that the MC_t^{δ} complexity's dependency on dimensions is in $\cong 100,000$ times stronger than the dependency on γ. At a given *n*, the complexity's grows is limited by a maximal $\gamma\to 1$ when the system decays[22].

In addition to the cooperative complexity, the model possesses the local $MC(t_i)$ and $MC(t_{i+1})$ complexities, determined by the macromovement *within and at the borders* of the time intervals t_i, t_{i+1} and the volumes $v_{io}^{-1}, v_{i+1,o}^{-1}$ −produced *during* these times:

$$MC(t_i)=(\alpha_{io}^t\, t_i)v_{io}^{-1}=\mathbf{a}_o(\gamma)\, v_{io}^{-1},\ MC(t_{i+1})=(\alpha_{i+1,o}^t\, t_{i+1})v_{i+1,o}^{-1}=\mathbf{a}_o(\gamma)\, v_{i+1,o}^{-1}, \quad (2.9)$$

$$MC_{\Sigma t_i}=MC(t_i)+MC(t_{i+1})=\mathbf{a}_o(\gamma)(v_{io}^{-1}+v_{i+1,o}^{-1}), \quad (2.10)$$

where the dynamics are initiated by the actions of regular controls $v_i(t_o)$. The complete model's complexity, generated by the cooperative macrodynamics of joining α_i^t to α_{i+1}^t, is the sum:

$$MC_{\Sigma t_i}+MC_t^{\delta}(n,\gamma)=MC_{\Sigma}(n,\gamma), \quad (2.11)$$

where both the invariants and the elementary volumes in (2.7-2.8, 2.10, 2.11) are determined by the basic model's parameters of dynamics (n,γ) and geometry (κ) [22]. Computer simulation [29] of the model's *adaptive* self-organizing process, considered in [22, 30], shows that at small *n*, a number of neighboring subsystems with a similar complexity does exist. With growing *n*, the number of close-complex neighboring subsystems decreases sharply. ●.

Comments 2.2. If the condition of decoupling the correlations at the punched windows [1, Corollary 4.1] is violated, then the invariants **a**, $\mathbf{a}_o$ in the complexity's measures (2.2), (2.7), (2.9) should be replaced with related information contribution considered in [1,(4.53b)].

In both cases, the complexity at the windows undergoes the *jump-vise changes*. ●.

Comments 2.3. Since $-\delta(\frac{\partial \Delta S_{ik}}{\Delta V_{ik} \partial t})$ is the increment per volume of information produced in this volume, which defines the *cooperative complexity* MC_{ik}^{δ}, this negentropy increment cannot be more than the corresponding internal increment of physical entropy $\delta(\frac{\partial \Delta S_{ik}^{\text{int}}}{\Delta V_{ik} \partial t})$ had been spent on the above negentropy production, according to the second thermodynamic law [26,27]. That's why the MC_{ik}^{δ} maximum is limited by the above maximal internal increment of entropy, admitted by the physical properties of this volume: $\max MC_{ik}^{\delta} \le \max | \delta(\frac{\partial \Delta S_{ik}^{\text{int}}}{\Delta V_{ik} \partial t}) |$. The analogous limitations are true also for the information complexity $MC_{ik} = -\frac{\partial \Delta S_{ik}}{\Delta V_{ik} \partial t}$: $\max MC_{ik} \le \max | \frac{\partial \Delta S_{ik}^{\text{int}}}{\Delta V_{ik} \partial t} |$, where $\frac{\partial \Delta S_{ik}^{\text{int}}}{\Delta V_{ik} \partial t}$ is the internal entropy production per the volume, limited by this volume (and $\gamma \to 1$). The maximal

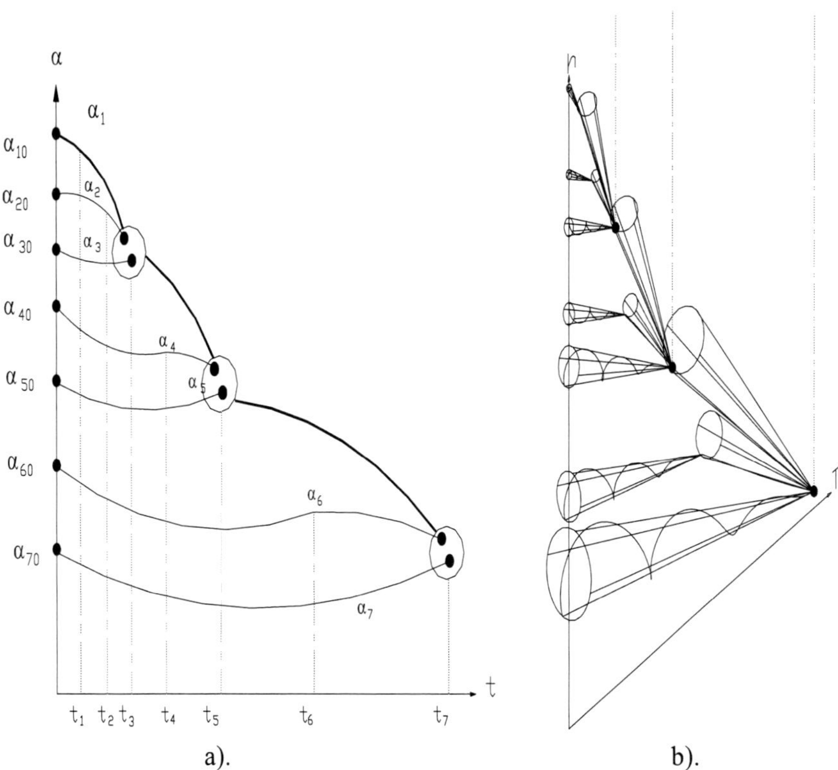

Figure 3a,b.The time-space cooperations with forming the triplet's structures: (a)-the time dynamics of the cooperating eigenvalues with the triplets, formed around the (t_1, t_2, t_3) locations; (b)- the structure of the hierarchical time-space distributed informational network (IN) (with the triplets' nodes), which allocates the distributed complexity.

complexity within each time interval t_i (at the preservation of invariants **a**, $\mathbf{a}_o$ and a system's maximal admissible entropy production within each t_i) can be reached, *if* this time interval acquires the *minimal* value, allowable by the VP during the evolutionary process. A minimal complexity corresponds to a minimal admissible $\gamma \to 0$, which limites a minimal system's structural uncertainty (γ=0.0072) [22]●.

The considered mechanism of cooperative dynamics reveals the MC origination and its connection to the dynamic's and physic's phenomena and parameters. The MC's invariant information measure emerges as an indicator of the cooperative phenomena, in particular, the contributions from different superimposing processes.

A sequential cooperation of all n eigenvalues for the model's operator spectrum (2.1a) generates the cooperative information hierarchical network (IN) (Fig.3), which we use for the information evaluation of a cooperative complexity in multi-dimensional evolutionary processes.

3. The IN's Formal Cooperative Mechanism with the MC Complexity Measure

The following questions arise: What is the general *mechanism coordinating* the formation of cooperative information *structures* in cooperative dynamics? What are an indicator and a measure of the *structural* cooperation? Below we show that the *general mechanism is an information cooperative network* (IN), initiated by the VP application with its MC function as the indicator and measure of the embedded cooperative dynamics. The IN's incoming information string of the model's eigenvalues is ranged automatically according to this mechanism, which assigns the information measure following from the VP solution.

The IN's Basics and the Considered Complexities

The optimal *procedure* of cooperating the extremal's sections consists of joining them in *threes* with the following addition of a new *pair* to those, which were assembled before in threes. (In [22] is shown that this triple composition follows, in particular, from the three-dimensional space metric).

The IN includes a sequence of such an elementary information dynamic structure (*triplet's nodes)*(Fig.3a,b). Each triplet's discrete point is the result of a joint solution of the three local extremal's equations that could be nonlinear. Such points are *singular* with a possibility of all kinds of *chaotic dynamic* phenomena. The controllable cooperative dynamics (that involve the chaotic resonances) automatically builds the IN hierarchical dynamic and geometrical organization through the enclosed sequence of the triplet's structures. The very first IN's triplet is formed by the cooperative dynamics of a triple of starting model's eigenvalues $\alpha_{1o}^t, \alpha_{2o}^t, \alpha_{3o}^t$, which can be assigned by the parameters of multiplication $\alpha_{1o}^t / \alpha_{3o}^t = \gamma_1^\alpha, \alpha_{2o}^t / \alpha_{3o}^t = \gamma_2^\alpha$ and a fixed α_{1o}^t. The preservation of $\gamma_i^\alpha, i = 1,2$ along the IN's triplet's chain define the IN's common parameter of uncertainty

γ that allows us to calculate the invariant quantity of information $\mathbf{a}(\gamma)$ *transferred* to each following triplet's eigenvalue (at every t_i). The optimal dynamics (Fig.3a) of a triplet's *cooperation* determine the moment t_3 when a first triplet's eigenvalue joins its third eiegenvalue, generating the entropy $\mathbf{a}/\gamma_1^\alpha$. At the same moment t_3, the second triplet's eigenvalue cooperates with its third eigenvalue, generating the entropy $\mathbf{a}/\gamma_2^\alpha$, while the cooperating third eigenvalue generates the entropy **a** at the same moment t_3. Thus, the IN's node collects information

$$\mathbf{a}(\gamma)(1+(\gamma_1^\alpha)^{-1}+(\gamma_2^\alpha)^{-1})=g(\gamma), \tag{3.1}$$

which is preserved at fixed γ. The MC for the IN's current hierarchical level $l_i(m_i)$ depends on the triplet's number m_i enclosed into this triplet's level from the previous levels, while the differential MC^δ (2.2) evaluates an instant production of information at every $(i,i+1)$ cooperation. Cooperation with each following node sequentially changes the IN's current MC(m_i,γ). Complexity of a total IN is determined by MC(m_n,γ), at $m_n=n/2-1$ of a final IN's node, according to formulas (2.9-2.11) and at preserving γ within the IN.

The information measure of the starting eigenvalues' string together with the values of $\alpha_{1o}^t, \gamma_i^\alpha$, γ, and $\mathbf{a}(\gamma)$ allow us to calculate the IN's dimension (n), restore the complete cooperative dynamics with the IN's space-time hierarchical information structure (Fig3b), including the sequence of cooperative resonance frequencies, the MC-function at each IN's hierarchical level, and for a whole IN.

Let us find the *limitations* on the IN's cooperative dynamics, its parameters, and the complexity.

Proposition 3.1(P3.1). The model's eigenvalues' sequence ($\ldots.\alpha_{i+1,o}^t, \alpha_{io}^t, \alpha_{i-1,o}^t$), satisfying the triplet's formation, is *limited* by the boundaries for $\gamma \in (0 \to 1)$, which at $\gamma \to 0$ (a)- forms a geometrical progression with $(\alpha_{io}^t)^2 = \alpha_{i-1,o}^t \alpha_{i+1,o}^t$, representing a geometric "gold section" at $\alpha_{io}^t \cong 0.618\,\alpha_{i-1,o}^t$ and the ratio $G=\dfrac{\alpha_{i+1,o}^t}{\alpha_{io}^t} \cong 0.618$; and at $\gamma \to 1$ (b) the sequence $\alpha_{io}^t, \alpha_{i+1,o}^t, \alpha_{i+2,o}^t, \ldots$ forms the Fibonacci series, where the ratio $\alpha_{i+1,o}^t / \alpha_{i+2,o}^t = \gamma_2^\alpha$ determines a "divine proportion" $PHI \cong 1.618$, satisfying $PHI \cong G+1$.

Proof. The model's invariant relation $\mathbf{a}(\gamma)\gamma_i^\alpha(\gamma)=inv$ is fulfilled for a triplet with a zero relative error $\varepsilon(\gamma)=|1-\dfrac{\alpha_{i-1,o}^t \alpha_{i+1,o}^t}{(\alpha_{io}^t)^2}| \to 0$, while the ratio $\dfrac{\alpha_{i-1,o}^t \alpha_{i+1,o}^t}{(\alpha_{io}^t)^2} = \dfrac{t_i^2}{t_{i-1}t_{i+1}}$

determines the corresponding discrete intervals, where the cooperation of these eigenvalues into a triplet is accomplished [1]. The above error evaluates the cooperative ability of the initial eigenvalues to form a triplet. At $\gamma \to 0$, $\varepsilon(0) \to 0$ if the nearest triplet's eigenvalues sequence $\alpha^t_{i-1,o}, \alpha^t_{io}, \alpha^t_{i+1,o}$ satisfies the relation $(\alpha^t_{io})^2 = \alpha^t_{i-1,o}\alpha^t_{i+1,o}$, which forms a geometrical progression $(....\alpha^t_{i+1,o}, \alpha^t_{io}, \alpha^t_{i-1,o})$. Such α^t_{io} represents an average geometrical value of two others, or a geometric "gold section" with $\alpha^t_{io} \cong 0.618\alpha^t_{i-1,o}$, which proves P3.1a. With growing γ, error $\varepsilon(\gamma)$ increases, at $\gamma \to 1$ the eigenvalues' triple looses its ability to cooperate, and the triplet disintegrates. In this situation, the sequence $\alpha^t_{io}, \alpha^t_{i+1,o}, \alpha^t_{i+2,o}, ...$ at the ratio $\alpha^t_{i+1,o} / \alpha^t_{i+2,o} = \gamma_2^\alpha$ for $\gamma \to 1$ forms the Fibonacci series with a "divine proportion" $PHI \cong 1.618$, satisfying $PHI \cong G + 1$ in P3.1b. ●

Comments 3.1. With a decay of the cooperative model (at $\gamma \to 1$), its current dimension n_i is changed. A minimal potential's growth (from the n_i dimension) corresponds to a possibility of adding the two cooperating eigenvalues (doublet) to the last triplet's eigenvalue of the n_i dimensional system, forming a new system of n_{i+1} dimension, and so on. Thus, we get the sequence of the feasible systems' dimensions n_i, n_{i+1}, n_{i+2} (with the different admissible $\gamma \in (0, 0.8)$, which satisfy a simple relation $n_{i+1} = n_i + 2$. If the cooperative dynamics of two previously composed subsystems n_i, n_{i+1} can provide the negentropy $N_{i+2} = n_{i+2} / 2$ necessary for generating the following subsystem n_{i+2}, then these dimensions become connected by relation $n_{i+2} = n_i + n_{i+1}$, which forms the Fibonacci sequence. Because the final IN's node encapsulates its total negentropy $N_i = n_i / 2, (\gamma = 0.5)$ and the ending eigenvalue is capable of transferring it toward other subsystems, both the considered eigenvalues and the dimensions form the Fibonacci sequence. The proportion PHI (at $\gamma \to 1$) expresses a completion of the cooperative dynamics in one subsystem and a potential start of the cooperation in a subsequent subsystem, exhibiting a border between the evolved subsystems. This also follows from the considered evolutionary dynamics [30] of a cyclic process, which consecutively creates the ordered dynamics (at $\gamma \to 0$) from the chaotic dynamics (at $\gamma \to 1$) and vice versa, or transfers the dynamic order into the disorder under the model's lifetime. Because the triplet's eigenvalues sequence forms a triplet's code, the ability of generating this code [31,1] is also limited by the conditions of P3.1a,b.

The VP minimax principle has the following meaning for the macrodynamics with above γ: *mini minimax* for $\gamma \to 0$, and *max minimmax* for $\gamma \to 1$. Therefore, the VP's law generates both well-known numbers G and PHI, which become not only a mutually connected but are also responsible for regularities of cooperative dynamics and the creation of new systems.

Let us study the elementary cooperation of two processes (i,k), represented by the corresponding IN_i, IN_k with *different* γ_i, γ_k and determine the cooperative MC.

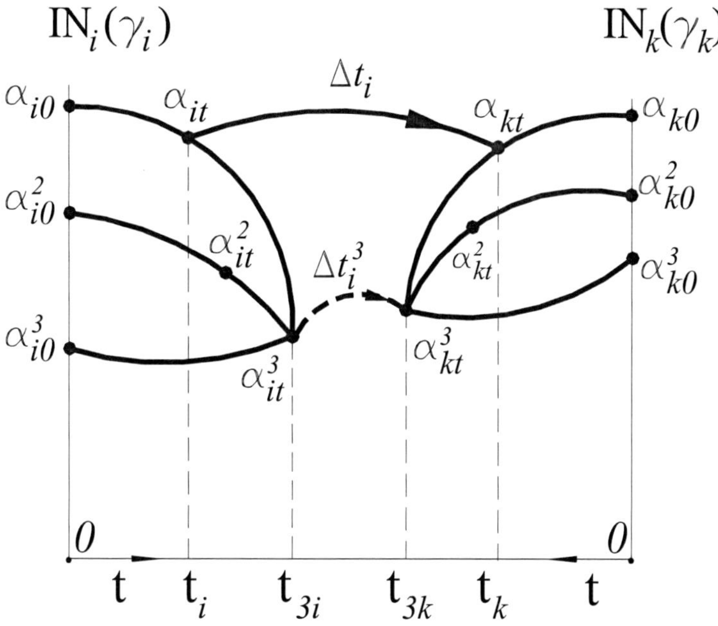

Figure 4. The cooperative schema for joining of two subsystems (i,k), represented by corresponding IN_i, IN_k with different γ_i, γ_k.

Proposition 3.2. Suppose the first eigenvalue α^1_{it} of the IN_i joins the first eigenvalue α^1_{kt} of the IN_k during a time interval Δt_i (Fig.4) according to the equation (2.2a,2.4), and assume that both $IN_i(\gamma_i)$, $IN_k(\gamma_k)$ are given, their dimensions n_i, n_k and $\alpha^1_{it}(\gamma_i, n_i)$, $\alpha^1_{kt}(\gamma_k, n_k)$ are known, as well as all other networks' eigenvalues.

Then the complexity's increment, reduced to the section's time t_i and related to the corresponding volume $v_i(t_i)$, acquires the form

$$MC^{\delta}_{ikt} = \mathbf{a}_o(\gamma_i)^2 \rho^1_{ik}\, \mathbf{a}(\gamma_i)/\,\mathbf{a}_o(\gamma_i)\mathbf{a}(\gamma_k)/\mathbf{a}_o(\gamma_k)$$

$$[1+ \rho^1_{ik}\, \mathbf{a}(\gamma_k)/\mathbf{a}_o(\gamma_k)\,[\mathbf{a}(\gamma_i)/\,\mathbf{a}_o(\gamma_i)\,]^{-1}]/v_i(t_i), \qquad (3.2)$$

determined by the invariants

$$\mathbf{a}(\gamma_i)=\alpha^1_{it}\, t_i,\ \mathbf{a}(\gamma_k)=\alpha^1_{kt} t_k\ ;\ \mathbf{a}_o(\gamma_i)=\alpha^1_{io} t_i,\ \mathbf{a}_o(\gamma_k)=\alpha^1_{ko} t_k\ , \qquad (3.2a)$$

and the relations

$$\alpha^1_{it} p_i = \alpha^1_{kt}, p_i = \alpha^1_{it}\exp(\alpha^1_{it}\Delta t_i)(2-\exp\alpha^1_{it}\Delta t_i)^{-1}, \ \alpha^1_{ko} = \alpha^1_{io}\rho^1_{\mathrm{ik}} \tag{3.2b}$$

Proof. From (3.2a,b) follow

$$\alpha^1_{io}\ p_i \mathbf{a}(\gamma_i)/\ \mathbf{a}_{\mathrm{o}}(\gamma_i) = \alpha^1_{ko}\,\mathbf{a}(\gamma_k)/\mathbf{a}_{\mathrm{o}}(\gamma_k), p_i \mathbf{a}(\gamma_i)/\ \mathbf{a}_{\mathrm{o}}(\gamma_i) = \rho^1_{\mathrm{ik}}\,\mathbf{a}(\gamma_k)/\mathbf{a}_{\mathrm{o}}(\gamma_k), \tag{3.3}$$

from which we have

$$p_i = \rho^1_{\mathrm{ik}}\,\mathbf{a}(\gamma_k)/\mathbf{a}_{\mathrm{o}}(\gamma_k)\,[\mathbf{a}(\gamma_i)/\ \mathbf{a}_{\mathrm{o}}(\gamma_i)\,]^{-1}. \tag{3.3a}$$

Applying (2.4) to equality

$$\dot{\alpha}^1_{it} = (\alpha^1_{io})^2\,(\mathbf{a}(\gamma_i)/\ \mathbf{a}_{\mathrm{o}}(\gamma_i)\,)^2\,(p_i + p_i^2), \tag{3.4}$$

and using (3.3), we get the Hamiltonian's derivation in the forms:

$$\begin{aligned}\dot{H}^1_{ik} &= (\alpha^1_{io})^2\ \rho^1_{\mathrm{ik}}\,\mathbf{a}(\gamma_i)/\ \mathbf{a}_{\mathrm{o}}(\gamma_i)\mathbf{a}(\gamma_k)/\mathbf{a}_{\mathrm{o}}(\gamma_k)\\ &[1+\ \rho^1_{\mathrm{ik}}\,\mathbf{a}(\gamma_k)/\mathbf{a}_{\mathrm{o}}(\gamma_k)\,[\mathbf{a}(\gamma_i)/\ \mathbf{a}_{\mathrm{o}}(\gamma_i)\,]^{-1}],\end{aligned} \tag{3.5}$$

$$\begin{aligned}\dot{H}^1_{ik} &= \mathbf{a}_o(\gamma_i)^2\,/\,t_i^2\ \rho^1_{\mathrm{ik}}\,\mathbf{a}(\gamma_i)/\ \mathbf{a}_{\mathrm{o}}(\gamma_i)\mathbf{a}(\gamma_k)/\mathbf{a}_{\mathrm{o}}(\gamma_k)\\ &[1+\ \rho^1_{\mathrm{ik}}\ \mathbf{a}(\gamma_k)/\mathbf{a}_{\mathrm{o}}(\gamma_k)\,[\mathbf{a}(\gamma_i)/\ \mathbf{a}_{\mathrm{o}}(\gamma_i)\,]^{-1}].\end{aligned} \tag{3.6}$$

and finally (3.2). ●

Comments 3.2. The obtained relations allow us to evaluate both the condition of cooperation (3.3a) and the cooperative complexity (3.2) by the INs' information invariants and the ratio ρ^1_{ik} of the starting cooperated eigenvalues. If the (i,k) cooperation is performed by joining the first IN_i - IN_k *triplets* during a time interval Δt_{3i} when the IN_i node's eigenvalue α^3_{it} cooperates with the IN_k node's eigenvalue α^3_{kt}, then the formula (3.3a) acquires the view

$$p_{3i}\,\mathbf{a}(\gamma_i)/\ \mathbf{a}_{\mathrm{o}}(\gamma_i) = \rho^3_{\mathrm{ik}}\,\mathbf{a}(\gamma_k)/\mathbf{a}_{\mathrm{o}}(\gamma_k). \tag{3.7}$$

Using the ratio of starting triplets' eigenvalues and the p_{3i}, expressed through Δt_{3i}, we get

$$\rho^3_{\mathrm{ik}} = \alpha_{3ko}\,/\,\alpha_{3io}\,,\ p_{3i} = \alpha^3_{it}\exp(\alpha^3_{it}\Delta t_{3i})(2-\exp\alpha^3_{it}\Delta t_{3i})^{-1}. \tag{3.7a}$$

For the triplets, by exchanging ρ^3_{ik} with ρ^1_{ik}, we come to formula for the MC complexity, analogous to (3.2). The related formulas follow from joining any other corresponding IN_i - IN_k nodes.

From eqs. (3.7),(3.3) we receive

$$p_{3i} / p_i = \rho^3_{ik} / \rho^1_{ik}, \tag{3.8}$$

which shows the proportionality between the time's parameters and the ratio's of cooperating eigenvalues. Within each IN, the ranged eigenvalues' ratios $\gamma^\alpha_{13}(\gamma_i) = \alpha^1_{i0} / \alpha^3_{i0}, \gamma^\alpha_{13}(\gamma_k) = \alpha^1_{k0} / \alpha^3_{k0}$, depending on a fixed $\gamma(\gamma_k, \gamma_i)$, are preserved. Because of that, the ratio $\rho^1_{ik} / \rho^3_{ik} = \gamma^\alpha_2(\gamma_k) = \gamma^\alpha_2(\gamma_i) = \alpha^i_{2t} / \alpha^i_{3t} = \alpha^{i=k}_{4t} / \alpha^{i=k}_{5t}(\gamma_i)$ (on the right side of (3.8)) is also fixed for both the INs' ranged cooperating eigenvalues. This leads to the fixed ratio (on the left side of (3.8)), or to the possibility of an *automatic* cooperation of all interacting IN_i - IN_k eigenvalues if the cooperation of any two eigenvalues of these networks has occurred. Actually, the automatic cooperation requires the sequential supply of produced information from each previous cooperating eigenvalue to each following cooperation. Such a supply is possible for any ranged cooperating sequence satisfying the VP, which leads to the automatic creation of each IN. If the ratio on the right (3.8) side is fixed for the IN_i - IN_k cooperation, then the MC complexity also gets fixed and stays the same for all cooperating eigenvalues. This means that the MC is changed only with the cooperation of eigenvalue having a distinctive γ. ●

Proposition 3.3. Let us consider the cooperative dynamics for a set of *triplets* with different γ, assuming that this set forms an IN's where the cooperation of a triplet's i eigenvalue α^i_{3t} with the triplet's k eigenvalue α^k_{3t} satisfies the relation (3.2a,b) and

$$\alpha^i_{3t} p^{ik}_{\Delta t} = \alpha^k_{3t}, p^{ik}_{\Delta t} = \exp(\alpha^i_{3t} \Delta t_{ik})(2 - \exp(\alpha^i_{3t} \Delta t_{ik})^{-1}. \tag{3.9}$$

Then, the entropy's consumption for the triplet k, corresponding (3.1), gets the invariant form:

$$\begin{aligned} g_k(\gamma_k, \gamma_i) &= 2\mathbf{a}(\gamma_i) / \gamma^\alpha_1(\gamma_k) + \mathbf{a}(\gamma_k) / \gamma^\alpha_2(\gamma_k) + \mathbf{a}(\gamma_k) \\ &= \mathbf{a}(\gamma_k) [2\mathbf{a}(\gamma_i) / \mathbf{a}(\gamma_k) (\gamma^\alpha_1(\gamma_k))^{-1} + (\gamma^\alpha_2(\gamma_k))^{-1} + 1], \end{aligned} \tag{3.10}$$

where the $g_k(\gamma_k, \gamma_i)$ series establish the information connection between any cooperating i , k triplets, modeling the IN's cooperative phenomena.

Proof. For such an IN, the α^k_{3t} is obtained by the cooperating pair ($\alpha^k_{4t}, \alpha^k_{5t}$) that forms the doublet

$$\alpha_{3t}^k \Rightarrow \alpha_{5t}^k = \alpha_{4o}^k \mathbf{a}(\gamma_k)/\mathbf{a}_o(\gamma_k)\ p_2^k = \alpha_{5o}^k \mathbf{a}(\gamma_k)/\mathbf{a}_o(\gamma_k),$$
$$p_2^k = \alpha_{5o}^k / \alpha_{4o}^k = \alpha_{5t}^k / \alpha_{4t}^k = (\gamma_2^\alpha(\gamma_k))^{-1}, \tag{3.11}$$

and the sequence of the cooperating triplet's eigenvalues $\alpha_{3t}^i(\alpha_{1t}^i, \alpha_{2t}^i), \alpha_{3t}^k(\alpha_{4t}^k, \alpha_{5t}^k)$, which is characterized by the ratios

$$\alpha_{3t}^k / \alpha_{3t}^i = \rho_3^{ki} = (\gamma_1^\alpha(\gamma_k))^{-1},\ \gamma_1^\alpha(\gamma_i) = \alpha_{1t}^i / \alpha_{3t}^i, \gamma_1^\alpha(\gamma_k) = \alpha_{3t}^i / \alpha_{3t}^k . \tag{3.12}$$

In such a sequence, the ratios $p_{\Delta t}^{ik} = p_2^k$ (and therefore the interval Δt_{ik}) are connected to γ_k through the parameter of multiplication $\gamma_1^\alpha(\gamma_k)$ of the adjoint triplet. This brings (3.1) to the related eqs. (3.10). ●

Comments 3.3. The $\mathrm{g_k}(\gamma_\mathrm{k},\gamma_\mathrm{i})$ component $2\mathbf{a}(\gamma_i)/\gamma_1^\alpha(\gamma_k)$ (which depends on the triplet's invariant and the nearest ratio $\gamma_1^\alpha(\gamma_k)$) measures the information *contribution* of a merged *phenomenon (defined by the ratio)*.

A *complex system represents a composition of these phenomena, which indicate an emergence of complexity*. The IN's structure allows numerical evaluation of these phenomena by the amount of generated *quantity and quality* of information. Because each k-triplet accumulates information $\mathrm{g_k}(\gamma_\mathrm{k},\gamma_\mathrm{i})=2\mathbf{a}(\gamma_k)$, (3.10) acquires the view $\mathbf{a}(\gamma_k)\,[2\mathbf{a}(\gamma_i)/\mathbf{a}(\gamma_k)+(\gamma_2^\alpha(\gamma_k))^{-1}+1]=2\mathbf{a}(\gamma_k)$, and we come to

$$2\mathbf{a}(\gamma_i)/\mathbf{a}(\gamma_k)\ (\gamma_1^\alpha(\gamma_k))^{-1}+(\gamma_2^\alpha(\gamma_k))^{-1}=1, \text{ or}$$
$$(1\text{-}(\gamma_2^\alpha(\gamma_k))^{-1})\,\gamma_1^\alpha(\gamma_k)=2\mathbf{a}(\gamma_i)/\mathbf{a}(\gamma_k), \tag{3.13}$$

which determines the condition, connecting γ_i, γ_k of the nearest triplets. Application of this condition allows us to get a chain of cooperating triplets, forming a *harmonic ensemble* of the sequentially coordinating frequencies, defined by the eigenvalues ratios. The computation *example* brings the following sequence of cooperating $(\gamma_1 \to \gamma_2 \to \gamma_3 \to \gamma_4) \Rightarrow (0.72 \to 0.42 \to 0.62 \to 0.5)$, which corresponds to the ratio's sequence of information frequencies: (4.08, 2.29),(3.69, 2.13),(3.89,2.21), where the first one starts from the last frequency of the triplet with γ_1=0.72. The computations according to formula (3.13) show that for a range of the feasible $\gamma \to$(0.0072-0.8), only a limited collection of $\gamma_k \to (0.4-0.7)$ exists with the frequency's ratios (4.08-2.049), which satisfies (3.13). In spite of these limitations, the absolute range of actual cooperating frequencies is essentially wider, depending on the specific frequencies' band for each cooperating subsystems according to its particular dimension and the locations of ending frequency, as well as this frequency band. For example, even for $\gamma = 0.5$,

γ_1^α=3.895, γ_2^α=2.215, and n=22, the frequency band is between 0.00015-500, i.e. 1:3300000. At the subsystems' cooperation, this band extends more radically. ●

Comments 3.4. Each triplet encapsulates three symbols representing a *code* of the cooperating eigenvalues $(\alpha_1,\alpha_2,\alpha_3)$, which carry the information transferred by these eigenvalues, evaluated by quantity of information 3**a**(γ). The model's control adds one more symbol (v_1) to this code: $(\alpha_1,\alpha_2,\alpha_3,v_1)$, evaluated by a total quantity of information 4**a**(γ), which possesses 1.44 bits at $\gamma=0.5$ (this is precisely the quantity of information necessary to encode each codon or each 3 nucleotides of 20 aminoacides [11]).

If the IN's starting eigenvalues' string is ranged by the invariant ratios of the nearest evaluated by quantity of information 3**a**(γ)eigenvalues $\gamma_{12}^\alpha(\gamma)=\alpha_1/\alpha_2=\alpha_3/\alpha_4,....,\gamma_{13}^\alpha(\gamma)=\alpha_1/\alpha_3=\alpha_3/\alpha_5,....$, then this four letter's code's word preserves also its sequence: $(\alpha_1,\alpha_2,\alpha_3,v_1)\Rightarrow(\alpha_3,\alpha_4,\alpha_4,v_2)\Rightarrow(\alpha_{i-1},\alpha_i,\alpha_{i+1},...)$, while the last symbol of the previous code transfers the node's information to the first symbol of subsequent code, and so on. This code possesses a *hierarchical* complexity for both the code's dynamic and geometrical structures [31]. Such a code consists of a set of the curved quasi-quadrate cells, whose sequence forms a hyperbolic cellular structure located along the IN nodes in the information geometric space [31, 1].

The quantity of information $\mathbf{a}_o$, accumulated by each following triplet includes the structural information from the previous triplet, which also depends on the node's geometric location. The corresponding hierarchical information measure for each node has an increasing *quality* of information. Even though the triplet's code preserves both the code word's information and the code symbols' sequence, each IN's node's code consists of a *new* sequence of distinctive *frequencies* $f_{i-1}\sim\alpha_{i-1}, f_i\sim\alpha_i, f_{i+1}\sim\alpha_{i+1}$, which are capable of encoding all initial string symbols. A single triplet can generate various combinations of the four code symbols at different γ being analogous to the DNA triplet's code [22, 31]. A system's genetic information is encoded by a sequence of multiple combinations of these four symbols.

Comments 3.5. Let us define a subsystem as an IN's structure with a different but fixed γ for each IN. Then the considered IN's invariants (depending on γ) determine the parameters of any particular subsystem's triplet and its complexity. The subsystems' set depends on the collection of admissible $\gamma\rightarrow(0.0-0.8)$, which follows from the model's simulation [30] and is defined by the condition of a subsystems' stability, consistent with the preservation of the total model's entropy *S*. This leads us to the conclusion that we can rename a collection of subsystems into a *system*, which keeps the total *S* constant. Because *S* consists of two parts: imaginary *I* and real information *R*, while their ratio is $\gamma=I/R$ [22], we get $S=(1+\gamma)R$, or $\gamma=S/R-1$. At γ=0.8, *S/R*=1.8, and at γ=0, we come to *S/R*=1. Thus the R/S range is between 1 and 0.555, and the *I/S* range is between 0 and 0.445. At the minimal feasible γ=0.0072, the practical rage is narrower: *R/S*=0.9928 to 0.555, and *I/S*=0.0072 to 0.445. The number of diverse subsystems within a single system, distinguished by the γ of a starting eigenvalue's sequence, is determined by the ratio (0.8 − 0.0072)/0.0072=110.111..., while each subsystem, having the same fixed γ, might consist of a maximal element's number n=90 [22]. This means that each system may include any of 110^{90} different combinations of those 90 subsystems representing the maximal number of diverse systems, having the distinct information characteristics and are capable of producing specific codes. •

4. The Equations of the Space Information Cooperative Dynamics, Information Attraction, and Complexity

Here we examine the role and contribution of space information dynamics to cooperation and complexity.

The information cooperative dynamics include the equalization of the model's eigenvalues and the consolidation of the equal eigenvectors into a common information unit [1]. The regular and needle controls, applied as a function of time, provide the equalization of the eigenvalues, while the consolidation of eigenvectors requires a space movement. Let's analyze this process, considering the cooperation of two local, dynamic, space models (triplets) transformed into diagonal form:

$$dx_i / dt = \lambda_i x_i, \; dx_k / dt = \lambda_k x_k, \tag{4.1}$$

where x_i, x_k are the model's macrocoordinates, λ_i, λ_k are the eigenvalues corresponding to the matrices

$$A_i^l = \| \lambda_i \|_{i=1}^3, A_k^l = \| \lambda_k \|_{k=1}^3 . \tag{4.1a}$$

If the equalization of the model's eigenvalues had been reached at the moment t_i+o:

$$\lambda_i(t_i+o) = \lambda_k(t_i+o) \tag{4.2}$$

during the cooperative dynamics, then the cooperation is accomplished after the coincidence of the model's eigenvectors $\vec{x}_i$, $\vec{x}_k$, generally defined in Riemann's space R^3 with the invariant metric

$$ds^2 = g^3 d\vec{x}_i^T d\vec{x}_i, g^3 = \| g_{ij}^3 \|_{i,j=1}^3, \vec{x}_i = (x_1^i \overline{e}_1^i, x_2^i \overline{e}_2^i, x_3^i \overline{e}_1^i), \; g^3 \neq 0, \tag{4.3}$$

where the vectors $\overline{e}_i = (\overline{e}_1^i, \overline{e}_2^i, \overline{e}_3^i), \overline{e}_k = (\overline{e}_1^k, \overline{e}_2^k, \overline{e}_3^k)$, represent the local orths (fundamental orthogonal triad's vectors), defined in the considered Riemann's space R^3, whereas each of $\vec{e}_i, \vec{e}_k$ is a tangent vector to a curve –trajectory of the eigenvector space movement in R^3.

(We assume that the phase speed's vector and the corresponding state's vector are defined at each R^3 point).

Definition 4.1. Let's move vector $\vec{x}_i$ toward equalization with vector $\vec{x}_k$, considering an increment $d\vec{x}_i$ on the path interval $ds_i = ds$ between $\vec{x}_i$, $\vec{x}_k$. Then, at the VP fulfillment, the metric (4.3) in the cooperative motion is described by the equation of a geodesic line in R^3, written in the traditional coordinate form [32]:

$$\frac{d^2 x_j^i}{ds^2} = -K_{mn}^j \frac{dx_m^i}{ds}\frac{dx_n^i}{ds}, j, m, n = 1,2,3, \tag{4.4}$$

where K_{mn}^j is a Gaussian curvature at a $\delta\vec{x}_i(\delta s_{ik})$-locality, which along the geodesic is defined by the vector equation

$$d\vec{x}_i / ds = -K_i\vec{x}_i \tag{4.5}$$

K_i is the operator's representation of the K_{mn}^j components. At a positive cooperative speed $d\vec{x}_i / ds$>0 and $\vec{x}_i$>0, the space gets a negative curvature K_i<0 corresponding an attraction; at K_i>0, we get repulsion. That's why the vector speed in (4.5) defines an *intensity of information attraction (or a repulsion)*, determined by the space curvature K_i. (The vector's derivations here and below mean that the derivations for each vector's components are similar to (4.4). Along the geodesic line (4.4), the vector undergoes a parallel displacement, where the vector increment according to (4.5) depends linearly and homogeneously upon the increments of the related coordinates, and only the first vector of curvature is significant). ●

Comments 4.1. The space curvature, defining the cooperative speed, is a fundamental source of both the complexity and the cooperative dynamics. At the VP execution, the K_i satisfies the following expressions via the fundamental tensor g^3:

$$K_i = 1/2(g^3)^{-1} dg^3 / ds\ . \tag{4.6}$$

The formula for a scalar curvature K^i is expressed through the determinant of the fundamental tensor in this eigenvector's space:

$$K^i = (\sqrt{g}\ )^{-1}\partial\sqrt{g}\ /\partial s\ , g\ = \det \| g_{ij}^3 \|, (\sqrt{g}\ )^{-1} = | x_1^i x_2^i x_3^i |. \tag{4.7}$$

●

Proposition 4.1. Assume that the eigenvalues of matrix $A_i^l = \| \lambda_i \|_{i,j=1}^3$ (4.1a) are simple with the distinctive roots, and the related eigenvectors $(x_1^i \overline{e}_1^i, x_2^i \overline{e}_2^i, x_3^i \overline{e}_1^i)$ are mutually orthogonal (the matrix satisfies the conditions for a Vandermonde matrix [33]). Then the curvature in the considered geometrical phase space (4.7) has the form

$$K^i = -3\frac{\alpha_o^i d\alpha_o^i}{ds}, \tag{4.8}$$

where $\alpha_o^i = Re\lambda_o^i, \lambda_1^i = \lambda_2^i = \lambda_3^i = \lambda_o^i$.

Proof. The above matrix's conditions allow expressing the space components (x_1, x_2, x_3) for each eigenvector in (4.3) through the eigenvalues:

$$x_1 = \{1, \lambda_1^i, (\lambda_1^i)^2\}, x_2 = \{1, \lambda_2^i, (\lambda_2^i)^2\}, x_3 = \{1, \lambda_3^i, (\lambda_3^i)^2\},\ \lambda_{j=1,2,3}^i \neq 0, \tag{4.9}$$

According to the cooperative dynamics, the eigenvectors $(\vec{x}_1, \vec{x}_2, \vec{x}_3)$ underwent the consolidation into a single eigenvector $(\vec{x}_1, \vec{x}_2, \vec{x}_3) \rightarrow \vec{x}_o, \vec{x}_o = (1, \alpha_o, \alpha_o^2)\vec{e}_o$ with the real eigenvalues $\alpha_o = Re\lambda_o$. The eigenvector's space coordinates are formed in the process of the equalization and transformation of the corresponding eigenvalues $\{(1, \lambda_1^i, (\lambda_1^i)^2), (1, \lambda_2^i, (\lambda_2^i)^2), (1, \lambda_3^i, (\lambda_3^i)^2)\} \rightarrow (1, \alpha_o^i, (\alpha_o^i)^2)$, which specifically take the form: $\{(1, \lambda_1^i, (\lambda_1^i)^2) = (1, \lambda_2^i, (\lambda_2^i)^2) = (1, \lambda_3^i, (\lambda_3^i)^2)\} \rightarrow (1, \alpha_o^i, (\alpha_o^i)^2)$. Applying (4.7) we get

$$(\sqrt{g})^{-1} = | 1\alpha_o^i (\alpha_o^i)^2 |, \partial\sqrt{g} / \partial s = \partial(\alpha_o^i)^{-3} / \partial s = -3(\alpha_o^i)^{-2} \partial(\alpha_o^i) / \partial s. \tag{4.9a}$$

Assuming that vector $\vec{x}_o^i = (1, \alpha_o^i, (\alpha_o^i)^2)\vec{e}_o^i$ is cooperating with the analogous eigenvector $\vec{x}_o^k = (1, \alpha_o^k, (\alpha_o^k)^2)\vec{e}_o^k$, at $| 3\alpha_o^i | = | \alpha_o^k |$, the curvature (4.7) gets the form (4.8). •

Comments 4.5. Unification of the cooperating eigenvectors produces an *information cooperative mass* M_i, which is defined by the information value of an elementary phase volume υ forming at the cooperation:

$$M_i = \upsilon 3 div S_i, \tag{4.10}$$

where $3divS_i$ is the instant quantity of information generated at joining of the above three eigenvectors.

Using the relations $cdivS_i = -\partial S_i / \partial t, H^i = -\partial S_i / \partial t, \dot{H}^i / \upsilon = MC_t^\delta$ (at $\dot{H} / \dot{V}^\delta = MC_t^\delta, \dot{V}^\delta = \upsilon$), space speed $c = ds / dt$, and (4.10), we write curvature (4.8) in the form

$$K^i = -MC_t^\delta M_i, \quad (4.11)$$

where $$M_i = 3H^i c^{-1}\upsilon, M_k = H^k c^{-1}\upsilon, H^k = \alpha_o^k = 3\alpha_o^i = 3H^i, \quad (4.11a)$$

and α_o^k, H^k are the model's eigenvalue and Hamiltonian accordingly after the cooperation at $\overline{x}_k$.

The information, localized in the space (through $H^k = \alpha_o^k = -\partial S_k / \partial t,\ d\alpha_o^i / ds = dH^i / ds$), generates an increment of fundamental metric's tensor (in the forms (4.6, or 4.7)), analogous to the "metric's relative density", which changes the curvature. It is clearly seen that the curvature K^i turns to zero if any of $\partial S_i / \partial t = -\alpha_o^i, \partial^2 S_i / (\partial t)^2 = -\dot{\alpha}_o^i$, equals to zero, or $c \to \infty$. Because $\partial S_i / \partial t$ and $\partial^2 S_i / (\partial t)^2$ are the entropy production and its derivative (or an amount of the entropy's acceleration) accordingly, memorized at the cooperation, the curvature conceals the corresponding *information of cooperation*, possessing the cooperative system's complexity and information mass. Moreover, the curvature is a result of both the cooperation and a memorized information mass, or a cooperated information mass, which generates complexity and the mass. The cooperation decreases uncertainty and increases information mass. Considering the negative entropy production $-\partial S_i / \partial t\ c_i = divS_i$, let us find a maximal admissible space speed c_i, carrying $divS_i$. Using the IN's invariant's relation $\partial S_i / \partial t = |\mathbf{a}_o| / t_i$, at the minimum admissible $t_i \approx 2.66 \times 10^{-16}$ (defined by the minimal time-interval of the light wave), and a maximum of the normalized $div^* S_i = divS_i / |\mathbf{a}_o| \approx 1/137$ [22], we get the maximal information speed

$$c_{\max}^i = (t_i div^* S_i)^{-1} \quad (4.12)$$

evaluated by $c_{\max}^i \approx 51.504 \times 10^{16}$. This maximum restricts the cooperative speed and a minimal information curvature at other equal conditions. From (4.12) also follows that a bound into space information ($div^* S_i$) limits the maximal speed of incoming information (beyond the light speed, even the information is a nonmaterial substance), imposing an *information* connection on the time and space.

The physical mass-energy that satisfies the law of preservation energy (following the Einstein's equation [32]), distinguishes from the information mass (4.10), which does not obey this law.

Accoding to [32] the multiplication of a mass on $\sqrt{g}$ defines the mass *density* M_i^*, which following (4.8) and (4.10), acquires the view

$$M_i^* = (\alpha_o^i)^{-3} \upsilon 3\alpha_o^i = 3\upsilon(\alpha_o^i)^{-2} . \tag{4.13}$$

In the simulated IN hierarchy [22, 31], the values of cooperating α_o^{kl} decrease with a growing number of the hierarchical level $l = 1,2,..,k,..,m$, which leads to an increasing of $M_i^*(l)$. •

Comments 4.5. According to equation (4.1), function $\alpha_o^i = (x_i)^{-1} \partial x_i / \partial t$ corresponds to a relative information flow. Consider x_i to be an elementary information charge q_i, whose flow $\partial q_i / \partial t = -c_i div q_i$ characterizes the charge's divergence from a volume $\upsilon_i = \upsilon$, we get relation $-q_i(div q_i)^{-1} = q_i^*$ describing a relative concentration of the charge within the volume. With growing the divergence, the relative information charge q_i^* decreases. Writing the mass density M_k^*(4.13) in the form $M_k^*(c_k)^3 = (q_k^*)^2 \upsilon_k$ and taking into account $\upsilon_k = \dot{V}_k = c_k f_k$, where f_k is an elementary cross section of volume υ_k, we get

$$M_k^*(c_k)^2 = (q_k^*)^2 f_k , \tag{4.14}$$

which is associated with an internal information charge's *power,* carried by an elementary space square.

With growing the divergence, a concentration of the power diminishes, indicating a decline of a *quality* of the information power. The information power at $c_k = c_{\max}^k$:

$$M_k^*(c_{\max}^k)^2 = M_{k c_k=1}^* 2.6526 \times 10^{35} ,\ M_{k c_k=1}^* = [(q_k^*)^2 f_k]_{c_k=1} \tag{4.14a}$$

reaches a maximum at a fixed $M_{k c_k=1}^*$,whose mass's density characterizes the information power's concentration at a fixed $c_k = 1$, or it describes a related power's "quality" (connected to a decrease of a uncertainty density). The power, corresponding to the information equivalent of speed of light $c_{ko} = t_{ko}^{-1} = 0.375 \times 10^{16}$:

$$M_{ko}^* c_{ko}^2 \approx 1.14133 \times 10^{32} M_{k c_k=1}^* \tag{4.15}$$

is in $(137)^{-2} \approx 5.328 \times 10^{-5}$ time less than the maximal $M_k^*(c_k^m)^2$.

The information power for a *relative speed* $c_i^* = c_k / c_{max}$:

$$M_k^*(c_k / c_{max})^2 = [(q_k^*)^2 f_k]_{c=c_k} 0.377 \times 10^{-35}, \tag{4.15a}$$

at $c_k / c_{max} = 1$ defines the related information mass $M_{kr}^* = M_k^*(c_k / c_{max} = 1)^2$, which is equal to $M_{k\,c_k=1}^* = [(q_k^*)^2 f_k]_{c_k=1}$. From that we get the ratio

$$p_{c_k}^{kn} = [(q_k^*)^2 f_k]_{c=c_k} / [(q_k^*)^2 f_k]_{c_k=1} = 2.6526 \times 10^{35}, \tag{4.15b}$$

or a *relative information power*, where, in general, both components are unequal:

$$[(q_k^*)^2]_{c=c_k} \neq [(q_k^*)^2]_{c_k=1}, [f_k]_{c=c_k} \neq [f_k]_{c_k=1}. \tag{4.15c}$$

Applying the IMD approach to equation (4.1), we come to its information analogy: $\dot{q}_i = \rho_{qi} X_{qi}$, where $X_{qi} = grad_{qi} S_i$ is an information force acting on a charge q_i within a flow $\dot{q}_i$, and ρ_{qi} is a medium's conductivity to transfer the flow. The IMD equation for the VP's natural constraint [22, 1] in the form $\partial X_i / \partial x_j = -2 X_i X_j$, at $X_i = \sigma_i \alpha_i, X_j = \sigma_j \alpha_j$, with the corresponding coefficients of a resistance: $\sigma_i = (\rho_{qi})^{-1}, \sigma_j = (\rho_{qj})^{-1}$ and the information flows accordingly (connected to the information mass (4.10)): $\alpha_i = M_{ki} c_i / \upsilon_i, \alpha_j = M_{kj} c_j / \upsilon_j$, lead to equations

$$F_{ij} = G_m M_{ki} M_{kj},\ F_{ij} = -1/2 \partial X_i / \partial x_j, \tag{4. 16}$$

where the parameter $G_m = \sigma_i^f / \sigma_i^f, \sigma_i^f = \sigma_i / f_i, \sigma_j^f = \sigma_j / f_j, f_i = \upsilon_i / c_i, f_j = \upsilon_j / c_j$ characterizes an ability of a medium to resist for a mutual exchange of the information flows at the interaction of the information mass M_{ki}, M_{kj} under the action of a force F_{ij}; σ_i^f, σ_i^f are the medium's relative coefficients of resistance (related to the elementary squares f_i, f_j accordingly). Equation (4.16) determines the force acting *between* the masses, which are generated by the corresponding space curvatures (4.10) in the interactive cooperative dynamics. The mass is formed by an influx of information charges $\{q_i\}$, and the force F_{qij}, controlling the flows *interaction*, where the force F_{qij} is a derivative of the elementary information force X_{qi}:

$$F_{qij} = -1/2 \partial X_{qi} / \partial q_j. \tag{4.16a}$$

If q_i represents an elementary information fermion (particularly a graviton), then X_{qi} corresponds to an elementary information boson, acting on q_i, which is a physical analogy to an elementary gravitation particle that carries a gravitation force. (A physical analogy of gravitation boson is unknown).

Physical matter's particle a boson affects a fermion like a "messenger", which transfers information. Physics know four different bosons, which communicate with four different fermions, transferring the messages in some information code. We may assume that the message code is based on the four symbols optimal information code [22,31], which can cover the four pairs of communicating particles.

The Nature has already invented the information four letters DNA code.

Thus, the information boson-fermion's pairs' communication code pretends to be a *united information language, which describes all known four elementary interactions, and information systems theory could serve as an information approach to a theory of everything.*

For example, such a code could be memorized by gluons, holding a silent in the early Universe. Gluons, as the elementary color exchange particle, underlying the particle interactions, are represented by eight pairs of bi-colored charges-octets (carrying three kinds of color and anti-color units). The generated color charges specify the binding exchange forces, acting between the interacting quarks and gluons (as an analogy of needle control). This means that a specific color force can be *encoded by any pair of such a three bits' information code* (whose binary digit handles eight messages), which is memorized in anticipation of interactions. The triplet's code can encode these *three* color-anticolor forces, which binds the quarks' color *triplet.*

Example. Let's find the curvature for an elementary triplet, formed by joining three of the cones' vertices with the fundamental tensor's coordinates [22]:

$$\begin{aligned} g_1 &= \cos\psi\cos(\varphi\sin\psi) \\ g_2 &= \sin\psi\cos\varphi\cos(\varphi\sin\psi) - \sin\varphi\sin(\varphi\sin\psi) \\ g_3 &= \sin\psi\sin\varphi\cos(\varphi\sin\psi) + \cos\varphi\sin(\varphi\sin\psi) \end{aligned} \tag{4.17}$$

where φ is the projection of the space angle for the spiral's radius vector ρ at the cone's base (Fig. 2c), ψ is the angle at the cone's vertex. The invariant metric of information space has the following expression via the model's information invariant, defined at the moments of cooperation t_i:

$$mes \,||\, ds \,||= mes \,||\, \partial\rho_i(t_i)\partial\varphi_i(t_i) \,||= mes \,||\, \mu(t_o) \,|| = mes \,||\, inv_{\inf} \,|| = \mathbf{a}, \tag{4.18}$$

where mes is the information measure of the metric, ρ_i is the radius vector at the i-cone's base: $\rho_i = b_i \sin(\varphi_i \sin\psi_i)$, b_i is the parameter for each cone's radius vector: $b_i = \mu_i / \varphi_i$, μ_i is a piece of the dynamic trajectory at the cone's surface. At the moment of cooperation, the above vector and the angles acquire the increments:

$$\delta\rho(t_i) = b(\mu(t_i)),\ \delta\ \varphi(t_i) = \pi\ ,\ \delta\ \psi(t_i) = \pi/6\ ,\ \delta\varphi(t_i)\sin\delta\psi(t_i) = \pi/2\ , \quad (4.18a)$$

and $mes\,||\,\mu(t_o)\,||$ obtains the invariant informational measure $mes\,||\,inv_{\inf}\,||$=**a** (4.18b). Each of the IN cell-code's [1] space metric ds_m, defined by the cell m diameter, acquires the invariant information measure $mes_{\inf}\,||\,ds_m\,||= \mathbf{a}_m(\mathbf{a},\psi_m)$, which depends upon both the invariant $\mathbf{a}\,(\gamma)$ and the cell space angle $\psi_m\ (\gamma)$ of its location within the IN.

Writing the formula (4.7) for the metric's components (4.17) at $\psi_i = \psi = const$, we have

$$K_{123} = 1/2\,|\,g_j\,|^{-1} \sum_{j=1}^{3} \partial g_j / \partial\varphi\ . \quad (4.19)$$

By taking the derivative $\partial g_j / \partial\varphi$ of $g_j = (g_1, g_2, g_3)$ and considering $\partial g_j / \partial\varphi$ at the condition (4.16), after the substitution to (4.19), we get $K_{123} \approx -0.4$. We come to the information attraction of the triplet's eigenvectors. Finding the information mass and complexity requires to know the triplet's cone phase volume [22]: $\dot{V}_i(t) = (c_i)^3 t^2 \pi^{-2} (tg\psi_i)^{-1}, t = \Delta t_i$, where the triplet's space speed c_i is defined by applying a natural constrain $\Delta S(\mu(t_i)) \to 0$ to the triplet's space geometry. This determines c_i=c_i(Δt_i). For instance, at $i = 3, \Delta t_3 = t_3 - t_1 = 1.2$, we get $(c_3)^{-1} \approx 0.0125$ and $\dot{V}_3(\Delta t_3) \approx 0.1957 \times 10^{-3}$. Using formula (4.13) in the form

$$M_i^*(\Delta t_3) \approx 3\,\dot{V}_3(\Delta t_3)\ (\mathbf{a}/\,\Delta t_3)^2\ (c_3)^{-1}$$

we get $$M_i^*(\Delta t_3) \approx 0.169 \times 10^{-3}.$$

The calculated complexity in (4.11) leads to $MC_{t_3}^{\delta} = -0.4435 \times 10^{-3}$ and the corresponding curvature is $K^i = -0.433$.

Therefore both formulas (4.7) and (4.11) for the curvature illustrate the comparative results with a relative (to their average) error $\pm 4\%$, even at a rough approximation.

The equations (4.8, 4.15) show how the curvature of the geometrical space R^3, determined by the space fundamental tensor, enfolds an inner uncertainty-entropy, becoming an attribute of cooperative dynamics.

The physical and informational attractions are determined by the same curvature of the geometrical space, which are both a source of attraction and a carrier of information. Since a physical space's curvature generates gravitational attraction, the gravitational and informational attractions are connected by an inner uncertainty, enclosed by both physical and informational fields. Because the cooperation requires a curvature, it causes both uncertainty and curvature. This means that the curve space is a *result* of the cooperative dynamics, whereas both uncertainty and curvature are inseparable and mutually dependable space's attributes. The curved space embraces the cooperated information structures and the IN's information codes, which the geometrical curvature enfolds. Therefore, the IN's code is a primary information structure of a curved space. Actual formal originator of cooperative dynamics is the variation principle [1], which also causes the formation of the curved space, the code, and governs the evolutionary law.

Acknowledgments

Author thanks Dr. Michael Talyanker for computer design of the cooperative network and the art works; Ms. Dina Treyger and Ms. Sasha Moldavsky for help in editing this paper.

References

[1] Lerner V.S. The evolutionary dynamics equations and the information law of evolution, *International Journal of Evolution Equations* , Vol.2, No.4, 2007.

[2] Kolmogorov A., N. Three approaches to the quantitative definition of information. *Problems Information. Transmission*, **1** (1):1-7, 1965.

[3] Kolmogorov A., N. *Information Theory and Theory of Algorithms*, Selected Works, Nauka, Moscow,1987.

[4] Chaitin G., J. Computational Complexity and Godel's incompleteness theorem, *SIGACT News*, **9**:11-12, 1971

[5] Chaitin G., J. Information-theoretic computational complexity. *IEEE Trans. Information Theory*, **IT-20**:10-15, 1974.

[6] Bennett C., H. Logical depth and physical complexity. In R. Herken, ed. *The Universal Turing Machine*: 227-258, Oxford University Press, 1988.

[7] Bennett C., H. How to define complexity in physics, and why. In W.H.Zurek, ed., *Complexity, Entropy and Physics of Information:* 137-148, Addison Wesley, 1991.

[8] Lopez-Ruiz R., Mancini H.,L. and Calbet, X. A statistical measure of complexity, *Phys. Letters A*, **209**: 321, 1995.

[9] Calbet X. and Lopez-Ruiz R. Tendency toward maximum complexity in an isolated non-equilibrium system, *Phys. Review E* **6**, 066116(9), 2001.

[10] Solomonoff R.J. Complexity-based induction systems: comparisons and convergence theorems, *IEEE Transactions on Information Theory*, **24**: 422-432, 1978.

[11] Ebeling,W, Jimenez-Montano,M.,A. On Grammars,Complexity and Information Measures of Biological Macromolecules, *Journal Mathematical Bioscience*, **52**: 53-71, 1980.

[12] Huberman, B., A; Hogg, T. Complexity and Adaption, *Physica D*, **22**, 376-384, 1986.

[13] Traub,JF, Wasilkowski,GW, Wozniakowski,H. Information-Based Complexity, Academic Press, 1988, London.

[14] Crutchfield,JP Young,K, 1989 , Inferring Statistical Complexity, Physics *Review Letters*, **63**:105.

[15] Lopez,L., R Caufield,LJ, A Principle of Minimum Complexity in Evolution, *Journal Lecture Notes in Computer Science*, **496**: 405-409, 1991.

[16] Grassberger P. *Information and Complexity Measures in Dynamical Systems*, in Information Dynamics, Eds.Atmanspacher H, Scheingraber H. Plenum Press, New York, 15-33, 1991.

[17] Atmanspacher H, Kurths J, Scheingraber H, Wackerbauer R, Witt A. Complexity and meaning in nonlinear dynamical systems, *Open systems and Information Dynamics*, **1**, 269-289, 1992.

[18] Nicolis, G. and Prigogine, I.: *Exploring complexity*. New York: W.H. Freeman, 1989.

[19] Gell-Mann M. Lloyd S. Information Measures, Effective Complexity and Total Information, *Journal Complexity*, **2**, 44-52,1996 .

[20] Bar-Yam, Y. Multiscale complexity/entropy, *Advances in Complex Systems*, **7**(1):47–63, 2004.

[21] Lerner V.,S. Dynamic approximation of a random information functional, *J. Mathematical Analysis and Applications*, **327**:494-514, 2007.

[22] Lerner V.,S. *Variation Principle in Informational Macrodynamics*, Kluwer, 2003.

[23] Feynman R.,P. and Hibbs A.R. *Path Integral and Quantum Mechanics*, McGraw Hill, 1965.

[24] Prigogine I. *From being to becoming: time and complexity in physical sciences*, W.H. Freeman and Co., San Francisco, 1980.

[25] Luan Chang-Fu, Entropy of Baker's Transformation,*Chinese Phys. Lett.* **20:** 392-394, 2003.

[26] Stratonovich R., L. *Theory of information*, Sov. Radio, Moscow, 1975.

[27] Lin J. Divergence measures based on the Shannon entropy. *IEEE Trans. on Information Theory*, **37**(1):145-151,1991.

[28] De Groot S.,R. and Mazur P. *Non-equilibrium Thermodynamics*, N. Holland Publ. Co., Amsterdam, 1962.

[29] Lerner V.,S. Macrosystemic Modelling and Simulation, Journal *Systems Analysis-Modeling-Simulation*, **28**(5):149-184, 1997.

[30] Lerner V.,S. Information Functional Mechanism of Cyclic Functioning. *Journal of Biological Systems,* **9**(3):145-168, 2001.

[31] Lerner V.,S. Information Geometry and Encoding the Biosystems Organization, *Journal of Biological Systems*,**13**(2):1-41, 2005.

[32] Einstein A. *The meaning of relativity*, Princeton University Press, Princeton, 1921: 69-71.

[33] Ganthmacher F.,R. Theory of Matrices, *Nauka*, M, 1967:395-396.

In: Progress in Evolution Equations
Editor: Gaston M. N'Guerekata, pp. 65-90

ISBN: 978-1-60456-328-3

Chapter 4

ON MODIFIED LAGRANGE MULTIPLIER RULE AND ITS APPLICATION TO THE REGULARITY OF MAGNETIC FLUX FUNCTION IN NUCLEAR FUSION

Takayuki Yamauchi*
Department of Mathematics and Computer Science, Lincoln University,
1570 Baltimore Pike, Lincoln University, PA19352, USA

Abstract

A state of equilibrium of a plasma flow in a nuclear fusion reactor is achieved as a minimizer of the magnetic flux energy over a class of rearrangements of a prescribed profile magnetic flux function. It is known that a Lipschitz continuous minimizer exists when the cross section of the reactor is convex. The classical Lagrange Multiplier Rule with a functional constraint is modified and is applied to the Kruskal-Kulsrud Principle to derive a non-degenerate Euler-Lagrange equation satisfied by a minimizer of the energy functional that arises in magnetohydrodynamics (MHD). A sufficient condition for the existence of a smooth minimizer is presented in this work. Under this sufficient condition, an explicit Euler-Lagrange equation that is compatible with numerical algorithms is obtained within the framework of the Lagrange Multiplier Rule.

Key Words: Magnetohydrodynamics, Lagrange Multiplier Rule, Rearrangement of functions, Partial differential equations of elliptic type, Optimization.

AMS subject class: 76W05,35Q72,35J35,35J60,49K30,49K20

1. Introduction

In 1958, Kruskal and Kulsrud [29] characterized an equilibrium solution to axi-symmetric magneto-hydro-dynamics (MHD) problems as an energy minimizer (possibly a stationary point) of the energy functional subject to two sets of continuously many constraints associated with the preservation of the total mass inside flux tubes and the total magnetic

*E-mail address: tyamauchi@lincoln.edu

flux.

The study of an equilibrium solution to axi-symmetric MHD problems and other fluid dynamics problems has been closely tied with the development of the theory of rearrangement of functions [1] [2] [10] [11] [12] [13] [14] [15] [16] [18] [30] [31] [32] [33] [34] [35] [36][37] [38] [40] [42] [43] [44] [46] [45] [52]. Statistical equilibrium models of an equilibrium solution to axi-symmetric MHD problems have been studied in [19] [9] [27] [28].

Based on the variational principle of Kruskal and Kulsrud, in 1990, Turkington, Eydeland, and Spruck [21] and Turkington, Eydeland, Lifshitz, and Spruck [22] showed that the slow evolution problem for an axi-symmetric plasma-vacuum system in MHD can be reduced to a quasi-equilibrium problem at each instant of time, and that it has a natural variational structure in which the total magnetic flux and mass are conserved within every magnetic surface.

A state of equilibrium of an axi-symmetric plasma flow in a nuclear fusion reactor is achieved as a minimizer of the magnetic flux energy over a class of rearrangements of a prescribed smooth profile magnetic flux function. The prototype of this problem has the following form (cf: [21]).

$$(P)_\infty : E(u) \equiv \frac{1}{2}\int_\Omega |\nabla u|^2 dx \to \min, u \in M_\infty(\overline{u}) \tag{1}$$

where $\Omega \subset \mathbf{R}^2$ is open and bounded, $u \in H_0^1(\Omega)$, $\overline{u}$ is a smooth prescribed function defined on Ω, $0 \le u(x) \le b, \forall x \in \Omega$, $0 \le \overline{u}(x) \le b, \forall x \in \Omega$ and $M_\infty(\overline{u})$ is a class of functions equimeasurable with $\overline{u}$.

They discretized the range $[0,b]$ into n finite elements $0 = \sigma_0 < \sigma_1 < \cdots < \sigma_{n-1} < \sigma_n = b$, and defined n constraints

$$F_i(u) = \frac{1}{2}\int_\Omega [(u-\sigma_i)_+^2 - (u-\sigma_{i+1})_+^2]dx, 1 \le i \le n. \tag{2}$$

They minimized the energy integral $E(u)$ over the space

$$M_n(\overline{u}) = \{u \in B_X | F_i(u) = F_i(\overline{u}), 1 \le i \le n\}. \tag{3}$$

They obtained the following minimization problem with finitely many constraints.

$$(P)_n : E(u) \equiv \frac{1}{2}\int_\Omega |\nabla u|^2 dx \to \min, u \in M_n(\overline{u}) \tag{4}$$

They then obtained the following regularity result.

Proposition 1.1. *(Theorem 2.2 in [21]) If $\overline{u}$ and $0 = \sigma_0 < \sigma_1 < \cdots < \sigma_{n-1} < \sigma_n = b$ are given so that $F_i(\overline{u}) > 0, \forall i \in \{1, \cdots, n\}$,*

then there exists a minimizer u of (4) satisfying

$$(a) u \in C^{2,\theta}(\overline{\Omega}), \forall \theta \in (0,1] \tag{5}$$

$$(b)u(x) \geq 0, x \in \Omega \tag{6}$$

$$(c)\Delta u = \sum_{i=0}^{n-1} \lambda_i f_i'(u), \lambda_i \in \mathbf{R}. \tag{7}$$

In [21], the authors raised the following questions.

Question 1: (1)What happens to the $C^{2,\theta}(\bar{\Omega})$ regularity (5) of the minimizer u and the Euler-Lagrange equation (7) of the discretized problem (4) in the limit as the mesh size approaches to 0?

In the work of Rakotoson and Serre [46], a similar problem is considered:

$$J(u) = \frac{1}{2}\int_\Omega |\nabla u|^2 dx - \int_\Omega fudx \to \inf, u \in K(h)$$

where

$$K(h) = \left\{ v \in H_0^1(\Omega) \middle| \int_\Omega \phi(v)dx \leq \int_\Omega \phi(h)dx, \forall \text{ convex Lipschitz function } \phi : \mathbf{R} \to \mathbf{R} \right\}$$

for some prescribed $f \in L^\infty(\Omega)$ and $h \in L^1(\Omega)$. **They obtained a regularity result for a weak solution to this problem, and showed that it is in $W^{2,\infty}(\Omega)$.** The present work is strongly motivated by their result.

A similar problem arises in the study of vortices [11] [12] [13] [14] [15] [16]. In particular, Burton and McLeod [16] studied the problem:

$$E(u) = \int_\Omega |\nabla u|^2 dx \to max, \text{ subject to } -\Delta u \sim f$$

for a prescribed $f \in L^2(\Omega)$. Alvino, Trombetti, and Lions [2] worked on the same problem for the case $f \in L_+^p(\Omega), p \geq 2N/(N+2)$ if $N \geq 3$ and $p > 1$ if $N = 2$, where N is the dimension of the domain Ω. Burton [12] and Alvino, Trombetti, and Lions [2] independently established the existence and uniqueness of a maximizer. But **the regularity properties of the maximizer are still unknown** .

As A. Cianchi and N. Fusco (cf: [18]) pointed out, **very little is known about the effect of rearrangements on functions of bounded variation (BV functions)**. They described some regularizing effects of the symmetric rearrangement in connection with fine properties of functions of bounded variation in $\mathbf{R}^n$. For $n \geq 2$, they showed that there exists BV function u whose rearrangement $u^\star$ is smooth on a.e. level surface.

It is known that $E(u^\star) \leq E(u)$ [10], where $u^\star$ is the Schwarz symmetrization of u. But whether there is any other rearrangement v of u which satisfies $E(v) \leq E(u^\star)$ or not is still unknown .

The approach developed by Laurence and Stredulinsky is described in [30] [31] [32] [33] [34] [35] [36] [37] [38], and its complete proofs are shown in [33]. In the model problem used in [33] [35] [36] [37] [38] , the axi-symmetric torus is replaced by a periodic cylinder. p, ψ and ρ represent the pressure, stream function, and density, respectively, assuming an adiabatic equation of state. Here, ψ^* is the non-decreasing rearrangement of ψ defined by

$$\psi^*(s) = \sup\{t \in \mathbf{R} : |\{x \in \Omega : \psi(x) < t\}| < s\}. \tag{8}$$

The energy functional considered in [33] [35] is

$$J(\psi) = \int_\Omega |\nabla\psi|^2 dx + \int_0^{|\Omega|} ((\psi^*)'(s))^2 ds \tag{9}$$

where the functional J is minimized over the space

$$W = \{\psi \in H_0^1(\Omega) : \psi^* \in H^1(0, |\Omega|), \psi^*(0) = 0, \psi^*(|\Omega|) = 1\}. \tag{10}$$

Then, they obtained a gradient bound for certain minimizers, not for all minimizers, of J on W when Ω is convex. They approximated the variational problem by a sequence of free boundary problems associated with n-layers

$$A_i^{(n)} = \{x \in \Omega : t_{i-1}^{(n)} \le \psi_n < t_i^{(n)}\}, 1 \le i \le n-1, A_n^{(n)} = \{x \in \Omega : \psi_n = t_{n-1}^{(n)}\}.$$

They then obtained a uniform gradient bound on minimizers of the approximating problems. The uniform bound was then combined with a proof that the minimizers of the approximating problems converge to a minimizer of the variational problem. By making a range variation on J, they obtained the identity

$$(\psi^*)' + \int_{\{\psi=t\}} |\nabla\psi| dx = \inf J \tag{11}$$

showing that $(\psi^*)'$ is bounded by the minimum energy. Using the argument of Mossino [40] and Payne and Stakgold [41],

$$P(\psi) = |\nabla\psi|^2(x) + \int_0^{\psi(x)} (\psi^*)'' dt \tag{12}$$

which shows that its maximum where $\nabla\psi = 0$, they obtained the main result which says,

$$|\nabla\psi|(x) \le \inf J, a.e. x \in \Omega. \tag{13}$$

They also showed in [36] that current sheets cannot occur in ideal magnetohydrodynamics by showing that when $\psi \in H^1(\Omega)$ such that $(\psi_x - i\psi_y)^2$ is holomporphic on Ω, ψ is locally Lipschcitz continuous on Ω.

It is known that, in general, **classical differentiability is not preserved by rearrangements**. As described by F. Brock at the end of Section 2 of [10] (On page 170 with corresponding graphs shown in Fig. 3 of page 171), the Lipschitz continuity is preserved under a continuous symmetrization of a smooth function (cf: on page 163 of [10], the symmetrization of a measurable function u is defined as a symmetric non-increasing rearrangement of u. The concept of *continuous (Steiner) symmetrization* of a function is given in Definition 2.4 on p168 of [10]). As shown on this Fig. 3, the resultant graph is piecewise smooth having several non-differentiable points. This description seems to be in agreement with the result obtained in [32]. It should be noted that:

(Note 1) Continuous (Steiner) symmetrization is only one kind of rearrangement of a function. The results obtained in [10] have not ruled out a possibility that the differentiability of the prescribed function is preserved under a certain rearrangement.

(Note 2) The main result of [34] says that **there exists a Lipschitz continuous minimizer** when Ω is convex and the energy functional is minimized over the space

$$W = \{\psi \in H_0^1(\Omega) : \psi^* \in H^1(0,|\Omega|), \psi^*(0) = 0, \psi^*(|\Omega|) = 1\}. \tag{14}$$

where ψ^* is the non-decreasing rearrangement of ψ (cf: [17]).

(Note 3) On the other hand, in Section 3 of [31], the authors showed that a minimizer is analytic C^ω when Ω is an open disk $B(a;r)$ of radius r centered at $a \in \mathbf{R}^2$. They also conjectured that (on page 151 of [31]) an analytic minimizer (as a minimizer of the modified variational problem that can be obtained as a limit of free boundary problems) should exist when the cross section Ω of the nuclear fusion reactor is convex with C^2 boundary.

This generates the following question which still remains open.

Question 2: Does the differentiability of the minimizer that is obtained when Ω is an open disk break down and only the Lipschitz continuity of the minimizer is preserved if the domain Ω is deformed to a convex set, for example, an ellipse?

(Note 4) In Lemma 2.2 in [37], the authors showed that a Lipschitz continuous minimizer can be obtained if the energy functional is minimized over a function space

$$SC_\Omega = \{u \in H_0^1(\Omega) \bigcap C(\mathbf{R}^2) | u \geq 0, \{u \leq t\}, \{u > t\} \text{ are connected for } t \geq 0\} \tag{15}$$

(Note 5) As far as the author knows, the concept of *smooth rearrangement* is not precisely defined yet.

The actual minimization problem that arises in axi-symmetric magneto-hydro-dynamics (MHD) is given by

$$E(\psi, f, g) \equiv \frac{1}{2}\int_\Omega \left[|r^{-1}\nabla\psi|^2 + r^{-1} \cdot f^2 + r \cdot g^2\right] drdz \to \text{min}, \tag{16}$$

subject to

$$\int_\Omega r^{-1} \cdot f drdz = F_0^* \tag{17}$$

$$\int_{\{\Psi>\sigma\}} r^{-1} \cdot f \cdot drdz = F^*(\sigma), \forall \sigma \in [\sigma_0, b] \tag{18}$$

$$\int_{\{\Psi>\sigma\}} r \cdot g^{\frac{2}{\gamma}} \cdot drdz = G^*(\sigma), \forall \sigma \in [\sigma_0, b] \tag{19}$$

where $\sigma_0 > 0, b > \sigma_0, F_0^*$ are known constants, $F^*(\sigma), G^*(\sigma)$ are prescribed smooth functions, r is the horizontal axis, z is the vertical axis, Ω is an open bounded set on the rz-plane, $\Psi = \psi + \bar{\psi}$ is the total flux function. $\gamma = 5/3$. $b > 0$ is the maximum of the flux function ψ. $\bar{\psi}$ is a prescribed smooth flux function, and $\psi \in H_0^1(\Omega)$ is the flux function induced by the current density supported in the plasma region, and is an unknown. $f \in L^2(\Omega)$ is the toroidal flux, $g = (2/(\gamma-1))^{1/2} p^{1/2}$, and p is the pressure, which are also unknown (cf: see [22] for more detail). (17) , (18), (19) represent the conservation of toroidal flux, the conservation of mass, and the conservation of the total toroidal flux, respectively.

(Note 6) These constraints are equivalent to Eq (3.2) of [37]. The authors showed in [37] that a Lipschitz continuous minimizer is still obtained when topological constraints are prescribed. Since the constraints (18)(19) are only right continuous with respect to σ, they cannot be used to create a smooth variation structure that is necessary in using the fundamental idea of calculus of variation. The authors discuss cases of complex level set topology in the introduction. In particular, they discuss the case of a "volcanic crater" in which the direction of the magnetic field ψ is reversed at the ridge of the crater. Based on this observation, they consider sums of the form $\psi = \sum_i f_i(\psi_i)$ where ψ_i is of simple type, and f_i is Lipschitz continuous with $f(0) = 0, |f'| = 1$ so that f_i folds the graph of ψ_i, making the graph of ψ an oscillating graph. In this case, a vertical cross section of the ridge of the crater is a peaked curve such that the curve can be non-differentiable at the peak.

The present work was motivated by these observations. The main result Theorem 4.1 answers the following question.

Question: Is there any condition under which a smooth minimizer ψ is obtained?.

The approach taken in this work is fundamentally different from them approaches of the previous works described above in that it is purely functional analytic. The novelty of the present work is the introduction of a smooth variation structure to the constraints via Lemma 2.1. Lemma 2.1 converts the right continuous constraints (18)(19) into C^{k-1} constraints (as functions of σ). However, this conversion alone is not sufficient to obtain a non-degenerate Euler-Lagrange equation satisfied by a minimizer of the energy functional with continuously many constraints.

When the number of constraints is finite, the non-degeneracy of the constraints $F_1(x), \cdots, F_n(x)$ can be established by showing that the derivatives $F_1'(x), \cdots, F_n'(x)$ of the constraints are linearly independent since $F'(x) = (F_1'(x), \cdots, F_n'(x))$ is onto $\mathbf{R}^n$ if and only if $F_1'(x), \cdots, F_n'(x)$ are linearly independent. However, when the number of constraints is continuously infinite as in the case of the energy minimization problem that arises in magnetohydrodynamics, the classical Lagrange Multiplier Rule (cf: [4][5][6][7][8][20][48][49]) does not provide any practical method for determining whether the constraint functional $F(x)$ is surjective or not because the idea of linear independence of $F'(x)$ has not been well-defined yet. This limitation of the classical Lagrange Multiplier Rule will be eliminated in Sections 2 and 3 as follows.

In Section 2, a class of equimeasurable functions is transformed into a level set having a smooth variational structure. This is a novel approach which is not used in [21] [22] [32]-[38].

In Section 3, the classical Lagrange Multiplier Rule is modified so that the idea of linear independence can be naturally generalized to the case of continuously infinitely many constraints. For this reason, Theorem 3.1 is called "Modified Lagrange Multiplier Rule". This result should be applicable to other optimization problems with continuously many constraints.

In Section 4, Modified Lagrange Multiplier Rule will be used to derive the Euler-Lagrange equation satisfied by a minimizer of the energy functional of the prototype problem of the original energy minimization problem. The main result Theorem 4.1 presents partial answers to the above-described questions in the sense that the regularity of the minimizer u of the discretized problem (4) is preserved when Condition (A-1) holds, in which case a continuous analog of the Euler-Lagrange equation (7) holds. At this stage, however, the physical meaning of condition (A-1) is unclear.

In Section 5, the method used in Section 4 is generalized to the original problem in MHD to obtain a sufficient condition for the existence of a smooth solution of the original plasma confinement problem is given (Theorem 5.2).

2. Construction of Smooth Variational Structures

Notations
(1) $\Omega \subset \mathbf{R}^2$ is an open bounded set.
(2) dx denotes the Lebesgue measure on Ω.
(3) For any Lebesgue measurable set $A \subset \Omega$, $|A| \equiv \int_A dx$.
(4) $\langle \cdot, \cdot \rangle$ denotes the inner product on $L^2(\Omega)$.
(5) $H_0^1(\Omega) = W_0^{1,2}(\Omega)$
(6) $H^{-1}(\Omega)$ is the dual of $H_0^1(\Omega)$ with respect to the pairing $\langle \cdot, \cdot \rangle$.
(7) For $s \in \mathbf{R}, s_+ \equiv \max\{0, s\}$.
(8) χ_A is the characteristic function of set $A \subset \Omega$.
(9) L is the Laplacian

$$L \equiv -r\frac{\partial}{\partial r}\left(\frac{1}{r}\frac{\partial}{\partial r}\right) - \frac{\partial^2}{\partial z^2}.$$

Definition 2.1. *Let $f : \Omega \to \mathbf{R}$ and $g : \Omega \to \mathbf{R}$ be measurable functions. f is said to be* **equimeasurable** *with g or f is said to be a* **rearrangement** *with g, if $|\{x \in \Omega : f(x) > \sigma\}| = |\{x \in \Omega : g(x) > \sigma\}|, \forall \sigma \in \mathbf{R}$. A set of equimeasurable functions forms an equivalence relation, which is denoted by $\sim$. That is,*

$$f \sim g \Longleftrightarrow |\{x \in \Omega : f(x) > \sigma\}| = |\{x \in \Omega : g(x) > \sigma\}|, \forall \sigma \in \mathbf{R}. \tag{20}$$

Hereafter,

$$d_f(\sigma) \equiv |f > \sigma| \equiv |\{x \in \Omega : f(x) > \sigma\}| \tag{21}$$

Using this definition, $M_\infty(\bar{u})$ in (1) is defined as

$$M_\infty(\bar{u}) \equiv \{u \in H_0^1(\Omega) : u \sim \bar{u}\}. \tag{22}$$

where $u, \bar{u}$ are assumed to be bounded and non-negative functions satisfying $0 \leq u(x) \leq b, \forall x \in \Omega$, $0 \leq \bar{u}(x) \leq b, \forall x \in \Omega$.

(Note 7) It is known that (21) is a right continuous function of σ (cf: [17]), and is discontinuous at those σ at which $|\{x \in \Omega : f(x) = \sigma\}| > 0$. Therefore, the function space (22) does not have a smooth variational structure in general. A smooth variational structure can be constructed in (22) using the following lemma.

Lemma 2.1. *Assume that $f, g \in L^1(\Omega)$ and $0 \leq f(x), g(x) \leq b, a.e. x \in \Omega$ for some $b > 0$. Then,*

$$f \sim g \Longleftrightarrow \int_\Omega (f-\sigma)_+^k dx = \int_\Omega (g-\sigma)_+^k dx, \forall \sigma \in \mathbf{R}, \forall k \in \mathbf{N}. \tag{23}$$

$$d_f(\sigma) \equiv |f > \sigma| = \frac{(-1)^k}{k!} \frac{d^k}{d\sigma^k} \int_\Omega (f-\sigma)_+^k dx. \tag{24}$$

Let $A_{k,f}(\sigma) = \int_\Omega (f-\sigma)_+^k dx$ and $A_{k,g}(\sigma) = \int_\Omega (g-\sigma)_+^k dx$. Then, $A_{k,f}(\sigma), A_{k,g}(\sigma) \in C^{k-1}(\mathbf{R})$ such that $A_{k,f}^{(k)}(\sigma), A_{k,g}^{(k)}(\sigma)$ are right continuous on $\mathbf{R}$.

Proof. ($\Rightarrow$) Suppose $f \sim g$. Since $(s-\sigma)_+^k$ is a Borel measurable function, $\int_\Omega (f-\sigma)_+^k dx = \int_\Omega (g-\sigma)_+^k dx$ holds (cf: Proposition 2.5, p 14 in [17]).

($\Leftarrow$) Suppose $\int_\Omega (f-\sigma)_+^k dx = \int_\Omega (g-\sigma)_+^k dx, \forall \sigma \in \mathbf{R}, \forall k \in \mathbf{N}$. The Dominated Convergence Theorem will be applied to the forward difference quotient of the integrand function $(g-\sigma)_+^k$ to obtain the right derivative of $\int_\Omega (g-\sigma)_+^k dx$ with respect to σ. The left derivative of $\int_\Omega (g-\sigma)_+^k dx$ with respect to σ will be obtained similarly.
For $k = 2$, let

$$A_{2,f}(\sigma) = \int_\Omega (f-\sigma)_+^2 dx. \tag{25}$$

For fixed $\sigma \in \mathbf{R}$ and $\tau > 0$, $\sigma \in (\sigma, \sigma+\tau)$, first consider the forward difference quotient

$$I_{\sigma,\tau}(s) = \frac{(s-\sigma)_+^2 - (s-(\sigma+\tau))_+^2}{\tau} \geq 0 \tag{26}$$

as a function of s. This simplifies to

$$I_{\sigma,\tau}(s) = (s-\sigma)_+^2 \chi_{[\sigma,\sigma+\tau]}(s) + 2(s-\sigma)\chi_{[\sigma+\tau,b]}(s) - \tau \cdot \chi_{[\sigma+\tau,b]}(s) \tag{27}$$

For $\sigma \in [\sigma, \sigma+\tau), s = \sigma + \theta(\tau)\cdot\tau, 0 \le \theta(\tau) < 1$ for some $\theta(\tau)$ such that $\theta(\tau) \to 0^+$ as $\tau \to 0^+$. Thus, $(s-\sigma)_+^2 = (\theta(\tau))^2 \cdot \tau^2$ on $[\sigma, \sigma+\tau)$.
So, (27) reduces to

$$(\theta(\tau))^2 \cdot \tau \cdot \chi_{[\sigma,\sigma+\tau]}(s) + 2(s-\sigma)\cdot\chi_{[\sigma+\tau,b]}(s) - \tau\cdot\chi_{[\sigma+\tau,b]}(s) \to 2(s-\sigma)_+ \quad (28)$$

as $\tau \to 0^+$. Since $f \in L^1(\Omega)$, $(f-\sigma)_+ \in L^1(\Omega)$. So, the Dominated Convergence Theorem yields

$$\lim_{\tau\to 0+} \frac{A_{2,f}(\sigma+\tau) - A_{2,f}(\sigma)}{\tau} = -2\int_\Omega (f-\sigma)_+ dx. \quad (29)$$

Similarly,

$$\lim_{\tau\to 0+} \frac{A_{2,f}(\sigma) - A_{2,f}(\sigma-\tau)}{\tau} = -2\int_\Omega (f-\sigma)_+ dx. \quad (30)$$

so that

$$A'_{2,f}(\sigma) = -2\int_\Omega (f-\sigma)_+ dx. \quad (31)$$

Applying the same argument to $A'_{2,f}(\sigma)$ yields

$$A''_2(\sigma) = 2|f > \sigma|. \quad (32)$$

Let

$$A_{2,g}(\sigma) = \int_\Omega (g-\sigma)_+^2 dx. \quad (33)$$

Applying the same argument to $A_{2,g}(\sigma)$ and $A'_{2,g}(\sigma)$ successively yields

$$A'_{2,g}(\sigma) = -2\int_\Omega (g-\sigma)_+ dx. \quad (34)$$

and

$$A''_{2,g}(\sigma) = 2|g > \sigma|. \quad (35)$$

Using the induction principle, one obtains the desired conclusion. □

Using this lemma, one can create a level set having a smooth variational structure as follows.

Corollary 2.1. *Let*

$$K_k(u;\sigma) \equiv \frac{1}{k}[A_k(\sigma) - B_k(\sigma)] = \frac{1}{k}\int [(u-\sigma)_+^k - (\bar{u}-\sigma)_+^k]dx \quad (36)$$

and

$$S_k \equiv K_k^{-1}(\{0\}). \quad (37)$$

Then, $K_k(u;\sigma)$ is Frechét C^{k-1} for $k \ge 2$ satisfying

$$K_k(u;b) = 0. \quad (38)$$

Hence, S_k is the level set of K_k through 0 having a Frechét C^{k-1} structure.

Proof. Lemma 2.1 is combined with the fact that $0 \leq \bar{u}(x) \leq b, a.e. x \in \Omega$ and $0 \leq u(x) \leq b, a.e. x \in \Omega$ to obtain

$$d_u(b) = d_{\bar{u}}(b) = 0 \tag{39}$$

where $d_u(\sigma)$ is defined by (21). (39) implies that $A_{k,u}(b) = A_{k,\bar{u}}(b) = 0$.

Hence, (38) follows.

□

The following definitions and propositions are needed to handle the minimization problem (1) within the framework of Lagrange Multiplier Rule.

The dual space of $C([a,b])$ is now defined.

Definition 2.2. *Let $BV([a,b])$ be the set of all functions of bounded variation on $[a,b]$. An equivalence relation $\sim$ can be defined on $BV([a,b])$ by*

$$\forall v_1, v_2 \in BV([a,b]), v_1 \sim v_2 \Longleftrightarrow \int_a^b x(t)dv_1(t) = \int_a^b x(t)dv_2(t), \forall x \in C([a,b]) \tag{40}$$

$$v \sim 0 \Longleftrightarrow v(a) = v(b), v(c-) = v(c+) = v(a), \forall c \in (a,b). \tag{41}$$

Thus,

$$v_1 \sim v_2 \Rightarrow v_1 + C \sim v_2, \forall C \in \mathbf{R}. \tag{42}$$

holds by the definition of Riemann-Stieltjes integral. This implies that $(C([a,b]))^*$ cannot be identified with $BV([a,b])$. In order to obtain a unique representation for all members of $(C([a,b]))^*$, the following definition is needed.

Definition 2.3. *$v \in BV([a,b])$ is said to be* **normalized** *if v is left continuous on $(a,b]$ and vanishes at b, i.e., $v(t-) = v(t), \forall t \in (0,b], v(b) = 0$. The set of all normalized functions of bounded variation on $[a,b]$ is denoted by $\widehat{BV}([a,b])$, i.e.,*

$$\widehat{BV}([a,b]) = \{v \in BV([a,b]) | v(t-) = v(t), \forall t \in (0,b], v(b) = 0\} \tag{43}$$

Then, $\widehat{BV}([a,b])$ is a closed proper subspace of $BV([a,b])$, and each equivalence class in $BV([a,b])$ can be represented by exactly one normalized function as follows: $\forall v \in BV([a,b])$, define $v^* \in BV([a,b])$ by $v^*(b) = 0, v^*(a) = v(a) - v(b), v^*(t) = v(t-) - v(b), \forall t \in (a,b]$.

The following proposition (cf: p198-200 [50]) plays a key role in obtaining an explicit Euler-Lagrange equation satisfied by a state of equilibrium.

Proposition 2.1. (Dual of $C([a,b])$)
Let $(C([a,b]))^*$ be the dual of $C([a,b])$. Then $(C([a,b]))^*$ is congruent with $\widehat{BV}([a,b])$ in the sense that $\forall \lambda \in (C([a,b]))^*, \exists! v \in \widehat{BV}([a,b])$ such that $\langle \lambda, h \rangle = \int_a^b h(t)dv(t), \forall h \in (C([a,b]))$, where $dv(t)$ is a signed measure (Riemann-Stieltjes integrator).

3. Modified Lagrange Multiplier Rule with a Functional Constraint

In this section, the classical Lagrange multiplier rule with a functional constraint (cf: [23] [26]) is modified using Ljusternik's Theorem (cf: [26]). This modified Lagrange multiplier rule with a functional constraint is needed to derive an Euler-Lagrange equation satisfied by a state of equilibrium. The obtained Euler-Lagrange equation then yields the optimal regularity of a minimizer.

The following version of Lagrange Multiplier Rule with a functional constraint is due to Flett [23].

Proposition 3.1. *Let X, Y be Banach Spaces. Let U be an open subset of X, and $f : U \to \mathbf{R}, F : U \to Y$ be Frechét C^1 at x_0.*
Assume that f has a local extremum at x_0 subject to $F(x) = 0$, and $Im(F'(x_0)) = Y$.
Then, $\exists! y_0^* \in Y^* \setminus \{0\}$ such that $f'(x_0) = (F'(x_0))^* y_0^*$.

Note: When the number of constraints is finite, say n, the constraint functional F becomes as mapping from X into $\mathbf{R}^n$, so that $F(x) = (F_1(x), \cdots, F_n(x)) \in \mathbf{R}^n$. In this case, $F'(x_0)$ **is onto $\mathbf{R}^n$ if and only if $F_1'(x), \cdots, F_n'(x)$ are linearly independent**, that is, $\sum_{i=1}^n F_i'(x_0)\lambda_i = 0 \Rightarrow \lambda_1 = \cdots, \lambda_n = 0$.

In order to modify this classical Lagrange Multiplier Rule, so as to include the concept of "**Generalized Linear Independence**", Ljusternik's Theorem is used.

Definition 3.1. *(***Tangent Vector, Tangent Cone, and Tangent Space in a Banach Space***)*
Let X be a Banach Space with norm $||\cdot||$ and $M \subset X$. (1) Then, $h \in$ is said to be **tangent to the set** *M* **at** *x_0 if $\exists \varepsilon > 0$ and a mapping $r : [0, \varepsilon] \to X$ such that $x_0 + th + r(t) \in M, \forall t \in [0, \varepsilon]$ and $\lim_{t \to 0}(||r(t)||/t) = 0$.*
(2) A set of vectors tangent to the set M is denoted by $T(M, x_0)$ and is called the **tangent cone to** *M* **at** *x_0. If this cone is a subspace, it is called the* **tangent space to** *M* **at** *x_0. In this case, $h \in T(M, x_0) \iff \exists \varepsilon > 0$ such that $x_0 + th + r(t) \in M, \forall r \in [-\varepsilon, \varepsilon]$ and $\lim_{t \to 0}(||r(t)||/t) = 0$.*

Proposition 3.2. *(***Ljusternik's Theorem***)*
Let X, Y be Banach Spaces with norms $||\cdot||_X, ||\cdot||_Y$, respectively. Let U be a neighborhood of a point x_0, and $F : U \to Y$ be Frechét C^1 at x_0.
Assume that $Im(F'(x_0)) = Y$ and $F(x_0) = 0$.
Then,
(1) $T(S, x_0) = \ker F'(x_0)$ where $S = \{x \in U | F(x) = 0\}$ is the level set of F through x_0.

(2) $\exists$ a smaller neighborhood $U' \subset U$ of x_0 and a constant $K > 0$, and a mapping $\xi \mapsto x(\xi)$ of U' into X such that $F(\xi + x(\xi)) = 0$ and $||x(\xi)||_X \leq K||F(\xi)||_Y$.

(3) $\forall h \in \ker F'(x_0) \setminus \{0\}, \exists \varepsilon > 0$ such that $|t| \leq \varepsilon \Rightarrow x_0 + th \in U'$ and $||F(x_0 + th) - F(x_0)||_Y = o(t), F(x_0 + th + r(t)) = F(x_0) = 0$, where $r(t) = x(x_0 + t\xi)$ and $||r(t)||_X \leq K||F(x_0 + th) - F(x_0)||_Y = o(t)$.

Proof. See [26].

Theorem 3.1. (**Modified Lagrange Multiplier Rule with a Functional Constraint**) *Let X, Y be Banach Spaces. Let U be an open subset of X, and $f : U \to \mathbf{R}, F : U \to Y$ be Frechét C^1 at x_0.*
Assume that: (1) f has a local extremum at x_0 subject to $F(x) = 0$, and $Im(F'(x_0))$ is closed in Y.
(2) F is **non-degenerate at** x_0 *in the sense that*

$$(F'(x_0))^* y_0^* = 0 \Rightarrow y_0^* = 0 \in Y^* \tag{44}$$

and $\ker F'(x_0) \setminus \{0\}$ *is non-empty.*
Then, $\exists ! y_0^* \in Y^* \setminus \{0\}$ such that

$$f'(x_0) = (F'(x_0))^* y_0^*. \tag{45}$$

Proof. In view of the classical version due to Flett [23], it suffices to show that the non-degeneracy of F at x_0 implies the surjectivity of $F'((x_0))$.
Suppose that $ImF'((x_0))$ is a proper closed subspace of Y. Then, the geometric Hahn-Banach Theorem implies that $\exists$ a closed hyperplane H through $0 \in Y$ containing $ImF'((x_0))$, i.e., $H = (y^*)^{-1}(\{0\})$ for some $y^* \in Y^* \setminus \{0\}$. Thus,

$$\forall h \in X, \langle y^*, F'(x_0)h \rangle = 0 = \langle ((F'(x_0))^* y^*, h \rangle. \tag{46}$$

This implies that

$$((F'(x_0))^* y^* = 0, y^* \in Y^* \setminus \{0\}. \tag{47}$$

Thus, the surjectivity of $F'(x_0)$ follows from the non-degeneracy by contraposition.

Since $F'(x_0)$ is onto Y, it follows from Ljusternik's Theorem that $T(S, x_0) = \ker F'(x_0)$ where $S = F^{-1}(\{0\}) = \{x \in X | F(x) = 0\}$. If $h \in \ker F'(x_0) \setminus \{0\}$, then, by the second assertion of Ljusternik's Theorem, $\exists \varepsilon > 0$ such that a C^1 path $x(t,h) = x_0 + th + r(t), |t| \leq \varepsilon$ lies on the level set S of F through x_0, i.e.,

$$x(t,h) \in S, ||r(t)|| = o(t), \forall t \in [-\varepsilon, \varepsilon], F(x(t,h)) = 0.$$

Then, $\phi(t) = f(x(t,h))$ attains a local extremum at $t = 0$. Hence,

$$\phi'(0) = f'(x_0)h = 0, \forall h \in \ker F'(x_0)$$

so that

$$f'(x_0) \in (\ker F'(x_0))^{\perp} = Im(F'(x_0))^*.$$

Thus, $f'(x_0) = (F'(x_0))^* y^*, \exists y^* \in Y^* \setminus \{0\}$.
□

Remark 3.1. *The non-degeneracy (44) is equivalent to the concept of* **Generalized Linear Independence** *in the sense that*

$$(((F'(x_0))^* y^* = 0 \Rightarrow y^* = 0) \Longleftrightarrow ((ImF'(x_0))^{\perp} = \{0\}). \tag{48}$$

4. Main Result: Regularity Result for Prototype Problem

Theorem 4.1. *Suppose that* $\Omega \subset \mathbf{R}^2$ *is open and bounded.*
Let $X = H_0^1(\Omega)$. *Let*

$$Y = C^{k-1}([0,b])_R = \{v \in C^{k-1}([0,b]) | v(b) = 0\}, \forall k \geq 2. \tag{49}$$

Let $\bar{u} \in C^\infty(\Omega)$ *such that* $|\bar{u} = 0| > 0, E_0 > 0$. *Let* $B_X = \{u \in X | ||u||_X \leq E_0\}$
Let

$$K_k(u;\sigma) = \int_\Omega [(u(x) - \sigma)_+^k - (\bar{u}(x) - \sigma)_+^k] dx \tag{50}$$

$$S_k = \{u \in B_X | K_k(u;\sigma) = 0\} \tag{51}$$

= *the level set of* K_k *through* $0 \in B_X$.

(A1) *Suppose that depending on* Ω *and* $\bar{u}$, *there exist a closed subspace* V *of* X *and a closed subspace* W *of* Y
such that $K_k(u_{op})V$ *is closed in* W.

Let

$$S_{k,V} \equiv S_k \bigcap V. \tag{52}$$

Consider the minimization problem

$$E(u) \equiv \frac{1}{2} \int_\Omega |\nabla u|^2 dx \to \min, u \in S_{k,V}. \tag{53}$$

Then,
(1)The variational problem (53) has a solution $u_{op,k,V}$.

(2) K_k *is non-degenerate at* $u_{op,k,V}$ *for* $k \geq 2$, *in the sense of Modified Lagrange Multiplier Rule, and*

(3) the minimizer $u_{op,k,V}$ *satisfies the Euler-Lagrange equation*

$$\Delta u_{op,k,V}(x) = \int_0^b (u_{op,k,V}(x) - \sigma)_+^{k-1} d\lambda_{k,V}(\sigma) \tag{54}$$

for some unique signed measure (Riemann-Stieltjes integrator) $\lambda_{k,V} \in \widehat{BV}([0,b])$ *satisfying*

$$||\lambda_{k,V}|| \leq \frac{2E_0}{r_{op,k,V}} \tag{55}$$

where $||\lambda_{k,V}||$ *is the total variation of* $\lambda_{k,V}$, *and* $r_{op,k,V}$ *is the radius of the largest ball centered at* $0 \in C([0,b])$ *contained in* $K'_{k,V}(u_{op,k,V})(B_X)$

and

(4)

$$u_{op,k,V} \in C^{k,1}(\Omega). \tag{56}$$

Remark 4.1. *(54) is the continuous analog of the Euler-Lagrange equation (7) satisfied by the discretized problem as stated in Proposition 1.1.*

Proof. (1) Since the energy $E(u)$ is bounded, there exists a minimizing sequence $\{u_j\}_{j\geq 1} \subset S_{k,V}$ such that $\lim_{j\to} E(u_j) = \inf\{E(u)|u \in S_{k,V}\}$. Since the sequence $\{u_j\}_{j\geq 1}$ is norm-bounded, it has a subsequence $\{u_{j(l)}\}_{l\geq 1}$ which converges weakly to a limit u_{op} (cf: [53] p126). Hence,

$$E(u_{op}) \leq \lim_{l\to\infty} E(u_{j(l)}) = \inf\{E(u)|u \in S_{k,V}\}$$

and

$$\int_\Omega (u_{op} - \sigma)_+^k dx = \lim_{l\to\infty} \int_\Omega (\Psi_{j(l)} - \sigma)_+^k dx = \int_\Omega (\bar{u} - \sigma)_+^k dx, \forall \sigma \in [0, b]$$

by the lower semicontinuity of E with respect to weak convergence in X, and the continuity of of each constraint with respect to strong convergence in $L^2(\Omega)$. Therefore, u_{op} is a solution.

(2) (54) will be established. Since Condition (A-1) is satisfied by the assumption, in view of the Modified Lagrange Multiplier Rule (Theorem 3.1), it suffices to show that the constraint functional K_k is non-degenerate at $u_{op,k,V}$ for $k \geq 2$ in the sense of Theorem 3.1, i.e.,

$$(K_k'(u_{op}))^* y^* = 0, y^* \in W^* \Rightarrow y^* = 0. \tag{57}$$

Let $h \in B_X$, and $y^* \in W$.
In the remainder of the proof, the subscript k, V will be omitted, and $u_{op,k,V}, K_k$ will be denoted by u_{op}, K, respectively.
Then,

$$\langle y^*, (K'(u_{op}))h\rangle = \int_0^b \left[\int_\Omega h(x) \cdot (u_{op}(x) - \sigma)_+^{k-1} dx\right] d\lambda(\sigma) \tag{58}$$

where λ is the Riemann-Stieltjes integrator that represents $y^* \in W^*$.
By Fubini's Theorem, one obtains

$$\int_0^b \left[\int_\Omega h(x) \cdot (u_{op}(x) - \sigma)_+^{k-1} dx\right] d\lambda(\sigma) = \int_\Omega \left[\int_0^b (u_{op}(x) - \sigma)_+^{k-1} d\lambda(\sigma)\right] h(x) dx \tag{59}$$

so that

$$\langle y^*, (K'(u_{op}))h\rangle = \langle (K'(u_{op}))^* y^*, h\rangle. \tag{60}$$

Thus,

$$(K'(u_{op}))^* y^* = \int_0^b (u_{op}(x) - \sigma)_+^{k-1} d\lambda(\sigma), a.e. x \in \Omega \tag{61}$$

Put $s = u_{op}(x)$. Then, s assumes almost all values in $[0, b]$ as x varies over Ω.

Suppose

$$\int_0^b (s - \sigma)_+^{k-1} d\lambda(\sigma) = 0, a.e. s \in [0, b] \tag{62}$$

It will be shown that

$$\lambda = 0 \in W^*. \tag{63}$$

Applying the Riemann-Stieltjes integration by parts formula (cf: [3] p144) to (62) yields

$$0 = \int_0^b (s-\sigma)_+^{k-1} d\lambda(\sigma) = (-\sigma+s)_+^{k-1}\lambda(\sigma)|_{\sigma=0}^{s} + (k-1)\int_0^s (\sigma+s)^{k-2}\lambda(\sigma)d\sigma. \tag{64}$$

Thus,

$$\int_0^s (-\sigma+s)^{k-2}\lambda(\sigma)d\sigma = \frac{\lambda(0)}{k-1}s^{k-1}, a.e.s \in [0,b]. \tag{65}$$

For $k = 2$, (65) becomes

$$\int_0^s \lambda(\sigma)d\sigma = \lambda(0)\cdot s, a.e.s \in [0,b]. \tag{66}$$

Taking the sequential derivative of (66),

$$\lambda(s) = \lambda(0), a.e.s \in (0,b]. \tag{67}$$

In particular, $\lambda(b) = 0 \Rightarrow \lambda(0) = 0$.
Thus, $\lambda(s) = 0, a.e.s \in [0,b]$.
The left continuity of λ on $(0,b]$ implies that

$$\lambda(\sigma) = 0, \forall\sigma \in [0,b] \tag{68}$$

(In fact, the set of points of discontinuity of λ is countable and nowhere dense in $[0,b]$.)

For $k \geq 2$, take the sequential left derivative of (65) $k-1$ times to obtain $\lambda = 0$.

(3) Next, (55) and (56) will be shown.

Let K and u_{op} denote K_k and $u_{o,k,Vp}$, respectively. Since K is non-degenerate at u_{op}, $K'(u_{op};\sigma)$ is a bounded linear operator from V onto W. Hence, it follows from the Open Mapping Theorem that $K'(u_{op};\sigma)$ is an open mapping.
So, $(K'(u_{op};\sigma))B_X$ contains balls centered at $0 \in W$.
Let r_{op} be the radius of the largest ball centered at $0 \in W$ contained in $(K'(u_{op};\sigma))B_X$. Then,

$$||y^*|| = ||\lambda|| = \sup\left\{ ||\langle\lambda(\sigma), (u_{op}-\sigma)_+^{k-1}h||_{C([0,b])}\frac{1}{r_{op}} : \left|\left|\int_\Omega (u_{op}-\sigma)_+^{k-1}hdx\right|\right|_{C([0,b])} = r_{op}\right\}$$

$$\leq \sup\{|\langle\Delta u_{op}, h\rangle| : h \in B_X\}\cdot\frac{1}{r_{op}} \leq \frac{1}{r_{op}}||\Delta u_{op}||_{L^2(\Omega)}||\Delta h||_{L^2(\Omega)} \leq 2E_0/r_{op}. \tag{69}$$

Since $0 \leq u_{op}(x) \leq b, a.e.x. \in \Omega$, (54) yields

$$|\Delta u_{op}(x)| \leq b^{k-1}||\lambda|| < \infty, a.e.x \in \Omega. \tag{70}$$

Thus, $u_{op} \in W^{2,\infty}(\Omega)$. By the Sobolev Imbedding Theorem, one has $u_{op} \in C^{1,1}(\Omega)$.

Next, it will be shown that $u_{op} \in C^{k-1,1}(\Omega)$ using Arzela-Osgood Theorem and the Interior Regularity Theorem (cf: [24] Lemma 6.10) with a bootstrap argument.

(4) Let

$$A(s) = \int_0^b (s-\sigma)_+^{k-1} d\lambda(\sigma), \lambda \in BV([0,b]). \tag{71}$$

It will be shown that $A \in C^{k-2,1}([0,b]), k \geq 2$.
First, consider the case $k=2$. Fix $s \in [0,b)$, and let $\tau > 0$ be sufficiently small. Then,

$$\frac{A(s+\tau)-A(s)}{\tau} = \int_0^b \frac{(-\sigma+s+\tau)_+ - (-\sigma+s)_+}{\tau} d\lambda(\sigma). \tag{72}$$

Note that

$$\begin{aligned}\frac{(-\sigma+s+\tau)_+ - (-\sigma+s)_+}{\tau} &= \chi_{[0,s)}(\sigma) + \frac{(-\sigma+s+\tau)_+}{\tau}\chi_{[s,s+\tau)}(\sigma) \\ &= \chi_{[0,s)}(\sigma) + \theta(\tau)\chi_{[s,s+\tau)}(\sigma)\end{aligned} \tag{73}$$

where $0 < \theta(\tau) \leq 1$ and $\theta(\tau) \to 0+$ as $\tau \to 0+$. So,

$$\frac{(-\sigma+s+\tau)_+ - (-\sigma+s)_+}{\tau} \leq 2, \forall \tau > 0 \tag{74}$$

and

$$\lim_{\tau \to 0+} \frac{(-\sigma+s+\tau)_+ - (-\sigma+s)_+}{\tau} = \chi_{[0,s)}(\sigma). \tag{75}$$

Applying the Arzela-Osgood Theorem (cf: [25] p71) to (75), one obtains

$$\lim_{\tau \to 0+} \frac{A(s+\tau)-A(s)}{\tau} = \int_0^b \chi_{,s)}(\sigma) d\lambda(\sigma). \tag{76}$$

Next, fix $s \in (0,b]$ and let $\tau > 0$ be sufficiently small. Then,

$$\frac{A(s)-A(s-\tau)}{\tau} = \int_0^b \frac{(-\sigma+s)_+ - (-\sigma+s-\tau)_+}{\tau} d\lambda(\sigma). \tag{77}$$

Note that

$$\begin{aligned}\frac{(-\sigma+s)_+ - (-\sigma+s-\tau)_+}{\tau} &= \chi_{[0,s-\tau]}(\sigma) + \frac{(-\sigma+s)_+}{\tau}\chi_{(s-\tau,s)}(\sigma) \\ &= \chi_{[0,s-\tau]}(\sigma) + \theta_1(\tau)\chi_{(s-\tau,s)}(\sigma)\end{aligned} \tag{78}$$

where $0 < \theta_1(\tau) \leq 1$ and $\theta_1(\tau) \to 0+$ as $\tau \to 0+$. So,

$$\frac{(-\sigma+s)_+ - (-\sigma+s-\tau)_+}{\tau} \leq 2, \forall \tau > 0 \tag{79}$$

and

$$\lim_{\tau\to 0+}\frac{(-\sigma+s)_+-(-\sigma+s-\tau)_+}{\tau}=\chi_{[0,s)}(\sigma). \tag{80}$$

Applying the Arzela-Osgood Theorem (cf: [25] p71) to (80), one obtains

$$\lim_{\tau\to 0+}\frac{A(s)-A(s-\tau)}{\tau}=\int_0^b \chi_{[0,s)}(\sigma)d\lambda(\sigma). \tag{81}$$

By (76)(81), $A(s)$ is differentiable on $[0,b]$ with

$$A'(s)=\int_0^b \chi_{[0,s)}(\sigma)d\lambda(\sigma)=\int_0^s d\lambda(\sigma)\in C([0,b]),\forall s\in[0,b], \tag{82}$$

which shows that $A\in C^1([0,b])$.
For $k\geq 2$, the same argument can be used to show that

$$A'(s)=(k-1)\int_0^b(-\sigma+s)_+^{k-2}d\lambda(\sigma),\forall s\in[0,b]. \tag{83}$$

Consider the difference quotient

$$\frac{(-\sigma+s)_+^{k-1}-(-\sigma+s-\tau)_+^{k-1}}{\tau} \tag{84}$$

for $\tau>0, 0<s\leq b$.

Expanding the numerator of (84) , one has

$$(k-1)(\sigma+s)_+^{k-2}\chi_{[0,s-\tau)}(\sigma)+\sum_{l=2}^{k-1}{}_{k-1}C_l(-\sigma+s)_+^{k-l-1}(-\tau)^{l-1}\chi_{[0,s)}(\sigma)$$

$$+\frac{(-\sigma+s)_+^{k-1}}{\tau}\chi_{[s-\tau,s)}(\sigma). \tag{85}$$

Note that $\forall\sigma\in[s-\tau,s),\sigma=s-\tau+\theta(\tau)\tau$,
where $0<\theta(\tau)\leq 1$ and $\lim_{\tau\to 0+}\theta(\tau)=1$.

Thus, the middle term of (85) becomes

$$(1-\theta(\tau))^{k-1}\tau^{k-2}\chi_{[s-\tau,b)}(\sigma). \tag{86}$$

(84) (85) (86) yield

$$0\leq\frac{(-\sigma+s)_+^{k-1}-(-\sigma+s-\tau)_+^{k-1}}{\tau}\leq(k-1)b^{-2}+\sum_{l=2}^{k-1}{}_{k-1}C_l b^{k-l-1}\tau^{l-1}+\tau^{k-2}$$

$$\leq b^{k-2}\left[k+\sum_{l=2}^{k-1}{}_{k-1}C_l b^2\right] \tag{87}$$

(86) (87) then yield

$$\lim_{\tau\to 0+}\frac{(-\sigma+s)_+^{k-1}-(-\sigma+s-\tau)_+^{k-1}}{\tau}=(k-1)(\sigma+s)_+^{k-2}\chi_{[0,s-\tau)}(\sigma) \tag{88}$$

Then, by the Arzela-Osgood Theorem, one has

$$\lim_{\tau\to 0+}\frac{A(s)-A(s-\tau)}{\tau}=(k-1)\int_0^b(-\sigma+s)_+^{k-2}d\lambda(\sigma),\forall s\in(0,b] \tag{89}$$

Next, consider the difference quotient

$$\frac{(-\sigma+s+\tau)_+^{k-1}-(-\sigma+s)_+^{k-1}}{\tau} \tag{90}$$

for $\tau>0, 0\le s<b$.

Arguing similarly, one obtains

$$\lim_{\tau\to 0+}\frac{(-\sigma+s+\tau)_+^{k-1}-(-\sigma+s)_+^{k-1}}{\tau}=(k-1)(-\sigma+s)_+^{k-1}\chi_{[0,s)}(\sigma), \tag{91}$$

and

$$0\le(k-1)(-\sigma+s)_+^{k-1}\chi_{[0,s)}(\sigma)\le(k-1)b^{k-2}+\sum_{l=2}^{k-1}{}_{k-1}C_l b^{k-l-1}\tau^{l-1}+\tau^{k-2}$$

$$\le b^{k-2}\left[k+\sum_{l=2}^{k-1}{}_{k-1}C_l b^2\right] \tag{92}$$

By Arzela-Osgood Theorem,

$$\lim_{\tau\to 0+}\frac{A(s+\tau)-A(s)}{\tau}=(k-1)\int_0^b(-\sigma+s)_+^{k-2}d\lambda(\sigma),\forall s\in[0,b) \tag{93}$$

and

$$\lim_{\tau\to 0+}\frac{A(s)-A(s-\tau)}{\tau}=(k-1)\int_0^b(-\sigma+s)_+^{k-2}d\lambda(\sigma),\forall s\in(0,b] \tag{94}$$

Thus, $A(s)$ is differentiable on $[0,b]$ with

$$A'(s)=(k-1)\int_0^b(-\sigma+s)_+^{k-2}d\lambda(\sigma),\forall s\in[0,b] \tag{95}$$

Repeating this procedure $k-1$ times, one obtains

$$A^{(k-1)}(s)=(k-1)!\int_0^b\chi_{[0,s)}(\sigma)d\lambda(\sigma),\forall s\in[0,b] \tag{96}$$

by induction on k.

Since A is $k-1$ time differentiable on $[0,b]$, $A\in C^{k-2,1}([0,b]), k\ge 2$.

If $k=2$, then $A \in C^{0,1}([0,b])$. Since $u_{op,2} \in C^{1,1}(\Omega)$, $A(u_{op,2}) \in C^{0,1}(\Omega)$. The Interior Regularity Theorem then yields $u_{op,2} \in C^{2,1}(\Omega)$.

If $k=3$, then $A \in C^{1,1}([0,b])$. Since $u_{op,3} \in C^{2,1}(\Omega)$, one has $A(u_{op,3}) \in C^{1,1}(\Omega)$. The Interior Regularity Theorem then yields $u_{op,3} \in C^{3,1}(\Omega)$.

Inductively choosing $k = 2,3,4,\cdots$, and using this bootstrap argument, one obtains $u_{op,k} \in C^{k,1}(\Omega)$.

□

5. A State of Equilibrium of Plasma Flow in Nuclear Fusion

In this section, the variational problem defined by (17) , (18), (19) will be reformulated in the framework of the Lagrange Multiplier Rule with a Functional Constraint.

Let $\Omega \subset \mathbf{R}^2$ be an open bounded set and $x=(r,z) \in \Omega$, such that

$$R_1 \leq r \leq R_2, Z_1 \leq z \leq Z_2 \tag{97}$$

$$X = H_0^1(\Omega) \times L^2(\Omega) \times L^2(\Omega) \tag{98}$$

$$Y = \mathbf{R} \times C([\sigma_0,b])_R \times C([\sigma_0,b])_R \tag{99}$$

where $C([\sigma_0,b])_R \equiv \{v \in C([\sigma_0,b]) | v(b)=0\}$.
Let

$$u = (\psi, f, g) \in X, ||u||_X = ||\psi||_{H_0^1(\Omega)} + ||f||_{L^2(\Omega)} + ||g||_{L^2(\Omega)} \tag{100}$$

$$K_k(\psi,f,g) \equiv (K_{k,1}(f), K_{k,2}(\psi,f;\sigma), K_{k,3}(\psi,g;\sigma)) \in Y \tag{101}$$

where

$$K_{k,1}(f) = \int_\Omega r^{-1} \cdot f drdz - F_0^* \in \mathbf{R} \tag{102}$$

$$K_{k,2}(\psi,f;\sigma) = \frac{1}{k+1}\int_\Omega r^{-1} \cdot f \cdot (\Psi-\sigma)_+^{k+1} drdz - F^*(\sigma) \in C([\sigma_0,b]) \tag{103}$$

$$K_{k,3}(\psi,g;\sigma) = \frac{1}{k+1}\int_\Omega r \cdot g^{2/\gamma} \cdot (\Psi-\sigma)_+^{k+1} drdz - G^*(\sigma) \in C([\sigma_0,b]) \tag{104}$$

$$B_X = \{u \in X : ||u||_X \leq (2E_0)^{1/2}, f,g \in L^\infty(\Omega), \sigma_0 \leq \psi(x) \leq b, a.e. x \in \Omega\} \tag{105}$$

$$S_k \equiv K_k^{-1}(\{0\}) \bigcap B_X \tag{106}$$

With (98) –(106), the original variational problem (16) – (19) can be concisely stated as follows.

$$E(u) \to \min, u \in S_k \tag{107}$$

Theorem 5.1. *(Existence of a Minimizer) The variational problem (107) has a solution* $u_{op} = (\psi_{op}, f_{op}, g_{op})$.

Proof. Since the energy $E(u)$ is bounded, there exists a minimizing sequence $\{u_j\}_{j\geq 1} = \{(\psi_j, f_j, g_j\}_{j\geq 1} \subset S_k$ such that $\lim_{j\to} E(u_j) = \inf\{E(u)|u \in S_k\}$. Since the sequence $\{u_j\}_{j\geq 1}$ is norm-bounded, it has a subsequence $\{u_{j(l)}\}_{l\geq 1}$ which converges weakly to a limit u_{op} (cf: [53] p126). Hence,

$$E(u_{op}) \leq \lim_{l\to\infty} E(u_{j(l)}) = \inf\{E(u)|u \in S_k\}$$

and

$$\frac{1}{k+1}\int_\Omega r^{-1} \cdot f_{op} \cdot (\Psi_{op} - \sigma)_+^{k+1} drdz = \lim_{l\to\infty} \frac{1}{k+1}\int_\Omega r^{-1} \cdot f_{j(l)} \cdot (\Psi_{j(l)} - \sigma)_+^{k+1} drdz = F^*(\sigma)$$

and

$$\frac{1}{k+1}\int_\Omega r \cdot g_{op}^{2/\gamma} \cdot (\Psi_{op} - \sigma)_+^{k+1} drdz = \lim_{l\to\infty} \frac{1}{k+1}\int_\Omega r \cdot g_{j(l)}^{2/\gamma} \cdot (\Psi_{j(l)} - \sigma)_+^{k+1} drdz = G^*(\sigma)$$

by the lower semicontinuity of E with respect to weak convergence in X, and the continuity of of each constraint with respect to strong convergence in $L^2(\Omega)$, and the fact that f, g are bounded. Therefore, u_{op} is a solution. □

Now the main result will be stated.

Theorem 5.2. *(Sufficient Condition for the Existence of a Smooth Minimizer) (i) Suppose that* $F_0^*, F^*(\sigma), G^*(\sigma)$ *be given as described in (98) –(106).*
(A-2) *Suppose that, corresponding to these prescribed data* $F_0^*, F^*(\sigma), G^*(\sigma)$, *and* $\forall u \in B_X$, *there exist a closed subspace* V *of* X *and a closed subspace* W *of* Y *such that* $K_k'(u)V$ *is closed in* W .
Let

$$S_{k,V} \equiv S_k \bigcap V \tag{108}$$

Let $u_{op,k} \equiv (\psi_{op,k}, f_{op,k}, g_{op,k})$ *be a solution of the problem*

$$E(u) \to \min, u \in S_{k,V} \tag{109}$$

Then, $\psi_{op,k}, f_{op,k}, g_{op,k}$ *satisfy the following system of Euler-Lagrange equations.*

$$L\psi_{op,k} = f_{op,k}\int_{\sigma_o}^b (\psi_{op,k} + \bar{\psi} - \sigma)_+^k d\lambda_k(\sigma) + r^2(g_{op,k})^{2/\gamma}\int_{\sigma_o}^b (\psi_{op,k} + \bar{\psi} - \sigma)_+^k d\mu_k(\sigma) \tag{110}$$

for some signed measures $\lambda_k, \mu_k \in \widehat{BV}([\sigma_0, b])$,

$$f_{op,k} = c_k + \frac{1}{k+1}\int_{\sigma_o}^{b}(\psi_{op,k} + \bar{\psi} - \sigma)_+^{k+1} d\lambda_k(\sigma), \exists c_k \in \mathbf{R} \tag{111}$$

$$g_{op,k} = \left[\frac{1}{\gamma(k+1)}\int_{\sigma_0}^{b}(\psi_{op} + \bar{\psi} - \sigma)_+^{k+1} d\mu_k(\sigma)\right]^{\frac{\gamma}{2\gamma-2}}, g_{op}(\sigma_0) = 0. \tag{112}$$

(ii) Moreover, $\psi_{op,k}, f_{op,k}, g_{op,k} \in C^{k,1}(\Omega)$.

Proof.

(i) For the sake of notational simplicity, the subscript k will be dropped in the proof. E, K are Frechét C^1 at any $u \in X$ with

$$E'(u) = (r^{-1}L\psi, r^{-1}f, rg) \in H^{-1}(\Omega) \times L^2(\Omega) \times L^2(\Omega) \tag{113}$$

First, K_k will be shown to be non-degenerate at u_{op}, i.e.,
Let $h = (h_1, h_2, h_3) \in X$.

$$y^* K'(u_{op})h = 0, y^* \in Y^*, \forall h \in X \Rightarrow y^* = 0 \in Y^*. \tag{114}$$

Let

$$y^* = (c, \lambda, \mu) \in \mathbf{R} \times (C([\sigma_0, b])^* \times (C([\sigma_0, b])^*. \tag{115}$$

Then, by Fubini's Theorem,

$$\langle y^*, K'(u_{op})h\rangle = \langle (K'(u_{op}))^* y^*, h\rangle. \tag{116}$$

Suppose, $(K'(u_{op}))^* y^* = 0$. Then, this is equivalent to

$$r^{-1} f_{op}\int_{s0}^{b}(\Psi_{op} - \sigma)_+^k d\lambda(\sigma) + r(g_{op})^{2/\gamma}\int_{s0}^{b}(\Psi_{op} - \sigma)_+^k d\mu(\sigma) = 0 \tag{117}$$

$$c + \frac{1}{\gamma(k+1)}\int_{s0}^{b}(\Psi_{op} - \sigma)_+^{k+1} d\lambda(\sigma) = 0 \tag{118}$$

$$(g_{op})^{\frac{2}{\gamma}-1}\int_{s0}^{b}(\Psi_{op} - \sigma)_+^{k+1} d\mu(\sigma) = 0 \tag{119}$$

In the plasma region, the pressure g_{op} and toroidal flux f_{op} are both positive, and Ψ_{op} assumes almost all values in $[\sigma_0, b]$.
Put $s = \Psi_{op}$. Then, (119) becomes

$$I(s) \equiv \int_{s0}^{b}(\Psi_{op} - \sigma)_+^{k+1} d\mu(\sigma) = 0, a.e. s \in [\sigma_0, b], I(\sigma_0) = 0. \tag{120}$$

Using the argument of the main result of Section 3, one obtains

$$\mu(\sigma) = 0, \forall \sigma \in [\sigma_0, b]. \tag{121}$$

Then, (117) simplifies to

$$\int_{s_0}^{b} (\Psi_{op} - \sigma)_+^k d\lambda(\sigma) = 0. \tag{122}$$

The same argument yields

$$\lambda(\sigma) = 0, \forall \sigma \in [\sigma_0, b]. \tag{123}$$

Then, (118) yields $c = 0$. Hence, $y^* = (c, \lambda, \mu) = (0,0,0)$. Now, it follows from the Modified Lagrange Multiplier Rule that

$$E'(u_{op}) = (K'(u_{op}))^* y^*, \exists y^* \in Y^* \setminus \{(0,0,0)\},$$

which are (110)–(112).

(ii)
Let $||\lambda||, ||\mu||$ denote the total variations of $\lambda, \mu \in \widehat{BV}([\sigma_0, b])$, respectively. Then, (111)(112) yield

$$|f_{op}(x)| \leq |c| + \frac{(b-\sigma_0)^{k+1}}{k+1} ||\lambda|| \equiv C_1, a.e. x \in \Omega. \tag{124}$$

and

$$|g_{op}(x)| \leq \left[\frac{1}{\gamma(k+1)} (b-\sigma_0)^{k+1} ||\mu|| \right]^{\frac{\gamma}{2\gamma-2}} \equiv C_2, a.e. x \in \Omega. \tag{125}$$

Substituting (124)(125) into (110), one obtains

$$|L\psi_{op}(x)| \leq C_1 (b-\sigma_0)^k ||\lambda|| + C_2^{2/\gamma} \gamma^{\frac{\gamma}{2\gamma-2}} R_2^2 (b-\sigma_0)^k ||\mu||, a.e. x \in \Omega \tag{126}$$

Therefore, $\psi_{op} \in W^{2,\infty}(\Omega)$, and hence $\psi_{op} \in C^{1,1}(\Omega)$ by the Sobolev Imbedding Theorem. The argument used in the proof of the main Theorem in the previous section shows that f_{op}, g_{op} are $k+1$ times Frechét differentiable with respect to ψ_{op}. Hence, $f_{op}, g_{op} \in C^{1,1}(\Omega)$. When $k = 3$, (110)–(112) imply that f_{op}, g_{op} are 4 times Frechét differentiable with respect to ψ_{op}. Thus, $f_{op}, g_{op} \in C^{3,1}(\Omega)$. Since $\psi_{op} \in C^{2,1}(\Omega)$ already, the right hand side of (110) is in $C^{2,1}(\Omega)$. By Lemma 6.10 of [24], $\psi_{op} \in C^{4,1}(\Omega)$.
For $k \geq 4$, this bootstrap argument with Lemma 6.10 of [24] can be used repeatedly until one obtains $\psi_{op}, f_{op}, g_{op} \in C^{k,1}(\Omega)$.

References

[1] A. Alvino, V. Ferone, G. Trombetti,and P. L. Lions Convex symmetrization and applications, *Ann. Inst. Henri Pincaré*, **14** (1997), no. 2, 275-293.

[2] A. Alvino, G. Trombetti, and P. L. Lions, On Optimization Problems with Prescribed Rearrangements, *Nonlinear Analysis*, **13** (1989), no. 2, 185–220.

[3] T.M.Apostol, *Mathematical Analysis*, 2nd Ed. 1974, Addison-Wesley.

[4] A. V. Arutyunov, On Necessary Optimization Conditions in a Problem with Phase Constraints *Soviet Math. Dokl.*, **31** (1985), no. 1, 174-177.

[5] A. V. Arutyunov, On the Theory of Quadratic Mappings in Banach Spaces, *Soviet Math. Dokl.*, **42** (1991), no. 2, 621-623.

[6] A. V. Arutyunov, Higher Order Conditions in Anormal Extremal Problems with Constraints of Equality Type, *Soviet Math. Dokl.,* **42** (1991), no. 3, 799-804, .

[7] E. R. Avakov, Extremum Conditions for Smooth Anormal Problems with Equality and Inequality Type Constraints, *Math. Notes,* **47** (1990), 431-437.

[8] M.S. Bazaraa and J. J. Goode, Necessary Optimality Criteria in Mathematical Programming in Normed Linear Spaces, *Journal of Optimization Theory and Applications,* **111** (1973), no. 3, 235-244, .

[9] C. Boucher, R. Ellis, and B. Turkington Derivation of Maximum Entropy Principles in Two-Dimensional Turbulence via Large Deviation, *Journal of Statistical Physics*, **98** (2000), nos.5/6.

[10] Friedemann Brock, Continuous rearrangement and symmetry of solutions of elliptic problems, *Proc. Indian Acad. Sci. (Math. Sci.)*, **110** (2000), no.2, May, 157-204.

[11] G. R. Burton, Vortex Rings in a Cylinder adn Rearrangements, *Journal of Differential Equations*, **70** (1987), 333-348.

[12] G. R. Burton, Rearrangements of Functions, Maximization of Convex Functionals, and Vortex Rings, *Mathematische Annalen*, **276** (1987), 225-253.

[13] G. R. Burton, Steady Symmetric Vortex Pairs and Rearrangements, *Proceedings of the Royal Society of Edinburgh*, **108A** (1988), 269-290.

[14] G. R. Burton, Rearrangements and Steady Vortices, *A.I.H.P. Analyse Nonlineaire*, **6** (1989), 295-319.

[15] G. R. Burton, Rearrangements of Functions, Saddle Points, and Uncountable Families of Steady Configurations for a Vortex, *Acta Math.* **163** (1989), 291-309.

[16] G. R. Burton and J. B. McLeod Maximization and Minimization on Classes of Rearrangements *Proceedings of the Royal Society of Edinburgh*, **119A** (1991), 287-300.

[17] K.M.Chong and N.M.Rice, Equimeasurable Rearrangements of Functions, *Queens's Papers in Pure and Applied Mathematics* , no. **28**, 1971, Queen's University, Kingston, Ontario, Canada.

[18] A. Cianchi and N. Fusco, *Functions of bounded variation and rearrangements* , Preprint, 2005.

[19] M. Costeniuc, R.S. Ellis, H. Touchette, and B. Turkington *The Generalized Canonical Ensemble and Its Universal Equivalence with the Microcanonical Ensemble* , Preprint 2005.

[20] B. D. Craven, A Generalization of Lagrange Multipliers, *Bull. Australia. Math. Soc.,* **3** (1970), 353-362.

[21] A.Eydeland, J.Spruck, and B. Turkington, Multiconstrained Variational Problems of Nonlinear Eigenvalue Type: New Formulations and Algorithms, *Mathematics of Computation*, **55** (1990), no. 192 , 509-535.

[22] A.Eydeland, J.Spruck, B. Turkington, and A.Lifshitz, Multiconstrained Variational Problems in Magnetohydrodynamics, *Journal of Computational Physics*, **106** (1993), no. 2, 269-285.

[23] T. M. Flett, On Differentiation in Normed Vector Spaces, *Journal. London. Math. Soc.,* **42** (1967), 523-533.

[24] D. Gilbarg and N. Trudinger, *Elliptic Partial Differential Equations of Second Order*, Springer-Verlag, New York, 1982.

[25] T.H.Hildebrandt, *Introduction to the Theory of Integration*, 1963, Academic Press.

[26] A. D. Ioffe and V. M. Tihomirov, *Theory of Exremal Problems*, Elsevier North Holland, New York, 1979.

[27] R.Jordan A statistical equilibrium model of coherent structures in magnetohydrodynamics *Nonlinearity*, **8** (1995), 585-613

[28] R.Jordan and B. Turkington, Ideal magnetofluid turbulence in two dimensions, *Journal of Statistical Physics,* **87** (1997), 661-695

[29] M.Kruskal and R. Kulsrud, Equilibrium of a Magnetically Confined Plasma in a Toroid , *Physics of Fluids*, **1** (1958), no. 1, 265-274.

[30] Peter Laurence and E. Stredulinsky A New Approach to Queer Differential Equations, *Comm. Pure Appl. Math.*, **38** (1985), 333-355.

[31] P. Laurence and E. Stredulinsky, A survey of recent regularity results for second order queer differential equations, *Proceedings of the Conference on the Calculus of Variations and Nonlinear Partial Differential Equations* , in Honor of Hans Lewy, Trento, Italy, June 1986, Springer Verlag.

[32] P. Laurence and E. Stredulinsky, Existence of Regular Solutions with Convex Level Sets for Semilinear Elliptic Equations with Monotone L^1 nonlinearities, Part II: Passage to the Limit, *Indiana University Math. Journal*, **3**, no. 2 (1990), 485-489.

[33] P. Laurence and E. Stredulinsky, Existence of Regular Solutions with Convex Level Sets for Semilinear Elliptic Equations with Monotone L^1 nonlinearities, Part I: An Approximating Free Boundary Problems, *Indiana University Math. Journal,* **39** (1990), no. 4, 1081-1114.

[34] Peter Laurence and E. Stredulinsky A GRADIENT BOUND FOR THE GRAD-KRUSKAL-KULSRUD FUNCTIONAL, *Mathematical Research Letters,* **1** (1994), 377-387.

[35] Peter Laurence and E. Stredulinsky A Gradient Bound for the Grad Kruskal-Kulsrud Variational Problem, *Communications on Pure and Applied Math*, **49** (1996), no. 3, 237-286.

[36] Peter Laurence and E. Stredulinsky Variational problems with topological constraints, *Calculus of Variations*, **10** (2000), 197-212.

[37] Peter Laurence and E. Stredulinsky Two-Dimesional Magnetohydrodynamics Equilibria with Prescribed Topology *Communications on Pure and Applied Math,* **LIII** (2000), no.3, 1177-1200.

[38] Peter Laurence and E. Stredulinsky Optimal regularity in a variational problem for current sheets in ideal magnetohydrodynamics, *Journal of Mathematical Physics*, **43** (2002), no. 11, November, 5707-5719

[39] I.J.Maddox, *Elements of Functional Analysis*, 2nd ed., 1988, Cambridge University Press.

[40] Mossino, J, A priori estimates for a model Mercier Type in Plasma Physics, *Appl.Anal.*, **13** (1982), no. 3, 185-207.

[41] Payne, L.L. and Srakgold, I., On the mean value of t he fundamental mode in the fixed membrane problem, *Appl.Anal.*, **3** (1973), 295-303.

[42] J. M. Rakotoson, A Differentiability Result for the Relative Rearrangement, *Differential and Integral Equations*, **2** (1989), no. 3, 366-377.

[43] J. M. Rakotoson, Some Properties of the Relative Rearrangement, *Journal of Mathematical Analysis and Applications,* **135** (1988), 488-500.

[44] J. M. Rakotoson, Relative Rearrangement and interpolation inequalities, *Rev. R. Acad. Cien. Series A. Mat.*, **97** (2003), no.1, 133-145.

[45] J.M. Rakotoson and M.L. Seoane, Numerical approximations of the relative rearrangement: The piecewise linear case. Application to some Nonlocal Problems, p477, *Mathematical Modelling and Numerical Analysis,* **M2AN**, 34 (2000), no. 2, March/April.

[46] J.-M. Rakotoson and D. Serre, *Sur un probléme d'optimisation lie aux équations de Navier-Stokes*, Preprint.

[47] F. Riesz and B.Sz.Nagy, *Functonal Analysis*, Frederick Unger Publishing Co., New York 1978

[48] D. Schechter, Derivatives of Mappings with Applications to Nonlinear Differential Equations, *Transactions of the AMS*, **293** (1986), 53-69.

[49] M. Schechter, Non-uniqueness in the Equilibrium Shape of a Confined Plasma, *Communications in Partial Differential Equations*, **2** (1986), no. 6, 587-600.

[50] A. E. Taylor, *Introduction to Functional Analysis, 6th Ed.*, 1967, John Wiley & Sons, Inc.

[51] A. E. Taylor, *General Theory of Functions and Integration*, Blaisdell Publishing Company, 1965.

[52] R. Temam, Monotone Rearrangements of a Function and the Grad-Mercier Equation of Plasma Physics, *Proc. International Meeting on Recent Methods in Nonlinear Analysis*, Rome, May 1978, 83-98.

[53] K. Yoshida, *Functional Analysis*, Springer-Verlag, New York, 1974.

In: Progress in Evolution Equations
Editor: Gaston M. N'Guerekata, pp. 91-101
ISBN: 978-1-60456-328-3

Chapter 5

DECAY OF SOLUTIONS OF A NONLINEAR BBM-BURGERS SYSTEM

Jardel Morais Pereira*
Departamento de Matemática, Universidade Federal de Santa Catarina
CEP 88040-900, Florianópolis, S.C., Brasil

Abstract

We consider a BBM-Burgers system with a homogeneous nonlinearity. For this system we obtain decay estimates exploiting properties of the semigroup generated by the linear part and estimating the associated integral equation.

AMS Subject classification: 35Q53, 35B40

Key-words: decay estimates, BBM-Burgers system, interaction of long waves.

1. Introduction

In this paper we consider the following BBM-Burgers system

$$U_t - AU_{xxt} + D(\nabla H(U))_x - BU_{xx} = 0, \quad (x,t) \in \mathbf{R} \times \mathbf{R}^+, \tag{1.1}$$

with initial condition

$$U(x,0) = \varphi(x), \quad x \in \mathbf{R}. \tag{1.2}$$

Here, $U = (u,v)^T$, $\varphi = (\varphi_1, \varphi_2)^T$, B and D are 2×2 real diagonal matrices, say, $B = \text{diag}(\varepsilon_1, \varepsilon_2)$, $D = \text{diag}(d_1, d_2)$, with $\varepsilon_i, d_i > 0$, $i = 1,2$, and $A = (a_{ij})$ is a 2×2 real matrix such that $D^{-1}A$ is positive definite. $\nabla H = (\partial_1 H, \partial_2 H)^T$ denotes the gradient of a C^3 homogeneous function $H : \mathbf{R}^2 \to \mathbf{R}$ of degree $p+2$, where $p \geq 1$ is any integer. An example of such a system is

*E-mail address: jardel@mtm.ufsc.br

$$\begin{aligned} &u_t - u_{xxt} - a_3 v_{xxt} + u^p u_x + a_2 (u^p v)_x + a_1 v^p v_x - \alpha_1 u_{xx} = 0 \\ &b_1 v_t - v_{xxt} - b_2 a_3 u_{xxt} + v^p v_x + b_2 a_2 u^p u_x + b_2 a_1 (u v^p)_x - \alpha_2 v_{xx} = 0 \end{aligned} \tag{1.3}$$

where $u = u(x,t)$, $v = v(x,t)$ are real-valued functions of the real variables x and t, the coefficients a_1, a_2, a_3, b_1, b_2, α_1, α_2 are real constants, with $\alpha_i > 0$, $b_i > 0$, $i = 1,2$, $a_3^2 b_2 < 1$, and p is a positive integer. System (1.3) is a BBM-Burgers version of the Gear and Grimshaw's model of strong interaction of long waves derived in [**6**]. Systems of coupled equations which model interaction of long waves in dispersive media have been studied by several authors (see for example [**1,2,4,5,8**] and the references therein). In this paper, we are interested in the study of asymptotic behavior as $t \to \infty$ of solutions of the initial-value problem (1.1)-(1.2) when φ belongs to a suitable class. In order to state our main result let us introduce some notations. First we observe that the norm of u in a Banach space X will be denoted by $||u||_X$ and we always consider the norm $||(u,v)||^2_{X\times Y} = ||u||^2_X + ||v||^2_Y$ in the Cartesian product $X \times Y$ of two Banach spaces. For $1 \le q \le \infty$, $L^q = L^q(\mathbf{R})$ denotes the space of qth-power integrable functions (essentially bounded functions if $q = \infty$) with the usual norm. For $s \in \mathbf{R}$, $H^s = H^s(\mathbf{R})$ denotes the Sobolev space of order s. Moreover, as in [**7**] we also consider the Banach space $L^2_1 = L^2_1(\mathbf{R})$ of all functions $w \in L^2$ such that $xw \in L^2$, equipped with the norm $||w||_{L^2_1} = \left(\int_{\mathbf{R}} (1+x^2)|w(x)|^2 dx\right)^{\frac{1}{2}}$. To simplify notation we will write $L_{2,q} = L^q \times L^q$, $H_{2,s} = H^s \times H^s$, and $L^1_{2,2} = L^2_1 \times L^2_1$. The existence and uniqueness of a global solution for the initial-value problem (1.1)-(1.2) in the space $H_{2,s}$ can be studied in the same way as in [**9**] by using the Fixed Point Theorem and energy estimates. In particular, under the hypotheses on A, B, D and H already stated, if $\varphi \in H_{2,2}$ then there exists a unique solution $U = U(\cdot,t)$ of (1.1)-(1.2) such that $U \in C^1(\mathbf{R}^+; H_{2,2})$. In the present paper we prove decay estimates for the solution of (1.1)-(1.2) in the norm of $L_{2,q}$, for $1 \le q \le \infty$, when $\varphi \in H_{2,2} \cap L^1_{2,2}$, and the matrices A and B satisfy

$$(\operatorname{adj} A) B = B (\operatorname{adj} A) \tag{1.4}$$

More precisely, we prove the following result.

Theorem 1.1. *Suppose that* (1.4) *holds and let* $p \ge 2$. *If* $\varphi \in H_{2,2} \cap L^1_{2,2}$, *then for each* $1 \le q \le \infty$ *there exists a positive constant* $C = C(q,\varphi)$ *such that*

$$||U(\cdot,t)||_{L_{2,q}} \le C(1+t)^{-\frac{1}{2}\left(1-\frac{1}{q}\right)}, \quad t \ge 0.$$

Theorem 1.1 is proved in section 3. The proof consists in estimate the integral equation associated to (1.1)-(1.2) (see equation (3.1)) using estimates of $||E(t)\varphi||_{L_{2,q}}$, where $\{E(t)\}_{t\ge 0}$ denotes the semigroup generated by the linear part of system (1.1). These linear estimates are studied in section 2. We observe that decay estimates for the solution of (1.1)-(1.2) in the norm $||\ ||_{L_{2,q}}$, when $2 \le q \le \infty$, $p \ge 3$ and φ belongs to a larger class can be proved without the hypothesis (1.4) (see [**9**]). Here we use it together with the hypothesis $\varphi \in H_{2,2} \cap L^1_{2,2}$ to find a bound to $||U(\cdot,t)||_{L_{2,1}}$, which is the start point to prove Theorem 1.1. Theorem 1.1 in particular shows that the rate of decay enjoyed by a single generalized BBM-Burgers equation is still valid for a system of BBM-Burgers equations even in the case that the equations are coupled through both dispersive and nonlinear effects (see example 3.1 in section 3). Some other notations used in this paper are as follows. The Fourier transform of a function f is defined by $\hat{f}(z) = \mathcal{F}[f](z) = (1/\sqrt{2\pi})\int_{\mathbf{R}} e^{-ixz} f(x)\,dx$. We denote by $||N|| = \left(\sum_{j,l=1}^{2} n_{jl}^2\right)^{1/2}$ the norm of a 2×2 matrix $N = (n_{jl})$. Finally, we denote by C a generic constant, which sometimes depend on φ and whose value may be different from a line or inequality to another.

2. Linear Estimates

Consider the linear problem associated to (1.1)-(1.2)

$$U_t - AU_{xxt} - BU_{xx} = 0, \qquad U(x,0) = \varphi(x), \tag{2.1}$$

where $x \in \mathbf{R}$ and $t \geq 0$. We assume that $\varphi \in H_{2,2}$ and A and B satisfy (1.4). The initial-value problem (2.1) has a unique solution $U \in C^1(\mathbf{R}^+; H_{2,2})$ given by $U(\cdot,t) = E(t)\varphi$, where $\{E(t)\}_{t\geq 0}$ denotes the semigroup of linear operators on $H_{2,2}$, generated by the operator $\mathcal{L} = M^{-1}B\frac{\partial^2}{\partial x^2}$. Here, M^{-1} denotes the inverse of the operator $M = I - A\frac{\partial^2}{\partial x^2}$, acting from $H_{2,2}$ to $L_{2,2}$. I is the 2×2 identity matrix. Taking the Fourier transform of (2.1) we can easily find an explicit formula for $E(t)\varphi$. In fact, we have

$$E(t)\varphi(x) = \frac{1}{\sqrt{2\pi}}\int_{\mathbf{R}} e^{ixz}\mathcal{A}(z,t)\hat{\varphi}(z)\,dz \tag{2.2}$$

where $\mathcal{A}(z,t) = \exp\big(-t\mathcal{A}(z)\big)$, with $\mathcal{A}(z)$ given by

$$\mathcal{A}(z) = \frac{z^2}{\sigma(z)}(I + z^2\mathrm{adj}\,A)B, \text{and } \sigma(z) = \det(I + z^2A) > 0.$$

Moreover, we observe that using (1.4) we can write

$$\mathcal{A}(z,t) = \sum_{j,l=1}^{2} M_{jl}\exp\Big(-\frac{\mu_{jl}(z)}{\sigma(z)}z^2t\Big), \tag{2.3}$$

where $M_{j,l}$ are 2×2 matrices whose entries depend only on adj A, and $\mu_{jl}(z) = \varepsilon_j + \lambda_l z^2$, $j,l = 1,2$, where λ_l are the positive eigenvalues of $(\mathrm{adj}\,A)B$.

Our aim in this section is to find decay estimates for $E(t)\varphi$ in the norms $||\ ||_{L_{2,q}}$, $1 \leq q \leq \infty$, using (2.2) and (2.3). As in [7] we consider a function $\xi_1 \in C^\infty(\mathbf{R})$ such that $0 \leq \xi_1(z) \leq 1$, $\xi_1(z) = 1$, if $|z| \leq 1/2$ and $\xi_1(z) = 0$, if $|z| \geq 1$. Let $E(t)\varphi = E_1(t)\varphi + E_2(t)\varphi$, where

$$E_j(t)\varphi(x) = \frac{1}{\sqrt{2\pi}}\int_{\mathbf{R}} e^{ixz}\xi_j(z)\mathcal{A}(z,t)\hat{\varphi}(z)\,dz, \tag{2.4}$$

$j,l = 1,2$, and $\xi_2(z) = 1 - \xi_1(z)$. We will also use the following facts. If $\alpha > 0$ and $m \in \mathbf{N}$, then there exists a positive constant $C = C(\alpha,m)$ such that

$$\int_{-1}^{1} z^{2m}e^{-\alpha z^2 t}dz \leq C(1+t)^{-\frac{1}{2}-m}, \forall t \geq 0. \tag{2.5}$$

Concerning the space L_1^2 defined in the introduction, an application of Schwartz inequality shows that $L_1^2 \subset L^1$. Moreover, if $w \in L_1^2$, then there exists a positive constant C, independent of w, such that

$$||w||_{L^1} \leq C||w||_{L^2}^{1/2}||(\hat{w})'||_{L^2}^{1/2} \tag{2.6}$$

Inequality (2.6) is proved in [**7**, p.75], where the author used it in the study of asymptotic behavior of generalized BBM-Burgers equations. Some of the ideas developed in [**7**] are used in our proofs.

Lemma 2.1. *Suppose that* (1.4) *holds. If* $\varphi \in H_{2,2} \cap L^1_{2,2}$, *then there exists a positive constant* $C = C(\varphi)$ *such that* $||E(t)\varphi||_{L_{2,1}} \leq C$, *for all* $t \geq 0$.

Proof. Let $\omega_{jl}(z,t) = \exp\left(-\frac{\mu_{jl}(z)}{\sigma(z)}z^2 t\right)$ and

$$J_{jl}(x,t) = \frac{1}{\sqrt{2\pi}} \int_{\mathbf{R}} \xi_1(z) e^{ixz} \omega_{jl}(z,t)\, dz, \quad j,l = 1,2.$$

Using (2.4) and (2.3) we have that

$$\begin{aligned} E_1(t)\varphi(x) =& \frac{1}{\sqrt{2\pi}} \int_{\mathbf{R}} \left[e^{ixz} \xi_1(z) \Big(\sum_{j,l=1}^{2} M_{jl} \omega_{jl}(z,t) \Big) \hat{\varphi}(z)\right] dz \\ =& \sum_{j,l=1}^{2} M_{jl} \mathcal{F}^{-1}\big(\widehat{J_{jl}}(\cdot,t)\hat{\varphi}\big)(x) = \sum_{j,l=1}^{2} \frac{1}{\sqrt{2\pi}} M_{jl}\big(J_{jl}(\cdot,t) * \varphi\big)(x), \end{aligned} \tag{2.7}$$

where $J_{jl}(\cdot,t) * \varphi$ denotes the vector whose components are $J_{jl}(\cdot,t) * \varphi_1$, $J_{jl}(\cdot,t) * \varphi_2$, and $*$ denotes spatial convolution. Observe that if $|z| \leq 1$, then $|\omega_{jl}(z,t)| \leq \exp(-\alpha_0 z^2 t)$, for all $j,l = 1,2$ and $t \geq 0$, where α_0 is a positive constant which depends only on ε_j, λ_l and A. Thus, using Parseval's identity and (2.5) we obtain

$$\begin{aligned} ||J_{jl}(\cdot,t)||_{L^2}^2 = ||\widehat{J_{jl}}(\cdot,t)||_{L^2}^2 &\leq \int_{-1}^{1} |\xi_1(z)\omega_{jl}(z,t)|^2 dz \\ &\leq C \int_{-1}^{1} e^{-2\alpha_0 z^2 t} dz \leq C(1+t)^{-1/2}, \end{aligned} \tag{2.8}$$

for all $j,l = 1,2$ and $t \geq 0$. In addition, using the notation $\beta_{jl}(z) = \frac{\mu_{jl}(z)}{\sigma(z)} z^2$ and (2.5) with $m = 0$ and $m = 1$, we also have

$$\begin{aligned} ||\partial_z \widehat{J_{jl}}(\cdot,t)||_{L^2}^2 &\leq \int_{-\infty}^{\infty} 2|\xi_1'(z)\omega_{jl}(z,t)|^2 dz + \int_{-\infty}^{\infty} 2t^2 |\xi_1(z)\omega_{jl}(z,t)\beta_{jl}'(z)|^2 dz \\ &\leq C(1+t)^{-1/2} + Ct^2(1+t)^{-1/2} \leq C(1+t)^{1/2}, \end{aligned} \tag{2.9}$$

for all $j,l = 1,2$ and $t \geq 0$. By (2.8), (2.9) and (2.6) it follows that

$$||J_{jl}(\cdot,t)||_{L^1} \leq C, \quad \forall j,l = 1,2 \text{ and } t \geq 0. \tag{2.10}$$

Therefore, applying Young's inequality in each component of the vector $M_{jl}\left(J_{jl}(\cdot,t)*\varphi\right)$ in (2.7) and using (2.10) we deduce that

$$||E_1(t)\varphi||_{L_{2,1}} \leq C\Big(\sum_{j,l=1}^{2} ||M_{jl}||\Big)||\varphi||_{L_{2,1}}. \tag{2.11}$$

Next, let us estimate $||E_2(t)\varphi||_{L_{2,1}}$. By (2.4)

$$||E_2(t)\varphi||_{L_{2,1}} \leq C\sum_{j,l=1}^{2} ||M_{jl}||\,||\mathcal{F}^{-1}\left(\xi_2\omega_{jl}(\cdot,t)\hat{\varphi}\right)||_{L_{2,1}}. \tag{2.12}$$

We estimate the right-hand side of (2.12) using (2.6) and (2.4) as follows. First observe that if $|z|\geq 1/2$, then $|\beta_{jl}(z,t)|\geq\beta_0$, $\forall j,l=1,2$ and $t\geq 0$, where the positive constant β_0 depends only on ε_j, λ_l and A. Then, using Parseval's inequality again we have that

$$||\xi_2\omega_{jl}(\cdot,t)\hat{\varphi}||^2_{L_{2,2}} = ||\xi_2\omega_{jl}(\cdot,t)\hat{\varphi}_1||^2_{L^2} + ||\xi_2\omega_{jl}(\cdot,t)\hat{\varphi}_2||^2_{L^2} \leq Ce^{-2\beta_0 t}||\varphi||^2_{L_{2,2}}. \tag{2.13}$$

Also

$$||\partial_z\left(\xi_2\omega_{jl}(\cdot,t)\hat{\varphi}\right)||^2_{L_{2,2}} = ||\partial_z\left(\xi_2\omega_{jl}(\cdot,t)\hat{\varphi}_1\right)||^2_{L^2} + ||\partial_z\left(\xi_2\omega_{jl}(\cdot,t)\hat{\varphi}_2\right)||^2_{L^2}. \tag{2.14}$$

Estimating each term in (2.14) we find

$$\begin{aligned}
||\xi_2'\omega_{jl}(\cdot,t)\hat{\varphi}_m||^2_{L^2} &\leq Ce^{-2\beta_0 t}\,||\hat{\varphi}_m||^2_{L^2} \leq Ce^{-2\beta_0 t}\,||\varphi||^2_{L_{2,2}}\\
||\xi_2\partial_z\omega_{jl}(\cdot,t)\hat{\varphi}_m||^2_{L^2} &\leq Ct^2\,e^{-2\beta_0 t}\,||\hat{\varphi}_m||^2_{L^2} \leq Ct^2\,e^{-\beta_0 t}\,||\varphi||^2_{L_{2,2}}\\
||\xi_2\omega_{jl}(\cdot,t)(\hat{\varphi}_m)'||^2_{L^2} &\leq Ce^{-2\beta_0 t}\,||(\hat{\varphi}_m)'||^2_{L^2} \leq Ce^{-2\beta_0 t}\,||\varphi||^2_{L^1_{2,2}}
\end{aligned}$$

for all $j,l,m=1,2$ and $t\geq 0$. Replacing these estimates into (2.14) we obtain

$$||\partial_z\left(\xi_2\omega_{jl}(\cdot,t)\hat{\varphi}\right)||^2_{L_{2,2}} \leq Ce^{-\beta_0 t}\left(||\varphi||^2_{L_{2,2}} + ||\varphi||^2_{L^1_{2,2}}\right), \quad \forall t\geq 0. \tag{2.15}$$

Therefore, from (2.12),(2.13),(2.15) and (2.6) it follows that

$$||E_2(t)\varphi||_{L_{2,1}} \leq Ce^{-\frac{1}{4}\beta_0 t}, \quad \forall t\geq 0, \tag{2.16}$$

where C is a positive constant that depends on $||\varphi||_{L_{2,2}}$ and $||\varphi||_{L^1_{2,2}}$. Now, the conclusion of lemma 2.1 follows from (2.11) and (2.16). □

Lemma 2.2. *Suppose that* (1.4) *holds. If* $\varphi\in H_{2,2}\cap L^1_{2,2}$, *then there exists a positive constant* $C=C(\varphi)$ *such that*

$$||E(t)\varphi||_{L_{2,\infty}} \leq C(1+t)^{-\frac{1}{2}}, \quad t\geq 0.$$

Proof. We will use the notation of lemma 2.1. Applying the Young's inequality in each component of the vector $J_{jl}(\cdot,t)*\varphi$ and using (2.5) we have that

$$\begin{aligned}
||M_{jl}\left(J_{jl}(\cdot,t)*\varphi\right)||_{L_{2,\infty}} &\leq ||M_{jl}||\,||J_{jl}(\cdot,t)*\varphi||_{L_{2,\infty}} \leq ||M_{jl}||\,||J_{jl}(\cdot,t)||_{L^\infty}||\varphi||_{L_{2,1}}\\
&\leq C\Big(\int_{-1}^{1} e^{-\alpha_0 z^2 t}\,dz\Big)||M_{jl}||\,||\varphi||_{L_{2,1}} \leq C(1+t)^{-\frac{1}{2}}||M_{jl}||\,||\varphi||_{L_{2,1}}.
\end{aligned}$$

Then, by (2.7) it follows that

$$||E_1(t)\varphi||_{L_{2,\infty}} \leq C\Big(\sum_{j,l=1}^{2} ||M_{jl}||\Big)||\varphi||_{L_{2,1}}(1+t)^{-\frac{1}{2}}, \quad \forall t \geq 0. \tag{2.17}$$

Moreover, using (2.4) and (2.3) we also have that

$$\begin{aligned}
||E_2(t)\varphi||_{L_{2,\infty}} &\leq \frac{1}{\sqrt{\pi}} \int_{|z|\geq 1/2} |\xi_2(z)| \Big(\sum_{j,l=1}^{2} ||M_{jl}||\,|\omega_{jl}(z,t)|\Big)|\hat{\varphi}|dz \\
&\leq \Big(\sum_{j,l=1}^{2} ||M_{jl}||\Big)e^{-\beta_0 t} \int_{|z|\geq 1/2} |\xi_2(z)|\,|\hat{\varphi}(z)|\,dz \\
&\leq C\Big(\sum_{j,l=1}^{2} ||M_{jl}||\Big)||\varphi||_{H_{2,1}} e^{-\beta_0 t}, \quad \forall t \geq 0.
\end{aligned}$$

Therefore, from this estimate and (2.17) we conclude the proof of lemma 2.2. □

From the above lemmas and the interpolation inequality $||w||_{L^q} \leq ||w||_{L^1}^{1/q} ||w||_{L^\infty}^{1-1/q}$, $1 \leq q \leq \infty$, we deduce the following result

Lemma 2.3. *Suppose that* (1.4) *holds. If* $\varphi \in H_{2,2} \cap L^1_{2,2}$, *then for each* $1 \leq q \leq \infty$ *there exists a positive constant* $C = C(q,\varphi)$ *such that*

$$||E(t)\varphi||_{L_{2,q}} \leq C(1+t)^{-\frac{1}{2}(1-\frac{1}{q})}, \quad t \geq 0.$$

3. Proof of Theorem 1.1

Let $U = U(\cdot,t)$ denote the global solution of (1.1)-(1.2), with $\varphi \in H_{2,2}$, which also satisfies the integral equation

$$U(\cdot,t) = E(t)\varphi - \int_0^t E(t-\tau)M^{-1}\partial_x F(U(\cdot,\tau))\,d\tau, \quad t \geq 0 \tag{3.1}$$

in $H_{2,2}$, where $F(U) = D\nabla H(U)$. In order to prove Theorem 1.1 we will estimate the integral equation (3.1) using lemma 2.3. The proof will be a consequence of lemmas 3.3 and 3.6 below. We will need the following lemma.

Lemma 3.1. *If* $\varphi \in H_{2,2}$ *then there exists a positive constant* $C = C(D^{-1}, A)$ *such that*
(a) $||U(\cdot,t)||_{H_{2,1}} \leq C||\varphi||_{H_{2,1}}, \quad t \geq 0,$
(b) $||U_x(\cdot,t)||^2_{L_{2,2}} \in L^1(\mathbf{R}^+)$.

Proof. Multiplying equation (1.1) by $U^T D^{-1}$ and integrating on $\mathbf{R}$ we obtain

$$\frac{d}{dt}\mathcal{E}(t) + \int_{\mathbf{R}} U_x^T D^{-1} B U_x\,dx = 0, \tag{3.2}$$

where

$$\mathcal{E}(t) = \int_{\mathbf{R}} \left(U^T D^{-1} U + U_x^T D^{-1} A U_x\right) dx.$$

Since $D^{-1}A$ is a positive definite matrix, then we can find positive constants $c_j = c_j(D^{-1}, A)$, $j = 1, 2$, such that

$$c_1 ||U(\cdot,t)||^2_{H_{2,1}} \leq \mathcal{E}(t) \leq c_2 ||U(\cdot,t)||^2_{H_{2,1}}, \quad t \geq 0. \tag{3.3}$$

From (3.2) and (3.3) we obtain

$$\mathcal{E}(t) + \min\{d_1^{-1}\varepsilon_1, d_2^{-1}\varepsilon_2\} \int_0^t ||U_x(\cdot,t)||^2_{L_{2,2}} d\tau \leq \mathcal{E}(0),$$

from which it follows (a) and (b). □

Besides equation (3.1) and lemma 3.1 we need a formula for $\widehat{M^{-1}g}$, whenever $g \in L_{2.2}$. Taking the Fourier transform in the equation $V - AV_{xx} = g$ we find $\hat{V}(z) = \frac{1}{\sigma(z)}\big(I + z^2 \operatorname{adj} A\big)\hat{g}(z)$. Therefore,

$$\widehat{M^{-1}g} = \frac{1}{\sigma(z)} R(z)\hat{g}(z), \tag{3.4}$$

where $R(z) = I + z^2 \operatorname{adj} A$.

Lemma 3.2. *For $j, l, m, n = 1, 2$, let*

$$k_{mn}^{jl}(x,t) = \frac{1}{\sqrt{2\pi}} \int_{\mathbf{R}} \frac{r_{mn}(z)}{\sigma(z)} \omega_{jl}(z,t) e^{ixz}\, dz$$

where $r_{mn}(z)$ denote the entries of the matrix $R(z)$.Then, there exists a positive constant C such that $||k_{mn}^{jl}(\cdot,t)||_{L^1} \leq C$, for all $t \geq 0$ and $j, l, m, n = 1, 2$.

Proof. The proof of this lemma consists in decompose $k_{mn}^{jl}(x,t)$ as

$$k_{mn}^{jl}(x,t) = \frac{1}{\sqrt{2\pi}} \int_{\mathbf{R}} \frac{r_{mn}(z)}{\sigma(z)} \xi_1(z) \omega_{jl}(z,t) e^{ixz}\, dz + \frac{1}{\sqrt{2\pi}} \int_{\mathbf{R}} \frac{r_{mn}(z)}{\sigma(z)} \xi_2(z) \omega_{jl}(z,t) e^{ixz}\, dz,$$

and estimate each integral using (2.6) as in the proof of lemma 2.1. □

Lemma 3.3. *Suppose that* (1.4) *holds and let $p \geq 2$. If $\varphi \in H_{2,2} \cap L^1_{2,2}$, then there exists a positive constant $C = C(\varphi)$ such that*

$$||U(\cdot,t)||_{L_{2,1}} \leq C, \quad t \geq 0.$$

Proof. Taking the norm $L_{2,1}$ of (3.1) we obtain

$$||U(\cdot,t)||_{L_{2,1}} \leq ||E(t)\varphi||_{L_{2,1}} + ||G(\cdot,t)||_{L_{2,1}}, \quad t \geq 0, \tag{3.5}$$

where

$$G(\cdot,t) = \int_0^t E(t-\tau)M^{-1}\partial_x F(U(\cdot,\tau))\, d\tau.$$

Let $g(x,t) = \partial_x F(U(x,t))$. To estimate $||G(\cdot,t)||_{L_{2,1}}$ first we observe that

$$G(x,t) = \frac{1}{\sqrt{2\pi}} \sum_{j,l=1}^{2} M_{jl} \int_0^t \int_{\mathbf{R}} e^{ixz}\, \omega_{jl}(z,t-\tau)\, M^{-1}\widehat{g(x,\tau)}\, dz d\tau.$$

Thus, denoting by $K^{jl}(\cdot,t)$ the matrix 2×2 whose entries are $k^{jl}_{mn}(\cdot,t), m,n = 1,2$, and using formula (3.4) we deduce that

$$||G(\cdot,t)||_{L_{2,1}} \leq C \sum_{j,l=1}^{2} \left\{ ||M_{jl}|| \left\| \int_0^t K^{jl}(\cdot,t-\tau) * \partial_x F(U(\cdot,\tau))\, d\tau \right\|_{L_{2,1}} \right\}. \tag{3.6}$$

Using the Young's inequality in each component of the vector $K^{jl}(\cdot,t-\tau) * \partial_x F(U(\cdot,\tau))$ we obtain

$$\left\| \int_0^t K^{jl}(\cdot,t-\tau) * \partial_x F(U(\cdot,t))\, d\tau \right\|_{L_{2,1}} \leq \sum_{m,n=1}^{2} \int_0^t \left(||k^{jl}_{mn}(\cdot,t-\tau)||_{L^1} ||\partial_x F_n(U(\cdot,\tau))||_{L^1} \right) d\tau \tag{3.7}$$

where $F_n(U)$, $n = 1,2$, are the componentes of $F(U)$. We estimate $||\partial_x F_n(U)||_{L^1}$ as follows. Observe that $\partial_x F_n(U) = \partial_1 F_n(U)u_x + \partial_2 F_n(U)v_x$, and $\partial_j F_n : \mathbf{R}^2 \to \mathbf{R}$, $j = 1,2$, are homogeneous functions of degree p. Therefore

$$\begin{aligned}
||\partial_x F_n(U)||_{L^1} &\leq ||\partial_1 F_n(U)||_{L^2}||u_x||_{L^2} + ||\partial_2 F_n(U)||_{L^2}||v_x||_{L^2} \\
&\leq 2^{\frac{p}{2}} [C_{1n}(||u^p||_{L^2} + ||v^p||_{L^2})||u_x||_{L^2} + C_{2n}(||u^p||_{L^2} + ||v^p||_{L^2})||v_x||_{L^2}] \\
&\leq 2^{\frac{1}{2}(p+1)} C_0(||u^p||_{L^2} + ||v^p||_{L^2})||U_x||_{L_{2,2}},
\end{aligned}$$

where $C_{jn} = \sup_{|\eta|=1} |\partial_j F_n(\eta)|$ and $C_0 = \max\{C_{jn}, j,n = 1,2\}$. Then, using the Gagliardo-Nirenberg's inequality $||w||_{L^4} \leq C||w_x||^{1/2}_{L^2}||w||^{1/2}_{L^1}$ valid for any $w \in H^1 \cap L^1$, lemma 3.1 and the hypotesis $p \geq 2$ we obtain

$$\begin{aligned}
||\partial_x F_n(U)||_{L^1} &\leq 2^{\frac{1}{2}(p+1)} C_0(||u||^{p-2}_{L^\infty}||u||^2_{L^4} + ||v||^{p-2}_{L^\infty}||v||^2_{L^4})||U_x||_{L_{2,2}} \\
&\leq C(||u_x||_{L^2}||u||_{L^1} + ||v_x||_{L^2}||v||_{L^1})||U_x||_{L_{2,2}} \\
&\leq C||U_x||^2_{L_{2,2}}||U||_{L_{2,1}}, \quad n = 1,2.
\end{aligned} \tag{3.8}$$

From (3.6),(3.7),(3.8) and lemma 3.2 we conclude that

$$||G(\cdot,t)||_{L_{2,1}} \leq C||U_x||^2_{L_{2,2}}||U||_{L_{2,1}}, \quad \forall t \geq 0.$$

Consequently, by (3.5) and lemma 2.3

$$||U(\cdot,t)||_{L_{2,1}} \leq C + C\int_0^t ||U_x(\cdot,\tau)||^2_{L_{2,2}}||U(\cdot,\tau)||_{L_{2,1}}\, d\tau, \quad 0 \leq t < \infty.$$

Therefore, using lemma 3.1(a) and Gronwall's inequality we obtain lemma 3.3. □

The next lemma is an extension to system (1.1) of part of Theorem 5.1 of [**3**].

Lemma 3.4. *Suppose that* (1.4) *holds and let* $p \geq 2$. *If* $\varphi \in H_{2,2} \cap L^1_{2,2}$, *then there exists a positive constant* $C = C(\varphi)$ *such that*

$$||U(\cdot,t)||_{L_{2,2}} \leq C(1+t)^{-\frac{1}{4}}$$
$$||U_x(\cdot,t)||_{L^2\times L^2} \leq C(1+t)^{-\frac{1}{4}}, \tag{3.9}$$

for all $t \geq 0$.

Proof. In view of (3.2) and lemma 3.3, using Parseval's identity we have

$$\begin{aligned}\frac{d}{dt}[(1+t)\mathcal{E}(t)] &\leq c_2||U(\cdot,t)||^2_{L_{2,2}} - c_3(1+t)||U_x(\cdot,t)||^2_{L_{2,2}} + c_2||U_x(\cdot,t)||^2_{L_{2,2}} \\ &\leq \int_{\Omega_t} |\hat{U}(z,t)|^2 dz + c_2||U_x(\cdot,t)||^2_{L_{2,2}} \\ &\leq C||U_x(\cdot,t)||^2_{L_{2,1}}(1+t)^{-1/2} + c_2||U_x(\cdot,t)||^2_{L_{2,2}},\end{aligned} \tag{3.10}$$

where $c_3 = \min\{d_1^{-1}\varepsilon_1, d_2^{-1}\varepsilon_2\}$ and $\Omega_t = \{z \in \mathbf{R}; |z| \leq [\frac{c_3}{c_2}(1+t)]^{-1/2}\}$. Integrating (3.10) on $\mathbf{R}$ and using (3.3) together with lemma 3.1(b) we obtain (3.9). □

Finally, to estimate $||U(\cdot,t)||_{L_{2,\infty}}$ we will use the following lemma (see [7, p. 74]).

Lemma 3.5. *Let* $\alpha, \beta \in (-1,0]$. *Then, there exists a positive constant C such that*

$$\int_0^t (1+t-\tau)^{\alpha}(1+\tau)^{\beta} d\tau \leq C(1+t)^{\alpha+\beta+1},$$

for all $t \geq 0$.

Lemma 3.6. *Suppose that* (1.4) *holds and let* $p \geq 2$. *If* $\varphi \in H_{2,2} \cap L^1_{2,2}$, *then there exists a positive constant* $C = C(\varphi)$ *such that*

$$||U(\cdot,t)||_{L_{2,\infty}} \leq C(1+t)^{-1/2}, \quad \forall t \geq 0. \tag{3.11}$$

Proof. We will use the notation of lemma 3.3 and also $\rho^{jl}_{mn}(z,t) = \frac{z}{\sigma(z)}\omega_{jl}(z,t)r_{mn}(z)$, $j,l,m,n = 1,2$. First we observe that

$$\begin{aligned}||G(\cdot,t)||_{L_{2,\infty}} &\leq C\sum_{j,l=1}^{2}\left\{||M_{jl}||\,\Big|\Big|\int_0^t \mathcal{F}^{-1}\left(\omega_{jl}(\cdot,t-\tau)\widehat{M^{-1}g(\cdot,\tau)}\right)d\tau\Big|\Big|_{L_{2,\infty}}\right\} \\ &\leq C\sum_{j,l=1}^{2}\left\{||M_{jl}||\sum_{m,n=1}^{2}\int_0^t \Big|\Big|\mathcal{F}^{-1}\left(i\rho^{jl}_{mn}(\cdot,t-\tau)\widehat{F_n(U(\cdot,\tau))}\right)\Big|\Big|_{L^\infty}d\tau\right\} \\ &\leq C\sum_{j,l=1}^{2}\left\{||M_{jl}||\sum_{m,n=1}^{2}\int_0^t ||\rho^{jl}_{mn}(\cdot,t-\tau)||_{L^2}||F_n(U(\cdot,\tau))||_{L^2}d\tau\right\}.\end{aligned} \tag{3.12}$$

Estimating $||\rho^{jl}_{mn}(\cdot,t)||_{L^2}$ as in lemma 2.1 we find

$$||\rho^{jl}_{mn}(\cdot,t)||_{L^2} \leq C(1+t)^{-3/4}, \quad \forall t \geq 0 \text{ and } j,l,m,n = 1,2. \tag{3.13}$$

We estimate $||F_n(U)||^2_{L^2}$ as follows.

$$\begin{aligned}||F_n(U)||^2_{L^2} &\le C_1(||u^{p+1}||^2_{L^2}+||v^{p+1}||^2_{L^2})\\ &\le C_1(||u||^{2p}_{L^\infty}||u||^2_{L^2}+||v||^{2p}_{L^\infty}||v||^2_{L^2})\\ &\le C_1||U||^{2p}_{L_{2,\infty}}||U||^2_{L_{2,2}},\end{aligned} \tag{3.14}$$

where $C_1 = 2^{p+1}\max\{\sup_{|\eta|=1}|F_1(\eta)|,\sup_{|\eta|=1}|F_2(\eta)|\}$, because $F_j:\mathbf{R}^2\to\mathbf{R}$, $j=1,2$, are homogeneous of degree $p+1$. Therefore, using lemma 3.4 and the hypothesis $p\ge 2$ we deduce from (3.14) that

$$||F_n(U)||_{L^2}\le C(1+t)^{-3/4},\quad \forall t\ge 0 \text{ and } n=1,2. \tag{3.15}$$

Now, taking the norm $L_{2,\infty}$ of (3.1) and using (3.13) together with (3.14),(3.15) and lemma 2.1 we conclude that

$$||U(\cdot,t)||_{L_{2,\infty}}\le C(1+t)^{-1/2}+C\int_0^t(1+t-\tau)^{-3/4}(1+\tau)^{-3/4}d\tau.$$

Finally, using lemma 3.5 with $\alpha=-\frac{3}{4}$ and $\beta=-\frac{3}{4}$ we obtain (3.12).

□

The conclusion of Theorem 1.1 now follows from lemmas 3.3 and 3.6 and interpolation.

Next, we give two applications of Theorem 1.1.

Example 3.1 Consider the system (1.3) with initial conditions $u(\cdot,0)=\varphi_1$, $v(\cdot,0)=\varphi_2$ in $H^2\cap L^1_2$. To apply Theorem 1.1 observe that it can be write in the form (1.1) with

$$A=\begin{pmatrix}1 & a_3\\ b_1^{-1}b_2a_3 & b_1^{-1}\end{pmatrix},\quad B=\begin{pmatrix}\alpha_1 & 0\\ 0 & b_1^{-1}\alpha_2\end{pmatrix},\quad D=\begin{pmatrix}b_2^{-1} & 0\\ 0 & b_1^{-1}\end{pmatrix},$$

and

$$H(U)=b_2\frac{u^{p+2}}{(p+1)(p+2)}+b_2a_2\frac{v^{p+2}}{(p+1)(p+2)}+b_2a_1\frac{u^{p+1}}{p+1}v+b_2u\frac{v^{p+1}}{p+1}.$$

A simple calculation shows that condition (1.4) holds if only if $a_3=0$ or $\alpha_2=\alpha_1b_1$. Therefore, under this hypothesis and $p\ge 2$, for all $1\le q\le\infty$ we have

$$||u(\cdot,t)||_{L^q}\le C(1+t)^{-\frac{1+q}{2q}},\quad ||v(\cdot,t)||_{L^q}\le C(1+t)^{-\frac{1+q}{2q}},\quad \forall t\ge 0. \tag{3.16}$$

Example 3.2 Consider the system

$$\begin{aligned}u_t-a_1u_{xxt}-a_2v_{xxt}+b_1(u^rv^{r+1})_x-\varepsilon u_{xx}&=0\\ v_t-a_2v_{xxt}-a_3u_{xxt}+b_1(u^{r+1}v^r)_x-\varepsilon v_{xx}&=0,\end{aligned} \tag{3.17}$$

with initial conditions as in example 3.1, where a_i, b_1, ε are real constants with $b_1>0, \varepsilon>0$, $a_1a_3>a_2^2$ and $r\ge 1$ is an integer. Here, $H(U)=b_1\frac{u^{r+1}v^{r+1}}{r+1}$ is homogeneous of degree $2r+2$. System (3.17) is a BBM-Burgers version of a coupled system studied in [**1**]. Applying Theorem 1.1 we conclude that for all $1\le q\le\infty$ and $r\ge 1$ the decay estimates (3.16) also hold for system (3.17).

References

[1] E. Alarcon, J. Angulo, J.F. Montenegro, Stability and instability of solitary waves for a nonlinear dispersive system, *Nonlinear Analysis* **36**, no.8, Ser.A: Theory Methods (1999), 1015–1035.

[2] J. Albert, F. Linares, Stability and symmetry of solitary-wave solutions to systems modeling interactions of long wave, *J. Math. Pures Appl.* **(9)** 79, no.3 (2000), 195–226.

[3] C.J. Amick, J.L. Bona, M.E. Schonbek, Decay of solutions of some nonlinear wave equations, *J. Diff. Equations* **81** (1989), 1–49.

[4] J.L. Bona, M. Chen, and J-C. Saut, Boussinesq equations and other systems for small-amplitude long waves in nonlinear dispersive media: II. The nonlinear theory, *Nonlinearity* **17** (2004), 925–957.

[5] E. Bisognin, V. Bisognin, and G. Perla Menzala, Exponential stabilization of a coupled system of Korteweg-de Vries equations with localized damping, *Advances in Differential Equations* **8**, nº4 (2003), 443–469.

[6] J.A. Gear, R. Grimshaw, Weak and strong interactions between internal solitary waves. *Studies in Appl. Math.* **70** (1984), 235–258.

[7] G. Karch, L^p-decay of solutions to dissipative-dispersive perturbations of conservations laws, *Ann. Polon. Math.* **67** (1997), 65-86.

[8] J.M. Pereira, Stability and instability of solitary waves for a system of coupled BBM equations, *Applicable Analysis,* **84**, nº8 (2005), 807–819.

[9] J.M. Pereira, Global existence and decay of solutions of a coupled system of BBM-Burgers equations. *Rev. Matemática Complutense* **18**, nº2 (2000), 423–443.

In: Progress in Evolution Equations
Editor: Gaston M. N'Guerekata, pp. 103-123
ISBN: 978-1-60456-328-3

Chapter 6

VARIATIONAL HYPERBOLIC INEQUALITY IN THE DOMAINS UNBOUNDED IN SPATIAL VARIABLES

***Serhiy Lavrenyuk*[1] *and Petro Pukach*[2]**
[1]Ivan Franko National University of Lviv, Lviv, Ukraine
[2]Lviv Polytechnic National University, Lviv, Ukraine

Abstract

The paper deals with the investigation of second order weakly nonlinear variational hyperbolic inequality in domain bounded in time variable and unbounded in spatial variables. We have obtained the conditions of existence and uniqueness of the solution of the inequality with initial data. These conditions do not include any restrictions as to a behavior at infinity of the solution, the nonhomogeneous term and the initial data. The classes of the existence and the uniqueness are spaces of locally integrable functions.

Key words: variational hyperbolic inequality, unilateral problem, unbounded domain, Galerkin method.

1. Introduction

Variational inequalities appear in physics and mechanics problems, namely in elasticity theory and plasticity theory [1, 2], as well as when modelling the cracks and shocks in solids [3] etc. The theory of nonlinear variational (particularly, hyperbolic) inequalities in bounded domains has been actively and widely developed. The papers [4–8] deal with investigating some unilateral problems for strictly hyperbolic operators

$$L_1 u \equiv a_0(u)u_{tt} - \sum_{i=1}^{n} (a_i(u_{x_i}))_{x_i} + f(x,t,u,u_t), \quad a_0 > 0,$$

$$L_2 u \equiv u_{tt} - a(u)\Delta u + f(x,t,u,u_t), \quad a > 0$$

in the bounded domain $Q_T = \Omega \times (0,T)$. In [9, 10], in the domain of the same form, unilateral problems are investigated for nonlocal quasi-linear hyperbolic operator

$$L_3 u \equiv u_{tt} - (-1)^m a \left(\int_\Omega |\nabla_x|^2 dx \right) \Delta_x^m u + f(x,t), \quad m \geq 1.$$

The paper [11] deals with studying a unilateral problem in a bounded domain for the system of quasi-linear hyperbolic inequalities of the form

$$u_{tt} - M(\|\nabla u\|^2_{L^2(\Omega)})\Delta u + \delta |u_t|^\rho u_t \geq \mu |u|^\alpha u,$$

$M(s) = a + bs, \quad a+b \geq 0, \quad b \geq 0, \quad \alpha > 0, \quad \rho \geq 0, \quad \delta > 0, \quad \mu \in \mathrm{R}.$

In [12], there has been proved a one-valued solvability of the unilateral problem in a bounded domain for so called strongly perturbed nonlinear hyperbolic third order operator

$$Lu \equiv k(u)u_{tt} - \Delta u - \sum_{i=1}^{n} (\omega_i(\nabla u_t))_{x_i} + f(u_t) - g(x,t),$$

where ω_i are nonlinear functions; as well as the solvability of the unilateral problem for the nonlinear equation

$$k(u_t)u_{tt} - \Delta u - \sum_{i=1}^{n} (\omega_i(\nabla u_t))_{x_i} + f(u_t) = g(x,t).$$

Similar questions are considered for unilateral problems with the singular hyperbolic operator $k(u_t)u_{tt} - u_{xx} = f(x,t)$ in $Q_T = (0,T) \times (0,1)$, $k(s) \geq 0$ (see [13]) as well as for high order variational inequalities (see [14]).

Some results on existence and uniqueness of solutions of hyperbolic variational inequalities have been obtained also for unbounded domains. Namely, in the paper [15], the unilateral problem in unbounded in t cylinder $(0,\infty) \times \Omega$ is considered for the equation $u_{tt} - \Delta u + a(u) = f$ with the constraint $|u(x,t)| \leq \varphi(t)$. Also, the paper [7] deals with the unilateral problem with the constraint $|u_{xt}| \leq \theta(t)$ for the equation $u_{tt} - u_{xx} + \alpha u_t + f(x,t,u) = 0$. Linear hyperbolic variational inequalities in the half-space $(x,t) \in \mathrm{R}^n_+ \times \mathrm{R}$, as well as their physical interpretation, are studied in [3].

The problems for weakly nonlinear hyperbolic equations and systems of the form $u_{tt} - \Delta u + A|u|^\rho = f$, $\rho > 0$ in unbounded domains have been considered by many authors (see [16-27]). It has been assumed that the degree of nonlinearity is considerably connected with the dimension of the domain in which the problem is being investigated, and that the coefficients of the elliptic operator are bounded. The results on the existence and uniqueness of the solutions in unbounded domains have been obtained in those papers under the assumption of a certain behavior of the solution, the initial data and the free term of the equation at infinity, or without such an assumption. In [27], a mixed problem for weakly nonlinear system of hyperbolic equations with two independent variables has been studied.

We investigate a weakly nonlinear variational hyperbolic inequality of the second order in the domain unbounded in spatial variables. We consider the case of continuous coefficients of the elliptic operator, which grow in a certain way at infinity. The degree of power nonlinearity depends on the dimension of domain. We obtain the conditions of existence and uniqueness of the generalized solution of the inequality with the initial condition, without any restrictions on solution behavior, the free term of the inequality and the initial data as $|x| \to \infty$. We also propose some examples of unilateral problems in the domain unbounded in spatial variables, which could be investigated by means of the main result obtained in the present paper.

It should be noted, that when proving the corresponding results for the variational inequalities solutions, we have used a cutoff function introduced in the paper [28].

2. Problem Definition. Main Results Statement

We suppose $\Omega \subset R^n$ to be an unbounded domain with the boundary $\partial\Omega \in C^1$, the set $\Omega_R = \Omega \cap \{x \in R^n : |x| < R\}$, for arbitrary $R > 1$, to be a domain with the regular by Calderon (see [29], p.45) boundary $\partial\Omega_R$, $\bigcup\limits_{R>1} \Omega_R = \Omega$.

Remark 1. In particular, the convex domain Ω satisfies all the conditions mentioned above (see [29], p.46, rem. 1.11).

Denote $Q_\tau = \Omega \times (0,\tau)$, $Q_{R,\tau} = \Omega_R \times (0,\tau)$, $\Omega_{R,\tau} = Q_{R,T} \cap \{t : t = \tau\}$, $\Omega_\tau = Q_T \cap \{t : t = \tau\}$, $\partial\Omega_R = \Gamma_1^R \cup \Gamma_2^R$, $\Gamma_1^R = \partial\Omega \cap \partial\Omega_R$, $\Gamma_2^R = \partial\Omega_R \backslash \Gamma_1^R$, $\tau \in (0,T]$.

We shall use the following functional spaces:

$$H^1_{0,\Gamma_1^R}(\Omega_R) = \left\{u \in H^1(\Omega_R) : u|_{\Gamma_1^R} = 0\right\},\ H^1_{0,loc}(\overline{\Omega}) = \left\{u : u \in H^1_{0,\Gamma_1^R}(\Omega_R)\ \forall R > 1\right\},$$

$$H^1_{loc}(\overline{\Omega}) = \left\{u : u \in H^1(\Omega_R)\ \forall R > 1\right\},\ H^2_{loc}(\overline{\Omega}) = \left\{u : u \in H^2(\Omega_R)\ \forall R > 1\right\},$$

$$L^r_{loc}(\overline{\Omega}) = \{u : u \in L^r(\Omega_R)\ \forall R > 1\},\ r \in (1,+\infty).$$

Let $V_1(\Omega_R)$ be a Hilbert space such that $H^1_{0,\Gamma_1^R}(\Omega_R) \subset V_1(\Omega_R) \subset H^1(\Omega_R)$ holds for all $R > 1$, and, for all R_1, R_2 $(1 < R_1 < R_2)$, the inclusion $v \in V_1(\Omega_{R_2})$ implies $v \in V_1(\Omega_{R_1})$. Let us introduce the space $V_{1,0}(\Omega_R) = \{v \in V_1(\Omega_R) : v|_{\Gamma_2^R} = 0\}$, and by $V_{1,loc}(\overline{\Omega})$ we denote the space of functions v such that $v \in V_1(\Omega_R)$ holds for all $R > 1$.

In the domain Q_T, consider the variational hyperbolic inequality

$$\int\limits_{Q_\tau} \left[v_t(v-u_t)\psi(x) + \sum_{i,j=1}^{n} a_{ij}(x) u_{x_i}((v-u_t)\psi(x))_{x_j} \right] dxdt +$$

$$+ \int\limits_{Q_\tau} \left[|u_t|^{p-2} u_t (v-u_t)\psi(x) - f(x,t)(v-u_t)\psi(x) \right] dxdt \geq$$

$$\geq \frac{1}{2}\int\limits_{\Omega_\tau} (v-u_t)^2 \psi(x) dx - \frac{1}{2}\int\limits_{\Omega_0} (v-u_1)^2 \psi(x) dx \tag{1}$$

with the initial condition

$$u(x,0) = u_0(x), \quad x \in \Omega, \tag{2}$$

where $\tau \in (0,T]$, $p > 2$, $\psi \in C_0^1(R^n)$ is an arbitrary function such that $\psi(x) \geq 0$, $x \in R^n$.

Let K be a convex closed cone such that $K \subset V_{1,loc}(\overline{\Omega})$, $\phi K \subset K$, where $\phi \in C^1(R^n)$, $\phi(x) \geq 0$, $x \in R^n$.

As to the coefficients, the free term (1) and the initial data, we assume the following conditions to hold:

(A) functions a_{ij} $(i,j=1,...,n)$ belong to space $C(\overline{\Omega})$, moreover $\sum_{i,j=1}^{n} a_{ij}(x)\xi_i\xi_j \geq a_0 |\xi|^2$, $a_0 > 0$ for any vector $\xi = (\xi_1,...,\xi_n) \in \mathrm{R}^n$ and for all $x \in \Omega$; $a_{ij}(x) = a_{ji}(x)$, $|a_{ij}(x)| \leq a_1 (1+|x|^{\alpha})$ as $|x| \to +\infty$, $a_1 > 0$, $0 \leq \alpha < 1 - \frac{n(p-2)}{2p}$ for any $i,j = 1,...,n$ and for all $x \in \Omega$; $a_{ij_{x_j}} \in L^{\infty}(\Omega)$ for $i,j = 1,...,n$ and for almost all $x \in \Omega$.

(F) $f \in L^{p'}((0,T);L^{p'}_{loc}(\overline{\Omega}))$, $p' = p/(p-1)$.

(U) $u_0 \in V_{1,loc}(\overline{\Omega})$, $u_1 \in L^2_{loc}(\overline{\Omega}) \cap \mathrm{K}$.

Definition 1. Let a function u belongs to $C([0,T];V_{1,loc}(\overline{\Omega}))$. Besides, for this function, suppose the following to hold:

$$u_t \in \mathrm{K} \text{ for almost all } t \in (0,T),$$

$$u_t \in C([0,T];V_{1,loc}(\overline{\Omega})) \cap L^p((0,T);L^p_{loc}(\overline{\Omega})).$$

Let u satisfy (1) for all $\tau \in (0,T]$, for all functions $\psi \in C^1_0(\mathrm{R}^n))$, $\psi(x) \geq 0$, $x \in \mathrm{R}^n$, and for all functions v such that $v \in L^p((0,T);L^p_{loc}(\overline{\Omega})) \cap L^2((0,T);V_{1,loc}(\overline{\Omega}))$, $v_t \in L^2((0,T);L^2_{loc}(\overline{\Omega}))$, $v \in \mathrm{K}$ for almost all $t \in (0,T)$. Then u is called a ***strong solution*** of the variational hyperbolic inequality (1) with the initial condition (2).

Remark 2. It is easily seen that a strong solution of inequality (1) also satisfies the initial condition

$$u_t(x,0) = u_1(x), \quad x \in \Omega.$$

Definition 2. Let $\{f^k\}$, $\{u^k_0\}$, $\{u^k_1\}(k=1,2,...)$ be some function sequences such that

$$f^k \to f \text{ in the space } L^{p'}((0,T);L^{p'}_{loc}(\overline{\Omega})),$$

$$u^k_0 \to u_0 \text{ in the space } V_{1,loc}(\overline{\Omega}),$$

$$u^k_1 \to u_1 \text{ in the space } L^2_{loc}(\overline{\Omega})$$

as $k \to \infty$. Let for each $k \in \mathrm{N}$, the function u^k be a strong solution of the inequality

$$\int_{Q_\tau} \left[v_t(v-u_t)\psi(x) + \sum_{i,j=1}^{n} a_{ij}(x)u_{x_i}((v-u_t)\psi(x))_{x_j} \right] dxdt +$$

$$+ \int_{Q_\tau} \left[|u_t|^{p-2} u_t(v-u_t)\psi(x) - f^k(v-u_t)\psi(x) \right] dxdt \geq$$

$$\geq \frac{1}{2} \int_{\Omega_\tau} (v-u_t)^2 \psi(x)dx - \frac{1}{2} \int_{\Omega_0} \left(v-u^k_1\right)^2 \psi(x)dx$$

with the initial condition

$$u(x,0) = u^k_0(x), \quad x \in \Omega$$

and under condition that $\operatorname{supp} \psi \in \Omega_{k+1}$. If

$$u^k \to u \text{ in the space } C([0,T];V_{1,loc}(\overline{\Omega})),$$

$$u_t^k \to u_t \text{ in the space } C([0,T]; L^2_{loc}(\overline{\Omega})) \cap L^p((0,T); L^p_{loc}(\overline{\Omega}))$$

at $k \to \infty$, then the function u is called a ***weak solution*** of the variational hyperbolic inequality (1) with the initial condition (2).

Theorem 1. *Let conditions **(A)**, **(F)**, **(U)** hold, $n < \frac{2p}{p-2}$. Then a weak solution u of inequality (1) with initial condition (2) is unique.*

Let $K_R = K \cap \Omega_R$, $V(\Omega_R)$ be a Banach space such that $V_1(\Omega_R)$ is continuously and compactly embedded into $V(\Omega_R)$ for any $R > 1$, $V^*(\Omega_R)$ be an adjoint space for $V(\Omega_R)$. By $\langle \cdot, \cdot \rangle$, denote a scalar product of spaces $V^*(\Omega_R)$ and $V(\Omega_R)$. Suppose that there exists a monotone, bounded, semi-continuous operator $\beta_R : V(\Omega_R) \to V^*(\Omega_R)$ for which

$$K_R = \{v : v \in V_1(\Omega_R), \beta_R(v) = 0\}.$$

We also suppose β_R to satisfy the condition

(B) $\int_0^\tau \langle \beta_R(u) - \beta_R(v), (u-v)\psi \rangle dt \geq 0, \quad \int_0^\tau \langle (\beta_R(u))_t, u_{tt} \rangle dt \geq 0$

for all $\tau \in (0,T]$, for all functions $u, v \in L^2((0,T); V_{1,0}(\Omega_R))$, $u_t, u_{tt} \in L^2(Q_{R,T})$, $\psi \in C^1(\mathrm{R}^n)$, $\psi(x) \geq 0$, $x \in \mathrm{R}^n$.

Let $R > 1$ be an arbitrary fixed number. Consider the function ψ_R defining one as follows: $\psi_R(x) = \frac{R^2 - |x|^2}{R}$, at $|x| \leq R$; $\psi_R(x) = 0$, when $|x| > R$. Since $\frac{\partial \psi_R(x)}{\partial x_i} = -\frac{2x_i}{R}$, $|x| \leq R$, $i = 1, \dots, n$, we conclude that, for any real number $\beta > 1$, there holds a following estimate:

$$\left| \frac{\partial [\psi_R(x)]^\beta}{\partial x_i} \right| \leq 2\beta [\psi_R(x)]^{\beta - 1}. \tag{3}$$

Moreover, one can easily obtain the following:

$$R - |x| \leq \psi_R(x) \leq 2(R - |x|). \tag{4}$$

Theorem 2. *Let condition **(A)** be fulfilled, $p > 2$, and, besides, $f^R, f_t^R \in L^2(Q_{R,T})$, $u_0^R \in H^2(\Omega_R) \cap V_{1,0}(\Omega_R)$, $u_1^R \in V_{1,0}(\Omega_R) \cap L^{2p-2}(\Omega_R) \cap K_R$. Then there exists a function $u \in C([0,T]; V_{1,0}(\Omega_R))$ such that $u_t \in C([0,T]; V_{1,0}(\Omega_R)) \cap L^p(Q_{R,T})$, $u_t \in K_R$ for almost all $t \in (0,T)$, which satisfies the inequality*

$$\int_{Q_{R,\tau}} \left[v_t(v-u_t)\psi(x) + \sum_{i,j=1}^n a_{ij}(x) u_{x_i} ((v-u_t)\psi(x))_{x_j} \right] dxdt +$$

$$+ \int_{Q_{R,\tau}} \left[|u_t|^{p-2} u_t (v-u_t)\psi(x) - f^R(x,t)(v-u_t)\psi(x) \right] dxdt \geq$$

$$\geq \frac{1}{2} \int_{\Omega_{R,\tau}} (v-u_t)^2 \psi(x) dx - \frac{1}{2} \int_{\Omega_{R,0}} \left(v - u_1^R\right)^2 \psi(x) dx$$

with the initial condition

$$u(x,0) = u_0^R(x), \quad x \in \Omega_R$$

for all $\tau \in (0,T]$, *for all* $\psi \in C^1(\mathbb{R}^n)$, $\psi(x) \geq 0$ *in* $\mathbb{R}^n$ *and for all* $v \in C([0,T];V_{1,0}(\Omega_R)) \cap L^p(Q_{R,T})$ *such that* $v \in \mathrm{K}_R$ *for almost all* $t \in (0,T)$.

Theorem 3. *Let conditions* **(A)**, **(F)**, **(U)** *be fulfilled,* $n < \frac{2p}{p-2}$. *Then there exists a weak solution* u *of inequality (1) with initial condition (2).*

3. Proof of the Uniqueness Theorem

Suppose that there exist two weak solutions $u^{(1)}$, $u^{(2)}$ of inequality (1) with initial condition (2). Then, by Definition 2, there exist sequences $\{u_1^k\}$, $\{u_2^k\}$ of strong solutions of inequality (1), corresponding to the free terms $\{f_1^{(k)}\}$, $\{f_2^{(k)}\}$ and the initial functions $\{u_{0,1}^{(k)}\}$, $\{u_{1,1}^{(k)}\}$, $\{u_{0,2}^{(k)}\}$, $\{u_{1,2}^{(k)}\}$, where $\{f_l^{(k)}\}$ converge to the function f in the space $L^{p'}((0,T);L^{p'}_{loc}(\overline{\Omega}))$, $\{u_{0,l}^{(k)}\}$ converge to the function u_0 in the space $V_{1,loc}(\overline{\Omega})$, $\{u_{1,l}^{(k)}\}$ converge to the function u_1 in the space $L^2_{loc}(\overline{\Omega})$, $l = 1,2$. Besides,

$$u_l^k \to u_l \text{ in the space } C([0,T];V_{1,loc}(\overline{\Omega})),$$

$$u_{lt}^k \to u_{lt} \text{ in the space } C([0,T];L^2_{loc}(\overline{\Omega})) \cap L^p((0,T);L^p_{loc}(\overline{\Omega}))$$

as $k \to \infty$.

Let $R > 1$ be an arbitrary fixed number. Then, for all functions $v \in L^p((0,T);L^p_{loc}(\overline{\Omega})) \cap L^2((0,T);V_{1,loc}(\overline{\Omega}))$ such that $v_t \in L^2((0,T);L^2_{loc}(\overline{\Omega}))$, and $v \in \mathrm{K}$ for almost all $t \in (0,T]$; for all functions $\psi \in C_0^1(\mathbb{R}^n)$, $\psi(x) \geq 0$, $x \in \mathbb{R}^n$, $\operatorname{supp}\psi \in \Omega_R$, the following inequalities hold:

$$\int_{Q_\tau} \left[v_t(v-u_{lt}^k)\psi(x) + \sum_{i,j=1}^n a_{ij}(x) u_{lx_i}^k ((v-u_{lt}^k)\psi(x))_{x_j} \right] dxdt -$$

$$+ \int_{Q_\tau} \left[|u_{lt}^k|^{p-2} u_{lt}^k (v-u_{lt}^k)\psi(x) - f_l^{(k)}(v-u_{lt}^k)\psi(x) \right] dxdt \geq$$

$$\geq \frac{1}{2} \int_{\Omega_\tau} \left(v-u_{lt}^k\right)^2 \psi(x)dx - \frac{1}{2} \int_{\Omega_0} \left(v-u_{1,l}^{(k)}\right)^2 \psi(x)dx, \tag{5}$$

where $l = 1,2$, $k > [R]+1$. Let us choose $\beta > \frac{2p}{p-2}$ in each of the inequalities (5) and assume $\psi = \psi_R^\beta$, $v = \frac{1}{2}\left(u_{1t}^k + u_{2t}^k\right)e^{-\gamma t}$, $\gamma > 0$, and finally sum up both inequalities. Denote $w = u_1^k - u_2^k$. Taking into account the initial conditions for u_l^k, $l = 1,2$ and

$$\int_{\Omega_\tau} \left[\left(v-u_{1t}^k\right)^2 + \left(v-u_{2t}^k\right)^2 \right] \psi(x)dx \geq 0,$$

we obtain the following inequality:

$$\int_{Q_\tau} \left[\sum_{i,j=1}^n a_{ij}(x) w_{x_i}(w_t\psi_R^\beta(x))_{x_j} + \frac{\gamma}{2}|w_t|^2\psi_R^\beta(x) + (|u_{1t}^k|^{p-2}u_{1t}^k - |u_{2t}^k|^{p-2}u_{2t}^k) w_t \psi_R^\beta(x) \right] \times$$

$$\times e^{-\gamma t}\,dxdt \le \int\limits_{Q_\tau}\left(f_1^{(k)}-f_2^{(k)}\right)w_t\psi_R^\beta(x)e^{-\gamma t}dxdt+\frac{1}{2}\int\limits_{\Omega_0}\left(u_{1,1}^{(k)}-u_{1,2}^{(k)}\right)^2\psi_R^\beta(x)dx. \quad (6)$$

Let us transform the terms of inequality (6). First of all, note that

$$\int\limits_{Q_\tau}\sum_{i,j=1}^n a_{ij}(x)w_{x_i}\left(w_t\psi_R^\beta(x)\right)_{x_j}e^{-\gamma t}dxdt=\int\limits_{Q_\tau}\sum_{i,j=1}^n a_{ij}(x)w_{x_i}w_{tx_j}\psi_R^\beta(x)e^{-\gamma t}dxdt+$$

$$+\int\limits_{Q_\tau}\sum_{i,j=1}^n a_{ij}(x)w_{x_i}w_t\left(\psi_R^\beta(x)\right)_{x_j}e^{-\gamma t}dxdt=I_1+I_2.$$

Using the condition **(A)** and the Hölder inequality, we can estimate the integrals obtained above:

$$I_1=\frac{1}{2}\int\limits_{Q_\tau}\frac{\partial}{\partial t}\left(\sum_{i,j=1}^n a_{ij}(x)w_{x_i}w_{x_j}\psi_R^\beta(x)\right)e^{-\gamma t}dxdt\ge\frac{a_0}{2}\int\limits_{\Omega_\tau}|\nabla w|^2\psi_R^\beta(x)e^{-\gamma\tau}dx+$$

$$+\frac{a_0\gamma}{2}\int\limits_{Q_\tau}|\nabla w|^2\psi_R^\beta(x)e^{-\gamma t}dx-\frac{C(R)}{2}\int\limits_{\Omega_0}\left|\nabla(u_{0,1}^{(k)}-u_{0,2}^{(k)})\right|^2\psi_R^\beta(x)dx;$$

$$I_2\le a_1\int\limits_{Q_\tau}\sum_{i,j=1}^n\frac{\left|w_{x_i}\psi_R^{\frac{\beta}{2}}(x)w_t\psi_R^{\frac{\beta}{p}}(x)\left(\psi_R^\beta(x)\right)_{x_j}\right|}{\psi_R^{\frac{\beta}{2}+\frac{\beta}{p}}(x)}e^{-\gamma t}dxdt\le$$

$$\le a_1\left(\int\limits_{Q_\tau}|\nabla w|^2\psi_R^\beta e^{-\gamma t}dxdt\right)^{1/2}\cdot\left(\int\limits_{Q_\tau}|w_t|^p\psi_R^\beta e^{-\gamma t}dxdt\right)^{1/p}\times$$

$$\times\left(\int\limits_{Q_\tau}\sum_{j=1}^n\left|\frac{\left(\psi_R^\beta\right)_{x_j}R^\alpha}{\psi_R^{\frac{\beta}{2}+\frac{\beta}{p}}}\right|^{p_1}e^{-\gamma t}dxdt\right)^{1/p_1},$$

where $\frac{1}{2}+\frac{1}{p}+\frac{1}{p_1}=1$. Using the Young inequality and taking into account (3), we obtain as follows:

$$I_2\le\frac{\delta_1}{2}\int\limits_{Q_\tau}|\nabla w|^2\psi_R^\beta e^{-\gamma t}dxdt+\frac{\delta_1}{2}\int\limits_{Q_\tau}|w_t|^p\psi_R^\beta e^{-\gamma t}dxdt+$$

$$+\frac{C(\delta_1)}{2}\int\limits_{Q_\tau}\frac{\beta\psi_R^{(\beta-1)p_1}R^{\alpha p_1}}{\psi_R^{\beta(p_1-1)}}e^{-\gamma t}dxdt\le\frac{\delta_1}{2}\int\limits_{Q_\tau}|\nabla w|^2\psi_R^\beta e^{-\gamma t}dxdt+$$

$$+\frac{\delta_1}{2}\int\limits_{Q_\tau}|w_t|^p\psi_R^\beta e^{-\gamma t}dxdt+\frac{C(\delta_1)}{2}R^{\beta+(\alpha-1)\frac{2p}{p-2}+n},\ \delta_1>0.$$

Since $p>2$, we conclude that

$$\int\limits_{Q_\tau}\left(|u_{1t}^k|^{p-2}u_{2t}^k-|u_{2t}^k|^{p-2}u_{2t}^k\right)\left(u_{1t}^k-u_{2t}^k\right)\psi_R^\beta e^{-\gamma t}dxdt\ge$$

$$\geq 2^{2-p}\int\limits_{Q_\tau}|w_t|^p\psi_R^\beta e^{-\gamma t}dxdt.$$

Moreover,

$$\int\limits_{Q_\tau}\left(f_1^{(k)}-f_2^{(k)}\right)w_t\psi_R^\beta(x)e^{-\gamma t}dxdt\leq\frac{\delta_2}{2}\int\limits_{Q_\tau}|w_t|^p\psi_R^\beta(x)e^{-\gamma t}dxdt+$$

$$+\frac{C(\delta_2)}{2}\int\limits_{Q_\tau}|f_1^{(k)}-f_2^{(k)}|^{p'}\psi_R^\beta(x)e^{-\gamma t}dxdt,\ \delta_2>0.$$

The above estimates of the terms of (6), yield the following inequality:

$$(\gamma a_0-\delta_1)\int\limits_{Q_\tau}|\nabla w|^2\psi_R^\beta(x)e^{-\gamma t}dxdt+(2^{3-p}-\delta_1-\delta_2)\int\limits_{Q_\tau}|w_t|^p\psi_R^\beta e^{-\gamma t}dxdt+$$

$$+a_0\int\limits_{\Omega_\tau}|\nabla w|^2\psi_R^\beta(x)e^{-\gamma\tau}dx+\gamma\int\limits_{Q_\tau}|w_t|^2\psi_R^\beta e^{-\gamma t}dxdt\leq$$

$$\leq C(\delta_1)R^{\beta+(\alpha-1)\frac{2p}{p-2}+n}+C(\delta_2)\int\limits_{Q_\tau}|f_1^{(k)}-f_2^{(k)}|^{p'}\psi_R^\beta(x)e^{-\gamma t}dxdt+$$

$$+C(R)\int\limits_{\Omega_0}\left|\nabla(u_{0,1}^{(k)}-u_{0,2}^{(k)})\right|^2\psi_R^\beta(x)dx+\int\limits_{\Omega_0}\left(u_{1,1}^{(k)}-u_{1,2}^{(k)}\right)^2\psi_R^\beta(x)dx.$$

Let $R_0>1$ be an arbitrary fixed number, $R>R_0$. Taking into account condition (4), and properly choosing the constants δ_1, δ_2, γ, we obtain the estimate

$$\int\limits_{\Omega_{R_0,\tau}}|\nabla w|^2dx+\int\limits_{Q_{R_0,T}}\left(|\nabla w|^2+|w_t|^p dxdt+|w_t|^2\right)dxdt\leq$$

$$\leq C_1\left(\frac{R}{R-R_0}\right)^\beta R^{n+(\alpha-1)\frac{2p}{p-2}}+C_1\left(\frac{R}{R-R_0}\right)^\beta\left[\int\limits_{Q_T}|f_1^{(k)}-f_2^{(k)}|^{p'}dxdt+\right.$$

$$\left.+\int\limits_{\Omega_0}\left(u_{1,1}^{(k)}-u_{1,2}^{(k)}\right)^2dx\right]+C(R)\int\limits_{\Omega_0}\left|\nabla(u_{0,1}^{(k)}-u_{0,2}^{(k)})\right|^2dx,\ \tau\in[0,T],$$

where C_1 does not depend on k, R_0, R. Let $\varepsilon>0$ be any arbitrarily small positive number. Since $\lim\limits_{R\to+\infty}\left(\frac{R}{R-R_0}\right)^\beta=1$ and, by theorem's conditions, $\lim\limits_{R\to+\infty}R^{n+(\alpha-1)\frac{2p}{p-2}}=0$, we conclude that there exists such $R_1>R_0$ that

$$C_1\left(\frac{R_1}{R_1-R_0}\right)^\beta R_1^{n+(\alpha-1)\frac{2p}{p-2}}<\frac{\varepsilon}{9}.$$

Taking into account the convergence of the sequences $\{f_l^{(k)}\}$, $\{u_{0,l}^{(k)}\}$, and $\{u_{1,l}^{(k)}\}$, $l = 1,2$, we can choose such $k_0 \in \mathrm{N}$, $k_0 > [R_1]+1$, that, for all $k > k_0$, there hold the estimates

$$C_1\left(\frac{R_1}{R_1-R_0}\right)^{\beta}\left[\int\limits_{Q_{R_1,T}} |f_1^{(k)}-f_2^{(k)}|^{p'}dxdt + \int\limits_{\Omega_{R_1,0}} \left(u_{1,1}^{(k)}-u_{1,2}^{(k)}\right)^2 dx\right] < \frac{\varepsilon}{9},$$

$$C(R_1)\int\limits_{\Omega_{R_1,0}} \left|\nabla(u_{0,1}^{(k)}-u_{0,2}^{(k)})\right|^2 dx < \frac{\varepsilon}{9}.$$

Consequently,

$$\int\limits_{\Omega_{R_0,\tau}} |\nabla w|^2\,dx + \int\limits_{Q_{R_0,T}} \left(|\nabla w|^2+|w_t|^p+|w_t|^2\right)dxdt < \frac{\varepsilon}{3}, \quad \tau\in[0,T].$$

By virtue of the convergence of the sequence $\{u_l^k\}$ in the space $C([0,T];V_{1,loc}(\overline{\Omega}))$ to the function u_l and of the convergence of the sequence $\{u_{lt}^k\}$ in the space $C([0,T];L^2_{loc}(\overline{\Omega}))\cap L^p((0,T);L^p_{loc}(\overline{\Omega}))$ to the function u_{lt}, there exists such $k_1\in\mathrm{N}$, $k_1\geq k_0$, that for all $k>k_1$ the following estimate holds:

$$\int\limits_{\Omega_{R_0,\tau}} \left|\nabla(u_l^k-u_l)\right|^2 dx + \int\limits_{Q_{R_0,T}} \left(\left|\nabla(u_l^k-u_l)\right|^2+\left|(u_l^k-u_l)_t\right|^p+\left|(u_l^k-u_l)_t\right|^2\right)dxdt < \frac{\varepsilon}{6},$$

$\tau\in[0,T]$, $l=1,2$. Using an elementary estimate for the square of sum of three terms, we obtain the inequality as follows:

$$\int\limits_{\Omega_{R_0,\tau}} |\nabla(u_1-u_2)|^2\,dx + \int\limits_{Q_{R_0,T}} \left(|\nabla(u_1-u_2)|^2+|(u_1-u_2)_t|^p+|(u_1-u_2)_t|^2\right)dxdt < \varepsilon,$$

$\tau\in[0,T]$. This implies that $u_1=u_2$ almost everywhere in $Q_{R_0,T}$. Taking into account the arbitrariness of the R_0, we obtain the statement of Theorem 1.

4. Proof of Theorem 2

Let $R>1$ be an arbitrary fixed number, $\{\omega^k\}$ be a "basis" in the function space $H^2(\Omega_R)\cap V_{1,0}(\Omega_R)\cap L^{2p-2}(\Omega_R)$, orthonormal in $L^2(\Omega_R)$. Consider the Galerkin approximations sequence $u^N(x,t)=\sum\limits_{k=1}^{N} C_k^N(t)\omega^k(x)$, $N=1,2,\ldots$, where $C_1^N,\ldots,C_N^N$ are solutions of the following Cauchy problem for a system of ordinary differential equations:

$$\int\limits_{\Omega_R}\left[u_{tt}^N\omega^k+\sum_{i,j=1}^{n} a_{ij}(x)u_{x_i}^N\omega_{x_j}^k+\left|u_t^N\right|^{p-2}u_t^N\omega^k - f^R\omega^k\right]dx +$$

$$+\left\langle\frac{1}{\varepsilon}\beta_R(u_t^N),\omega^k\right\rangle = 0, \qquad (7)$$

$$C_k^N(0) = u_{0,k}^{R,N}, \quad k = 1,\dots,N,$$

$$u_0^{R,N}(x) = \sum_{k=1}^{N} u_{0,k}^{R,N}\omega^k(x), \quad \left\| u_0^{R,N} - u_0^R \right\|_{H^2(\Omega_R)} \to 0, \quad N \to +\infty;$$

$$\left.\frac{\partial C_k^N}{\partial t}\right|_{t=0} = u_{1,k}^{R,N}, \quad k = 1,\dots,N,$$

$$u_1^{R,N}(x) = \sum_{k=1}^{N} u_{1,k}^{R,N}\omega^k(x), \quad \left\| u_1^{R,N} - u_1^R \right\|_{L^{2p-2}(\Omega_R)} \to 0, \quad N \to +\infty.$$

Owing to the Carathéodory theorem (see [30], p.54), there exists a continuous solution of such a Cauchy problem, which has an absolutely continuous derivative with respect to t on the interval $[0,t_0]$. The estimates obtained below imply $t_0 = T$. We multiply each equation of the system (7), by $C_{kt}^N e^{-\gamma t}$, then sum up all the equations over k from 1 to N, and finally integrate the result over the interval $[0,\tau]$, $0 < \tau \le T$. Completing of these operations yields the equality as follows:

$$\int_{Q_{R,\tau}} \left[u_{tt}^N u_t^N + \sum_{i,j=1}^{n} a_{ij}(x) u_{x_i}^N u_{tx_j}^N + |u_t^N|^p - f^R u_t^N \right] e^{-\gamma t} dxdt +$$

$$+\frac{1}{\varepsilon}\int_0^\tau \langle \beta_R(u_t^N), u_t^N \rangle e^{-\gamma t} dt = 0. \tag{8}$$

Taking into account the conditions of Theorem 2, we estimate the terms of equality (8):

$$\int_{Q_{R,\tau}} \sum_{i,j=1}^{n} a_{ij}(x) u_{x_i}^N u_{tx_j}^N e^{-\gamma t} dxdt = \frac{1}{2}\int_{\Omega_{R,\tau}} \sum_{i,j=1}^{n} a_{ij}(x) u_{x_i}^N u_{x_j}^N e^{-\gamma\tau} dxdt -$$

$$-\frac{1}{2}\int_{\Omega_{R,0}} \sum_{i,j=1}^{n} a_{ij}(x) (u_0^{R,N})_{x_i} (u_0^{R,N})_{x_j} dx + \frac{\gamma}{2}\int_{Q_{R,\tau}} \sum_{i,j=1}^{n} a_{ij}(x) u_{x_i}^N u_{x_j}^N e^{-\gamma t} dxdt \ge$$

$$\ge \frac{a_0}{2}\int_{\Omega_{R,\tau}} |\nabla u^N|^2 e^{-\gamma\tau} dx + \frac{\gamma a_0}{2}\int_{Q_{R,\tau}} |\nabla u^N|^2 e^{-\gamma t} dxdt - \frac{M_1}{2}\left\| u_0^{R,N} \right\|_{H^1(\Omega_R)}^2,$$

where $M_1 > 0$ is a certain constant which does not depend on N;

$$\int_{Q_{R,\tau}} u_{tt}^N u_t^N e^{-\gamma t} dxdt = \frac{1}{2}\int_{\Omega_{R,\tau}} |u_t^N|^2 e^{-\gamma\tau} dx - \frac{1}{2}\int_{\Omega_{R,0}} \left| u_1^{R,N} \right|^2 dx +$$

$$+\frac{\gamma}{2}\int_{Q_{R,\tau}} |u_t^N|^2 e^{-\gamma t} dxdt;$$

$$\int_{Q_{R,\tau}} f^R u_t^N e^{-\gamma t} dxdt \le \frac{1}{2}\int_{Q_{R,\tau}} |u_t^N|^2 e^{-\gamma t} dxdt + \frac{1}{2}\int_{Q_{R,\tau}} |f^R|^2 e^{-\gamma t} dxdt.$$

Using these estimates, from (8) we obtain the inequality

$$\int\limits_{\Omega_{R,\tau}} \left|u_t^N\right|^2 e^{-\gamma\tau} dx + (\gamma - 1) \int\limits_{Q_{R,\tau}} \left|u_t^N\right|^2 e^{-\gamma t} dxdt + a_0 \int\limits_{\Omega_{R,\tau}} \left|\nabla u^N\right|^2 e^{-\gamma\tau} dx +$$

$$+ \gamma a_0 \int\limits_{Q_{R,\tau}} \left|\nabla u^N\right|^2 e^{-\gamma t} dxdt + 2 \int\limits_{Q_{R,\tau}} \left|u_t^N\right|^p e^{-\gamma t} dxdt +$$

$$+ \frac{2}{\varepsilon} \int\limits_0^T \langle \beta_R(u_t^N), u_t^N \rangle e^{-\gamma t} dt \leq M_2 \Bigg(\left\|u_0^R\right\|^2_{H^1(\Omega_R)} +$$

$$+ \left\|u_1^R\right\|^2_{L^2(\Omega_R)} + \left\|f^R\right\|^2_{L^2(Q_{R,T})} \Bigg), \tag{9}$$

where $M_2 > 0$ does not depend on N. Note that inequality (9) implies the estimate

$$\int\limits_0^T \langle \beta_R(u_t^N), u_t^N \rangle e^{-\gamma t} dt \leq \varepsilon M_3. \tag{10}$$

Taking into account (9), (10), the boundedness of the operator β_R, and choosing $\gamma > 1$, one can conclude the following: there exists a sequence $\{u^{N_k}\} \subset \{u^N\}$ such that

$$u^{N_k} \to u^{R,\varepsilon} \quad * - \quad \text{weakly} \quad \text{in} \quad L^\infty((0,T); V_{1,0}(\Omega_R)),$$

$$u^{N_k} \to u^{R,\varepsilon} \quad \text{weakly} \quad \text{in} \quad L^2((0,T); V_{1,0}(\Omega_R)),$$

$$u_t^{N_k} \to u_t^{R,\varepsilon} \quad * - \quad \text{weakly} \quad \text{in} \quad L^\infty((0,T); L^2(\Omega_R)),$$

$$u_t^{N_k} \to u_t^{R,\varepsilon} \quad \text{weakly} \quad \text{in} \quad L^p((0,T); L^p(\Omega_R)),$$

$$\beta_R(u_t^{N_k}) \to z^{R,\varepsilon} \quad \text{weakly} \quad \text{in} \quad L^2((0,T); V_{1,0}^*(\Omega_R)).$$

We shall differentiate each equation of system (7) with respect to t, multiply by $C_{ktt}^N e^{-\gamma t}$, sum up all the equations over k from 1 to N, integrate the result with respect to t over the interval $[0,\tau]$, where $0 < \tau \leq T$:

$$\int\limits_{Q_{R,\tau}} \left[u_{ttt}^N u_{tt}^N + \sum_{i,j=1}^n a_{ij}(x) u_{x_i t}^N u_{x_j tt}^N + (p-1) \left|u_t^N\right|^{p-2} |u_{tt}^N|^2 - f_t^R u_{tt}^N \right] e^{-\gamma t} dxdt +$$

$$+ \frac{1}{\varepsilon} \int\limits_0^T \langle (\beta_R(u_t^N))_t, u_{tt}^N \rangle e^{-\gamma t} dt = 0. \tag{11}$$

From condition **(B)**, we obtain

$$\int_0^T \langle (\beta_R(u_t^N))_t, u_{tt}^N \rangle e^{-\gamma t} dt \geq 0. \tag{12}$$

Taking into account the conditions of Theorem 2, we transform and estimate the terms of equality (11). We obtain as follows:

$$\int_{Q_{R,\tau}} u_{ttt}^N u_{tt}^N e^{-\gamma t} dxdt = \frac{1}{2} \int_{\Omega_{R,\tau}} |u_{tt}^N|^2 e^{-\gamma\tau} dx -$$

$$-\frac{1}{2} \int_{\Omega_{R,0}} |u_{tt}^N(x,0)|^2 dx + \frac{\gamma}{2} \int_{Q_{R,\tau}} |u_{tt}^N|^2 e^{-\gamma t} dxdt;$$

$$\int_{Q_{R,\tau}} \sum_{i,j=1}^n a_{ij}(x) u_{x_it}^N u_{x_jtt}^N e^{-\gamma t} dxdt = \frac{1}{2} \int_{\Omega_{R,\tau}} \sum_{i,j=1}^n a_{ij}(x) u_{x_it}^N(x,\tau) u_{x_jt}^N(x,\tau) e^{-\gamma\tau} dx -$$

$$-\frac{1}{2} \int_{\Omega_{R,0}} \sum_{i,j=1}^n a_{ij}(x) u_{x_it}^N(x,0) u_{x_jt}^N(x,0) dx + \frac{\gamma}{2} \int_{Q_{R,\tau}} \sum_{i,j=1}^n a_{ij}(x) u_{x_it}^N u_{x_jt}^N e^{-\gamma t} dxdt \geq$$

$$\geq \frac{a_0}{2} \int_{\Omega_{R,\tau}} |\nabla u_t^N|^2 e^{-\gamma\tau} dx + \frac{\gamma a_0}{2} \int_{Q_{R,\tau}} |\nabla u_t^N|^2 e^{-\gamma t} dxdt - \frac{M_4}{2} \int_{\Omega_{R,0}} |\nabla u_{1t}^R|^2 dx,$$

where M_4 does not depend on N. Further, using the Cauchy inequality, we obtain

$$\int_{Q_{R,\tau}} f_t^R u_{tt}^N e^{-\gamma t} dxdt \leq \frac{1}{2} \int_{Q_{R,\tau}} |u_{tt}^N|^2 e^{-\gamma t} dxdt + \frac{1}{2} \int_{Q_{R,\tau}} |f_t^R|^2 e^{-\gamma t} dxdt.$$

Using the estimates above, taking into account (12), we obtain the inequality as follows:

$$\int_{\Omega_{R,\tau}} |u_{tt}^N(x,\tau)|^2 e^{-\gamma\tau} dx + a_0 \int_{\Omega_{R,\tau}} |\nabla u_t^N(x,\tau)|^2 e^{-\gamma\tau} dx +$$

$$+ \gamma a_0 \int_{Q_{R,\tau}} |\nabla u_t^N|^2 e^{-\gamma t} dxdt + (\gamma - 1) \int_{Q_{R,\tau}} |u_{tt}^N|^2 e^{-\gamma t} dxdt \leq$$

$$\leq \int_{Q_{R,\tau}} |f_t^R|^2 e^{-\gamma t} dxdt + \int_{\Omega_{R,0}} |u_{tt}^N(x,0)|^2 dx + M_4 \int_{\Omega_{R,0}} |\nabla u_{1t}^R|^2 dx. \tag{13}$$

Considering system (7) at $t = 0$, multiplying each its equation by $C_{ktt}^N(0)$ and summing up over k from 1 to N, we obtain the following equality:

$$\int_{\Omega_{R,0}} \left[|u_{tt}^N(x,0)|^2 - \sum_{i,j=1}^n (a_{ij}(x)) u_{x_i}^N(x,0))_{x_j} u_{tt}^N(x,0) \right] dx +$$

$$+\int_{\Omega_{R,0}} \left|u_t^N(x,0)\right|^{p-2} u_t^N(x,0)u_{tt}^N(x,0)dx -$$

$$-\int_{\Omega_{R,0}} f(x,0)u_{tt}^N(x,0)dx = 0. \quad (14)$$

Using the conditions of Theorem 2 as well as the Hölder inequality, from (14) one can easily obtain the estimate

$$\int_{\Omega_{R,0}} \left|u_{tt}^N(x,0)\right|^2 dx \leq M_5 \int_{\Omega_{R,0}} \left[\sum_{i=1}^{n} |u_{0,x_ix_i}^R|^2 + \left|\nabla u_0^R\right|^2 + \left|u_{1t}^R\right|^{2p-2} dx + |f(x,0)|^2\right] dx,$$

where M_5 does not depend on N. Owing to the restrictions as to the initial data and the free term, this implies the estimate as follows:

$$\int_{\Omega_{R,0}} \left|u_{tt}^N(x,0)\right|^2 dx \leq M_6, \quad (15)$$

where the constant M_6 does not depend on N. Taking into account inequalities (13), (15), we can state that there exists a subsequence of the sequence $\{u^N\}$ (without a generality limitation, we consider one to be $\{u^{N_k}\}$), such that

$$u_{tt}^{N_k} \to u_{tt}^{R,\varepsilon} \quad * - \text{weakly} \quad \text{in} \quad L^\infty((0,T);L^2(\Omega_R)),$$

$$u_t^{N_k} \to u_t^{R,\varepsilon} \quad * - \text{weakly} \quad \text{in} \quad L^\infty((0,T);V_{1,0}(\Omega_R)),$$

$$u_{tt}^{N_k} \to u_{tt}^{R,\varepsilon} \quad \text{weakly} \quad \text{in} \quad L^2((0,T);L^2(\Omega_R)),$$

$$u_t^{N_k} \to u_t^{R,\varepsilon} \quad \text{weakly} \quad \text{in} \quad L^2((0,T);V_{1,0}(\Omega_R)).$$

For the reason of the above obtained convergences of the sequence $\{u^{N_k}\}$, we can state that $u^{N_k} \to u^{R,\varepsilon}$ strongly in $L^2((0,T);H^1(\Omega_R))$ (see [1, p. 70, Theorem 5.1]), $\left|u_t^{N_k}\right|^{p-2} u_t^{N_k} \to \left|u_t^{R,\varepsilon}\right|^{p-2} u_t^{R,\varepsilon}$ weakly in the space $L^{p'}((0,T);L^{p'}\Omega_R))$ (see [1, p. 25, Lemma 1.3]. Moreover, $u^{R,\varepsilon} \in C([0,T];V_{1,0}(\Omega_R))$ (see [1, p. 20, Lemma 1.2]. Since $u^{N_k}(\cdot,0) \to u^{R,\varepsilon}(\cdot,0)$ weakly in $V_{1,0}(\Omega_R)$, $u_0^{R,N_k}(\cdot,0) \to u_0^R$ strongly in $H^1(\Omega_R)$, we conclude that $u^{R,\varepsilon}(x,0) = u_0^R(x)$, $x \in \Omega_R$. Since $u_t^{N_k} \to u_t^{R,\varepsilon}$ weakly in $L^2((0,T);V_{1,0}(\Omega_R))$, $u_{tt}^{N_k} \to u_{tt}^{R,\varepsilon}$ weakly in the space $L^2((0,T);L^2(\Omega_R))$, we conclude that $u_t^{N_k} \to u_t^{R,\varepsilon}$ strongly in $C([0;T];V_{1,0}(\Omega_R))$(see [1, p. 20]), which yields $u_t^{R,\varepsilon}(x,0) = u_1^R(x)$, $x \in \Omega_R$.

On the basis of the monotonicity of the operator β_R, similar to [1, p. 171], we prove that $z^{R,\varepsilon} = \beta_R(u_t^{R,\varepsilon})$. Using (7), we can easy obtain the equality as follows:

$$\int_{Q_{R,\tau}} \left[-u_t^{R,\varepsilon} v_t + \sum_{i,j=1}^{n} a_{ij}(x)u_{x_i}^{R,\varepsilon} v_{x_j} + \left|u_t^{R,\varepsilon}\right|^{p-2} u_t^{R,\varepsilon} v - f^R v\right] dxdt +$$

$$+\int\limits_{\Omega_{R,\tau}} u_t^{R,\varepsilon} v dx + \frac{1}{\varepsilon}\int\limits_0^{\tau}\left\langle \beta_R(u_t^{R,\varepsilon}), v\right\rangle dt = \int\limits_{\Omega_{R,0}} u_1^R v dx, \tag{16}$$

which holds for all functions $v \in L^2((0,T);V_{1,0}(\Omega_R))$ such that $v_t \in L^p(Q_{R,T})$, for all $\tau \in [0,T]$ and for all $\psi \in C^1(\mathrm{R}^n)$, $\psi(x) \geq 0$, $x \in \mathrm{R}^n$. Moreover,

$$\int\limits_{Q_{R,\tau}}\left[\sum_{i,j=1}^{n} a_{ij}(x) u_{x_i}^{R,\varepsilon}\left(u_t^{R,\varepsilon}\psi(x)\right)_{x_j} + \left|u_t^{R,\varepsilon}\right|^p \psi(x) - f^R u_t^{R,\varepsilon}\psi(x)\right] dxdt +$$

$$+\frac{1}{2}\int\limits_{\Omega_{R,\tau}}\left|u_t^{R,\varepsilon}\right|^2 \psi(x) dx + \frac{1}{\varepsilon}\int\limits_0^{\tau}\left\langle \beta_R(u_t^{R,\varepsilon}), u_t^{R,\varepsilon}\psi(x)\right\rangle dt = \frac{1}{2}\int\limits_{\Omega_{R,0}}\left|u_1^R\right|^2 \psi(x) dx \tag{17}$$

and $u^{R,\varepsilon}(x,0) = u_0^R(x)$. Note that the constants in estimates (9), (10), (13), (15) do not depend on ε. Therefore, there exists such a subsequence $\{u^{R,\varepsilon_k}\} \subset \{u^{R,\varepsilon}\}$ ($\lim\limits_{k\to+\infty}\varepsilon_k = 0$) that

$$\begin{aligned}
u^{R,\varepsilon_k} &\to u^R \quad * - \quad \text{weakly} \quad \text{in} \quad L^\infty((0,T);V_{1,0}(\Omega_R)),\\
u^{R,\varepsilon_k} &\to u^R \quad \text{weakly} \quad \text{in} \quad L^2((0,T);V_{1,0}(\Omega_R)),\\
u_t^{R,\varepsilon_k} &\to u_t^R \quad \text{weakly} \quad \text{in} \quad L^2((0,T);V_{1,0}(\Omega_R)),\\
u_t^{R,\varepsilon_k} &\to u_t^R \quad \text{weakly} \quad \text{in} \quad L^p((0,T);L^p(\Omega_R)),\\
u_{tt}^{R,\varepsilon_k} &\to u_{tt}^R \quad \text{weakly} \quad \text{in} \quad L^2((0,T);L^2(\Omega_R)),
\end{aligned}$$

$$\int\limits_0^T\left\langle \beta_R(u_t^{R,\varepsilon_k}), u_t^{R,\varepsilon_k}\right\rangle \to 0 \quad \text{as} \quad k \to +\infty.$$

Then, (16) implies

$$\int\limits_{Q_{R,T}}\langle \beta_R(u_t^{R,\varepsilon_k}), v\rangle dxdt \to 0$$

as $k \to +\infty$ for an arbitrary function $v \in L^2((0,T);V_{1,0}(\Omega_R))$.

Now consider the sequence $\{\eta_k\}$, where

$$\eta_k = \int\limits_0^T\left\langle\left(\beta_R(u_t^{R,\varepsilon_k}) - \beta_R(w)\right), \left(u_t^{R,\varepsilon_k} - w\right)\right\rangle dt,$$

w is an arbitrary function of the space $L^2((0,T);V_{1,0}(\Omega_R))$. By the monotonicity of the operator β_R, after the passage to limit as $k \to +\infty$, we conclude that

$$\int\limits_o^T\langle \beta_R(w), (u_t^R - w)\rangle dxdt \leq 0.$$

Since β_R is a semi-continuous operator, choosing $w = u_t^R - \lambda z$, $\lambda \in (0,1)$ (z is an arbitrary function from the same space as u_t^R), we obtain $\beta_R(u_t^R) = 0$, e.g. $u_t^R \in \mathrm{K}_R$.

Now we prove the limit function u^R to be a strong solution of the variational hyperbolic inequality in the domain $Q_{R,T}$. Let w be an arbitrary function of the space $L^2((0,T);V_{1,0}(\Omega))\cap L^p(Q_{R,T})$ such that $w_t\in L^2(Q_{R,T})$, $w\in \mathrm{K}_R$ for almost all $t\in(0,T]$. Let us write the equality (16) for the functions u^{R,ε_k}, $v(x,t)=-w(x,t)\psi(x)$, $\psi\in C^1(\mathrm{R}^n)$, $\psi(x)\geq 0, x\in\mathrm{R}^n$:

$$\int\limits_{Q_{R,\tau}}\left[-u_t^{R,\varepsilon_k}w_t\psi+\sum_{i,j=1}^{n}a_{ij}(x)u_{x_i}^{R,\varepsilon_k}(w\psi)_{x_j}\right]dxdt-$$

$$-\int\limits_{\Omega_{R,\tau}}u_t^{R,\varepsilon_k}w\psi dx-\int\limits_{Q_{R,\tau}}\left[\left|u_t^{R,\varepsilon_k}\right|^{p-2}u_t^{R,\varepsilon_k}-f^R\right]w\psi dxdt-$$

$$-\frac{1}{\varepsilon_k}\int\limits_{o}^{\tau}\left\langle\beta_R(u_t^{R,\varepsilon_k}),w\psi\right\rangle dxdt=-\int\limits_{\Omega_{R,0}}u_1^Rw\psi dx. \tag{18}$$

Let us write the equality (17) for the functions u^{R,ε_k}:

$$\int\limits_{Q_{R,\tau}}\left[\sum_{i,j=1}^{n}a_{ij}(x)u_{x_i}^{R,\varepsilon_k}\left(u_t^{R,\varepsilon_k}\psi\right)_{x_j}+\left(\left|u_t^{R,\varepsilon_k}\right|^{p}-f^Ru_t^{R,\varepsilon_k}\right)\psi\right]dxdt+$$

$$+\frac{1}{2}\int\limits_{\Omega_{R,\tau}}\left|u_t^{R,\varepsilon_k}\right|^2\psi dx+\frac{1}{\varepsilon_k}\int\limits_{o}^{\tau}\left\langle\beta_R(u_t^{R,\varepsilon_k}),u_t^{R,\varepsilon_k}\psi\right\rangle dxdt=\frac{1}{2}\int\limits_{\Omega_{R,0}}\left|u_1^R\right|^2\psi dx. \tag{19}$$

Besides, the equality below is obvious:

$$-\int\limits_{Q_{R,\tau}}ww_t\psi dxdt=-\frac{1}{2}\int\limits_{\Omega_{R,\tau}}w^2\psi dx+\frac{1}{2}\int\limits_{\Omega_{R,0}}w^2\psi dx, \tag{20}$$

where $\tau\in(0,T]$. Summing up (18), (19), (20) and taking into account condition **(B)**, we obtain the inequality as follows:

$$\int\limits_{Q_{R,\tau}}\left[w_t\left(w-u_t^{R,\varepsilon_k}\right)\psi+\sum_{i,j=1}^{n}a_{ij}(x)u_{x_i}^{R,\varepsilon_k}\left(\left(w-u_t^{R,\varepsilon_k}\right)\psi\right)_{x_j}\right]dxdt+$$

$$+\int\limits_{Q_{R,\tau}}\left[\left|u_t^{R,\varepsilon_k}\right|^{p-2}u_t^{R,\varepsilon_k}\left(w-u_t^{R,\varepsilon_k}\right)\psi-f^R\left(w-u_t^{R,\varepsilon_k}\right)\right]\psi dxdt-$$

$$\geq\frac{1}{2}\int\limits_{\Omega_{R,\tau}}\left(w-u_t^{R,\varepsilon_k}\right)^2\psi dx-\frac{1}{2}\int\limits_{\Omega_{R,0}}\left(u_1^R-w\right)^2\psi dx. \tag{21}$$

Note that inequality (21) holds also for the function $w=u^R$, therefore

$$u^{R,\varepsilon_k}\to u^R\quad\text{strongly in}\quad L^2((0,T);V_{1,0}(\Omega_R)),$$

$$u_t^{R,\varepsilon_k} \to u_t^R \quad \text{strongly} \quad \text{in} \quad L^p((0,T);L^p(\Omega_R)).$$

Taking into account $u^{R,\varepsilon_k} \to u^R$, $u_t^{R,\varepsilon_k} \to u_t^R$ in the respective functional spaces, and passing in (21) to limit as $k \to +\infty$, we obtain the inequality as follows:

$$\int\limits_{Q_{R,\tau}} \left[w_t \left(w - u_t^R\right) \psi(x) + \sum_{i,j=1}^{n} a_{ij}(x) u_{x_i}^R \left(\left(w - u_t^R\right) \psi(x)\right)_{x_j} \right] dxdt +$$

$$+ \int\limits_{Q_{R,\tau}} \left[|u_t^R|^{p-2} u_t^R \left(w - u_t^R\right) - f^R \left(w - u_t^R\right) \right] \psi(x) dxdt \geq$$

$$\geq \frac{1}{2} \int\limits_{\Omega_{R,\tau}} \left(w - u_t^R\right)^2 \psi(x) dx - \frac{1}{2} \int\limits_{\Omega_{R,0}} \left(u_1^R - w\right)^2 \psi(x) dx, \qquad (22)$$

that completes the proof of Theorem 2.

5. Proof of Theorem 3

Consider the sequences of functions $\{F_k(x,t)\}$, $\{U_{0,k}(x)\}$, $\{U_{1,k}(x)\}$ with the following properties: $F_k, F_{kt} \in L^2((0,T);L^2_{loc}(\overline{\Omega}))$, $F_k \to f$ in the space $L^{p'}((0,T);L^{p'}_{loc}(\overline{\Omega}))$, $U_{0,k} \in H^2_{loc}(\overline{\Omega}) \cap V_{1,loc}(\overline{\Omega})$, $U_{0,k} \to u_0$ in the space $V_{1,loc}(\overline{\Omega})$, $U_{1,k} \in V_{1,loc}(\overline{\Omega}) \cap L^{2p-2}_{loc}(\overline{\Omega}) \cap \mathrm{K}$, $U_{1,k} \to u_1$ in the space $L^2_{loc}(\overline{\Omega})$.

Let R sequentially possess the values $2, 3, 4, \ldots$. Consider the functions

$$f^k(x,t) = \begin{cases} F_k(x,t), & (x,t) \in Q_{k+1,T}, \\ 0, & (x,t) \in Q_T \setminus Q_{k+1,T}, \end{cases}$$

$u_{0,k}(x) = U_{0,k}(x)\zeta_k(x)$, $u_{1,k}(x) = U_{1,k}(x)\zeta_k(x)$, where $\zeta_k \in C^2(\mathrm{R}^n)$, $\zeta_k(x) = 1$ as $|x| \leq k-1$, $\zeta_k(x) = 0$ as $|x| \geq k$, $0 \leq \zeta_k(x) \leq 1$ as $k-1 < |x| < k$. Then, owing to Theorem 2, for each $k \in \mathrm{N}$, there exists a strong solution u^k of inequality (1) in the domain $Q_{k+1,T}$ with the initial functions $u_{0,k}$, $u_{1,k}$ and the free term f^k. We extend each of the functions u^k by zero into the domain $Q_T \setminus Q_{k+1,T}$ preserving its notation. Then there hold the following conditions:

$$u^k, u_t^k \in L^2((0,T);V_{1,loc}(\overline{\Omega})), \quad u_t^k \in L^p((0,T);L^p_{loc}(\overline{\Omega})), \quad u_{tt}^k \in L^2((0,T);L^2_{loc}(\overline{\Omega})).$$

Let $R_0 > 1$ be an arbitrary fixed number, and $R_1 > R_0$. Consider the sequence of natural numbers $R = [R_0] + m$, $m \in \mathrm{N}$. This sequence determines the function sequence $\{u^m\}$. Choose $m > [R_1] + 1$, $s > [R_1] + 1$; for each of the functions u^m and u^s, write the variational hyperbolic inequality (22) when $\psi(x) = \psi_{R_1}(x)$; and then sum up those inequalities. This yields the inequality as follows:

$$\int\limits_{Q_\tau} \left[\sum_{i,j=1}^{n} a_{ij}(x) \left(u_{x_i}^m ((w - u_t^m) \psi_{R_1}(x))_{x_j} + u_{x_i}^s ((w - u_t^s) \psi_{R_1}(x))_{x_j} \right) \right] dxdt +$$

$$+ \int\limits_{Q_\tau} \left[w_t (2w - u_t^m - u_t^s) + |u_t^m|^{p-2} u_t^m (w - u_t^m) \right] \psi_{R_1}(x) dxdt +$$

$$+\int\limits_{Q_\tau}\left[|u_t^s|^{p-2}u_t^s(w-u_t^s)-(f^m(w-u_t^m)-f^s(w-u_t^s))\right]\psi_{R_1}(x)dxdt\geq$$

$$\geq\frac{1}{2}\int\limits_{\Omega_\tau}(w-u_t^m)^2\psi_{R_1}(x)dx+\frac{1}{2}\int\limits_{\Omega_{R,\tau}}(w-u_t^s)^2\psi_{R_1}(x)e^{-\gamma\tau}dx-$$

$$-\frac{1}{2}\int\limits_{\Omega_0}\left[(u_{1,m}-w)^2+(u_{1,s}-w)^2\right]\psi_{R_1}(x)dx. \quad (23)$$

In (23), having chosen the function $w=\frac{1}{2}(u_t^m+u_t^s)e^{-\gamma t}$ and denoted $u^{m,s}=u^m-u^s$, we obtain the following inequality:

$$\frac{1}{2}\int\limits_{\Omega_\tau}\left|u_t^{m,s}\right|^2\psi_{R_1}(x)e^{-\gamma\tau}dx+\int\limits_{Q_\tau}\sum_{i,j=1}^{n}a_{ij}(x)u_{x_i}^{m,s}\left(u_t^{m,s}\psi_{R_1}(x)\right)_{x_j}e^{-\gamma t}dxdt+$$

$$+\int\limits_{Q_\tau}\left(|u_t^m|^{p-2}u_t^m-|u_t^s|^{p-2}u_t^s\right)u_t^{m,s}\psi_{R_1}(x)e^{-\gamma t}dxdt\leq$$

$$\leq\int\limits_{Q_\tau}(f^m-f^s)u_t^{m,s}\psi_{R_1}(x)e^{-\gamma t}dxdt+\frac{1}{2}\int\limits_{\Omega_0}(u_{1,m}-u_{1,s})^2\psi_{R_1}(x)dx. \quad (24)$$

From the inequality (24), thinking in the same way as when proving the uniqueness, one can conclude the following: for any arbitrarily small $\varepsilon>0$, there exist $R_1>R_0$ and such a natural number $n_0(\varepsilon)>[R_1]+1$, that, when $m,s>n_0(\varepsilon)$, there holds the estimate

$$\int\limits_{\Omega_{R_0,\tau}}\left(\left|u_t^{m,s}\right|^2+|\nabla u^{m,s}|^2\right)dx+\int\limits_{Q_{R_0,T}}\left(|\nabla u^{m,s}|^2+\left|u_t^{m,s}\right|^p\right)dxdt<\varepsilon,$$

where $\tau\in(0,T]$. Consequently, $\{u^m\}$, $\{u_t^m\}$ are fundamental sequences in the spaces $C([0;T];V_1(\Omega_{R_0}))$ and $C([0;T];L^2(\Omega_{R_0}))\cap L^p((0;T);L^p(\Omega_{R_0}))$ respectively. Therefore, taking into account the arbitrariness of R_0, as $m\to+\infty$, we obtain that $u^m\to u$ strongly in $C([0,T];V_{1,loc}(\overline{\Omega}))$, and that $u_t^m\to u_t$ strongly in $C([0,T];L^2_{loc}(\overline{\Omega}))\cap L^p((0,T);L^p_{loc}(\overline{\Omega}))$. Since $u_t^m\in\mathrm{K}$ and K is a closed set, $u_t\in\mathrm{K}$ for almost all $t\in(0,T]$. Moreover, $u(x,0)=u_0(x)$. Consequently, by Definition 2, the function u is a weak solution of inequality (1). This proves the theorem.

6. The Inequality Interpretation

Similarly to [1, p. 423–424], we can give examples of some problems in unbounded domains, which could be investigated by means of the results obtained in the present paper for the variational hyperbolic inequality.

Example 1. Let $V_{1,loc}(\overline{\Omega})=H^1_{0,loc}(\overline{\Omega})$, $\mathrm{K}=\{v:v\in H^1_{0,loc}(\overline{\Omega}),\ v\geq 0$ almost everywhere in $\Omega\}$. Consider the operator $\beta_R(v)=J(v-P_{\mathrm{K}_R}(v))$, where J is a unit operator, P_{K_R} is the operator of projecting onto K_R, at that $P_{K_R}(v) = \begin{cases} v(x),\ v(x) \geq 0, \\ 0,\ v(x) < 0 \end{cases}$. Inequality (1) with initial condition (2) is equivalent to the unilateral problem as follows:

$$u_t \geq 0 \quad \text{almost everywhere in} \quad Q_T,$$

$$u_{tt} - \sum_{i,j=1}^{n} (a_{ij}(x)u_{x_i})_{x_j} + |u_t|^{p-2} u_t - f \geq 0 \quad \text{almost everywhere in} \quad Q_T,$$

$$u_t \left(u_{tt} - \sum_{i,j=1}^{n} (a_{ij}(x)u_{x_i})_{x_j} + |u_t|^{p-2} u_t - f \right) = 0 \quad \text{almost everywhere in} \quad Q_T,$$

$$u(x,0) = u_0(x), \quad u_t(x,0) = u_1(x), \quad x \in \Omega.$$

Example 2. Let $V_{1,loc}(\overline{\Omega}) = H^1_{loc}(\overline{\Omega})$, $K = \{v : v \in H^1_{loc}(\overline{\Omega}),\ v \geq 0$ almost everywhere in $\partial\Omega\}$. Define the operator $\beta_R : V(\Omega_R) \to V^*(\Omega_R)$ as follows:

$$(\beta_R(u_t), v) = -\int_{\Gamma_1^R} u_t^- v dS,\ v \in V_{1,loc}(\overline{\Omega}),\ u_t^- = \begin{cases} -u_t,\ u_t < 0, \\ 0,\ u_t \geq 0. \end{cases}$$

Inequality (1) with initial condition (2) is equivalent to the unilateral problem as follows:

$$u_{tt} - \sum_{i,j=1}^{n} (a_{ij}(x)u_{x_i})_{x_j} + |u_t|^{p-2} u_t - f = 0 \ \text{ almost everywhere in } Q_T,\ u_t \geq 0 \text{ in} \partial\Omega \times (0,T),$$

$$\frac{\partial u}{\partial \nu_A} \geq 0 \text{ in } \quad \partial\Omega \times (0,T), \quad \frac{\partial u}{\partial t} \cdot \frac{\partial u}{\partial \nu_A} = 0 \text{ in } \quad \partial\Omega \times (0,T),$$

$$u(x,0) = u_0(x), \quad u_t(x,0) = u_1(x), \quad x \in \Omega,$$

where

$$\frac{\partial u}{\partial \nu_A} = \sum_{i,j=1}^{n} a_{ij}(x) u_{x_i} \cos(\nu, x_j),$$

ν is a unit vector of the external normal to the surface $\partial\Omega$.

Remark 3. Similar results could be obtained for the variational inequality

$$\int_{Q_\tau} \Big[v_t(v-u_t)\psi(x) + \sum_{i,j=1}^{n} a_{ij}(x,t)u_{x_i}((v-u_t)\psi(x))_{x_j} + \sum_{i=1}^{n} b_i(x,t)u_{x_i}(v-u_t)\psi(x) +$$

$$+ |u_t|^{p-2} u_t (v-u_t)\psi(x) - \frac{\gamma}{2}(v-u_t)^2\psi(x) + c(x,t)u(v-u_t)\psi(x) - f(v-u_t)\psi(x) \Big] \times$$

$$\times e^{-\gamma t} dxdt \geq \frac{1}{2}\int_{\Omega_\tau} (v-u_t)^2 \psi(x) e^{-\gamma\tau} dx - \frac{1}{2}\int_{\Omega_0} (v-u_1)^2 \psi(x) dx, \quad \gamma \geq \gamma_0,$$

with the initial condition

$$u(x,0) = u_0(x), \quad x \in \Omega.$$

References

[1] Lions J.L. *Some methods of the solving of nonlinear boundary value problems.* – Moscow: Mir, 1972. (in Russian)

[2] Duvaut G., Lions J.L. *Les inequations en mecanique et en physique.* – Paris: Dunod, 1972.

[3] Cooper R.G. Two variational inequality problems for the wave equation in a half – space// *J. Math. Anal. And Appl.*– **232.** – 1999.– P. 434–460.

[4] Maksudov F.G., Aliev A.B., Takhirov D.M. Unilateral problem for quasi-linear equation of hyperbolic type// *Dokl. Acad. of Sciences USSR.* – **258**, 4. – 1981. – P. 789–791. (in Russian)

[5] Aliev A.B. Global solvability of unilateral problems for quasi-linear operators of hyperbolic type// *Dokl. Acad. of Sciences USSR.* – **298**, 5. – 1988. – P. 1033–1036. (in Russian)

[6] Landes R. Quasi-linear hyperbolic variational inequalities// *Arch. Ration. Mech. Anal.* – **91**, 3. – 1986. – P. 267–282.

[7] Glazatov S.N. Some quasi-linear hyperbolic inequalities/ Preprint Russian Academy of Sciences. *Siberian Department. Inst. of Math.;* **2**. – Novosibirsk. – 1990. – 26 p. (in Russian)

[8] Lar'kin N.A. On global solutions of nonlinear hyperbolic inequalities// *Dokl. Acad. of Sciences USSR.* – **250,** 4. – 1980. – P. 806–809. (in Russian)

[9] Lar'kin N.A. Unilateral problem for nonlocal quasi-linear hyperbolic equation of elasticity theory// *Dokl. Acad. of Sciences USSR.* – **274,** 6. – 1984. – P. 1341–1344. (in Russian)

[10] Aliev A.B. Unilateral problems for quasi-linear hyperbolic operators in function spaces// *Dokl. Acad. of Sciences USSR.* – **297,** 2. – 1987. – P. 271–275. (in Russian)

[11] Park J. Y., Bae J.J. Variational inequality for quasi-linear wave equations with nonlinear damping terms// *Nonlin. Anal.* – **50.** – 2002. – P. 1065–1083.

[12] Glazatov S.N. Some problems for nonlinear third order equations/ Preprint Russian Academy of Sciences. *Siberian Department. Inst. of Math.;* **7**. – Novosibirsk. – 1992. – 22 p. (in Russian)

[13] Glazatov S.N. On some variational inequalities for nonclassical type operators// *Banach Center Publications.* -- **27**. – 1992. – P. 169–174.

[14] Glazatov S.N. *Variational inequality for some high order operator* // Actual problems of modern math.: digest of scientific transactions. – Novosibirsk: Novosibirsk State University Publishing House. – 1. – 1995. – P. 60-66. (in Russian)

[15] Glazatov S.N. On some approach to nonlinear wave equation without a priori estimates// Boundary value problems for nonclassical equations of mathematical physics: digest of scientific transactions. *Acad. of Sciences USSR*. Siberian Department. Inst. of Math. – Novosibirsk. – 1987. – P. 60-70. (in Russian)

[16] D'Ancona P., Manfrin R. A class of locally solvable semilinear equations of weakly hyperbolic type// *Ann. Mat. Pura Appl.* – **168**. – 1995. – P. 355–372.

[17] Agre K., Rammaha M.A. Global solutions to boundary value problems for a nonlinear wave equation in high space dimensions// *Diff. And Integr. Equat.* – **14**. – 2001. – P. 1315–1331.

[18] Georgiev V., Todorova G. Existence of a solution of the wave equation with nonlinear damping and sourse terms// *J. Diff. Equat.* – **109.** – 1994. – P. 295–308.

[19] Dragieva N.A. A hyperbolic equation with two space variables with strong nonlinearity// *Godishnik Vish. Uchebn. Zaved. Prilozhna Mat.* – **23**. – 1987, 4. – P. 95–106.

[20] Carpio A. Existence of global solutions to some nonlinear dissipative wave equations// *J. Math. Pures Appl.* – **73**. – 1994, 5. – P. 471–488.

[21] Vittilaro E. Global nonexistence theorems for a class of evolution equation with dissipation// *Arch. Ration. Mech. Anal.* – **149**. – 1999, 2. – P. 155–182.

[22] Rubino B. Weak solutions to quasilinear wave equations of Klein – Gordon or Sine – Gordon type and relaxation to reaction – diffusion equations// *Nonlinear Diff. Equat. And Appl.* – **48**. – 1997. – P. 439–457.

[23] Pecher H. Sharp existence results for self– similar solutions of semilinear wave equations// *Nonlinear Diff. Equat. And Appl.* – **7**. – 2000. – P. 323–341.

[24] Todorova G., Yordanov B. Critical exponent for a nonlinear wave equations with damping// *J. Diff. Equat.* – **174**. – 2001. – P. 464–489.

[25] Choquet – Bruhat Y. Global existence for solutions of $u_{tt} - \Delta u = A\,|u|^p$// *J. Diff. Equat.* – **82**. – 1989. – P. 98–108.

[26] Li M.– R., Tsai L. – Y. Existence and nonexistence of global solutions of some system of semilinear wave equations// *Nonlin. Anal.* – **54.** – 2003. – P. 1397–1415.

[27] Lavrenyuk S.P., Oliskevich M.O. Galerkin method for hyperbolic first order systems with two independent variables// *Ukrainian Math. Jourhal* – **54.** – 2002, 10. – P. 1356–1370. (in Ukrainian)

[28] Bernis F. Elliptic and parabolic semilinear problems without conditions at infinity // *Arch. Ration. Mech. Anal.* – **106**. – 1989, 3. – P. 217–241.

[29] Gajewski H., Gröger K., Zacharias K. *Nonlinear operator equations and operator differential equations*. – Moscow: Mir, 1978. (in Russian)

[30] Coddington E., Levinson N. *Theory of ordinary differential equations*. – Moscow: Izd. Inostr. Lit., 1958. (in Russian)

In: Progress in Evolution Equations
Editor: Gaston M. N'Guerekata, pp. 125-133
ISBN: 978-1-60456-328-3

Chapter 7

GLOBAL EXISTENCE AND BLOWUP TO A REACTION-DIFFUSION SYSTEM WITH GRADIENT TERM*

Jun Zhou[1†] ***and Chunlai Mu***[2‡]
[1]School of Mathematics and Statistics, Southwest University, Chongqing, 400715, P. R. China
[2]School of Mathematics and Physics, Chongqing University, Chongqing, 400044, P. R. China

Abstract

This paper deals with the initial problem $u_{it} = \Delta u_i + f_i(u_1,\cdots,u_m) + g_i(u_1,\cdots,u_m)|\nabla u_i|^2$ with $u_i|_{\partial\Omega} = 0$ and $u_i(x,0) = \phi_i(x)$, $i = 1,\cdots,m$, in a bounded domain $\Omega \subset R^n$. Under suitable assumptions on f_i, g_i, we proves that, if $\phi_i \geq (1+\varepsilon_0)\psi_i$ in $D_i \subset \Omega$ for some small $\varepsilon_0 > 0$, then the solutions blow up in finite time, where ψ_i is a positive solution of $\Delta\psi_i + f_i(\psi) + g_i(\psi)|\nabla\psi_i|^2 = 0$ in D_i with $\psi_i|_{\partial D_i} = 0$ for $i = 1,\cdots,m$. If $0 \leq \phi_i(x) \leq \lambda\psi_i(x)$ in Ω with $\lambda < 1$, then the solution exists for all $t > 0$ and decay exponentially in t, where ψ_i is a positive solution of $\Delta\psi_i + f_i(\psi) + g_i(\psi)|\nabla\psi_i|^2 = 0$ in Ω with $\psi_i|_{\partial\Omega} = 0$ for $i = 1,\cdots,m$.

2000 Mathematics Subject Classification: Primary 35K57, 35K45.

Key words: Blowup, Reaction-diffusion system, Gradient term.

1. Introduction

In this paper, we consider the following parabolic equations

$$\begin{cases} u_{it} = \Delta u_i + f_i(u_1,\ldots,u_m) + g_i(u_1,\ldots,u_m)|\nabla u_i|^2, & x \in \Omega, \quad t > 0, \\ u_i(x,t) = 0, & x \in \partial\Omega, \quad t > 0, \\ u_i(x,0) = \phi_i(x), & x \in \Omega, \end{cases} \tag{1}$$

*This work is supported in part by NNSF of China (10571126) and in part by Program for New Century Excellent Talents in University.
†E-mail address: zhoujun_math@hotmail.com
‡E-mail address: chunlaimu@yahoo.com.cn

for $i = 1, \cdots, m$, where $\Omega \subset R^n$ is a bounded domain with smooth boundary $\partial\Omega$. For convenience, we denote $u = (u_1, \cdots, u_m)$, $\psi = (\psi_1, \cdots, \psi_m)$.

We get from [8] that this sort of nonlinear PDE aries in stochastic optimal control theory. It is well know that, for some small initial values, the solution exists globally, while for some large initial values the solutions blow up in finite time (see [3,11,15,16]). A nature question is what kind of initial values produce solutions of (1) that blow up in finite time and exist globally.

If $m = 1$, many authors (see [7,9,13,17]) discuss the following problem

$$\begin{cases} u_t = \Delta u + |u|^{p-1}u - a|\nabla u|^q, & x \in \Omega, \quad t > 0, \\ u(x,t) = 0, & x \in \partial\Omega, \quad t > 0, \\ u(x,0) = \phi(x), & x \in \Omega. \end{cases} \tag{2}$$

Existence of global solution of (2) depends on the balance between the power of the damping term and the nonlinear source. In one dimension, Deng [7] and Fila [9] proved that under suitable assumption on p and q, the solution of (2) blow up in a finite time if $\phi(x) \geq \psi(x)$, $\forall x \in \Omega$, where ψ is the unique steady state of (2). Deng [7] also proved that if $\phi(x) < \psi(x)$, $\forall x \in \Omega$, the solution of (2) approach to zero as $t \to \infty$.

If $m = 1$ and $f(u) = u^p$, $g(u) = 0$ with $p > 1$, Levine [14] proved that solutions of (1) must blow up in finite time, as long as $\phi(x)$ is large enough in the sense that its *energy*

$$E(\phi) = \frac{1}{2}\| \nabla \phi\|_2^2 - \frac{1}{p+1}\|\phi\|_{p+1}^{p+1}$$

is negative. Tan [18] gave a sharp condition on $\phi(x)$ such that the solution of (1) blows up in finite time when p is a critical Sobolev exponent. Lacey [15] showed that for general $f(u)$, the solution blows up if $\phi(x)$ is greater than a nonminimal steady state and $f', f'' \geq 0$, $\int^{\infty} s/f(s)ds < \infty$. This is also a sharp result.

If $m = 2$ and $f(u) = ((1-u_2)\Lambda(u_1),\ (1-u_2)\Lambda(u_1))$, $g(u) = (0,0)$, with $\Lambda(u_1) = e^{u_1}$ Bebernes and Lacey [2] have shown that solutions of (1) blow up in finite time. Later, in [3], they extended their results to more general $\Lambda(u_1)$. If $f(u) = (u_1^{m_1}u_2^{n_1}, u_1^{m_2}u_2^{n_2})$, $g(u) = (0,0)$, Chen [4] proved that if $m_1 \leq 1$, $n_1 \leq 1$, and $m_2 n_1 \leq (1-m_1)(1-n_2)$ then all nonnegative solutons are global, while if $m_1 > 1$ or $n_1 > 1$ or $m_2 n_1 > (1-m_1)(1-n_2)$ then both global existence and finite time blowup coexist.

Gang and Sleeman [11] gave several sufficient conditions to obtain the blowup property in one-dimensional parabolic system with $m = 2$,

$$u_{it} = \alpha_i u_{ixx} + f_i(u_1,\ u_2), \quad -a < x < a, \ \alpha_i > 0, \ i = 1,2.$$

Pao [16] studied more general systems with two equations. He assumed that f_1 and f_2 are quasimonotone nondecreasing, $g_1(u) = g_2(u) = 0$,

$$f_i(x,u_1,u_2) \geq \sigma_i u_1^{m_i} u_2^{n_i} - c_i u_i, \quad i = 1,2,$$

for $u_i > 0$, and there exist positive constant $\mu \geq 1$ and $\nu \geq 1$ such that

$$\gamma = \min\{\mu(m_1-1) + \nu n_1,\ \mu m_2 + \nu(n_2-1)\} > 0.$$

Then for any $(\phi_1,\phi_2) \geq (\omega^\mu,\omega^\nu)$, the solution of (1) must blow up in finite time, where ω satisfies

$$\Delta\omega+\sigma(\omega)^{1+\gamma}-c_0\omega \geq 0 \text{ in } \Omega, \quad \omega = 0 \text{ on } \partial\Omega,$$

with $c_0 = \max\{c_1/\mu, c_2/\nu\}$ and $\sigma = \min\{\sigma_1/\mu, \sigma_2/\nu\}$.

If $m \geq 1$, Chen and Derric [5] considered the following system

$$\begin{cases} u_{it} = \Delta u_i + f_i(u_1,\cdots,u_m), & x\in\Omega, \quad t>0, \\ u_i(x,t)=0, & x\in\partial\Omega, \quad t>0, \\ u_i(x,0)=\phi_i(x), & x\in\Omega. \end{cases} \tag{3}$$

Under suitable assumption on the nonlinear term f_i, they proved that if $0 \leq \phi_i \leq \lambda\psi_i$ with $\lambda < 1$, the solutions are global, where if $\phi_i \geq \lambda\psi_i$ with $\lambda > 1$, the solutions blow up in a finite time, where ψ_i is positive solutions of $\Delta\psi_i + f_i(\psi_1,\cdots,\psi_m) = 0$ in Ω with $\psi_i|_{\partial\Omega} = 0$. In [6], Chen generalize the blowup result of [5] by removing the condition $\psi_i|_{\partial\Omega} = 0$. We assume that

(H_1) $g_i \in C^1(\overline{\Omega})$, $g_i \geq 0$ and $\frac{\partial g_i}{\partial u_j} \geq 0$ for $u_j > 0$ and $u_i\frac{\partial g_i}{\partial u_i}$ is bounded as $u_i \to 0^+$, $i,j = 1,\cdots,m$.

(H_2) $D_i \subset \Omega$ is an open domain in R^n with a closed, simple and smooth boundary ($\partial D_i \cap \partial\Omega$ may be nonempty) for $i = 1,\cdots,m$ and there exists a small ball B_r in Ω such that $\overline{B}_r \subset \bigcap_{i=1}^m D_i$, ψ_i is defined in Ω and is a positive solution of $\Delta\psi_i + f_i(\psi) + g_i(\psi)|\nabla\psi_i|^2 = 0$ in D_i with $\psi_i|_{\partial D_i} = 0$ for $i = 1,\cdots,m$. Furthermore, $\psi_i \leq 0$ on $D_j - D_i$ if $D_j - D_i \neq \emptyset$ and $i \neq j$.

$(H_{2'})$ ψ_i is a positive solution of $\Delta\psi_i + f_i(\psi) + g_i(\psi)|\nabla\psi_i|^2 = 0$ in Ω with $\psi_i|_{\partial\Omega} = 0$ for $i = 1,\cdots,m$.

(H_3) $\phi_i(x) \in C^{2+\alpha}(\overline{\Omega})$ for $\alpha \in (0,1)$ with $\phi_i|_{\partial\Omega} = 0$ for $i = 1,\cdots,m$.

(H_4) f_i is locally lipschitz continuous, $f_i(u) \geq 0$ if $u \geq 0$ and $\frac{f_i(u)}{u_i} > \frac{f_i(v)}{v_i}$ whenever $u > v$, $v > 0$, $i = 1,\cdots,m$.

(H_5) There exists an index k, $1 \leq k \leq m$, such that $\frac{f_k(u)}{u_k^\sigma} \geq C_0 > 0$ for some $\sigma > 1$, $u_k > 0$.

2. Blowup and Global Existence

In this section, we will give some conditions on the solutions of (1) blowup in finite time or global existence. Furthermore, we will find the solution of (1) is unstable though our results. We first give a lemma, which is important in our proofs.

Lemma 2.1. *Under the assumptions of H_1 H_2 and H_4, for any $i,j = 1,\cdots,m$, we have*

$$\int_{D_i} \frac{\partial g_i}{\partial u_j} \nabla u_j \cdot \nabla\psi_i dx \geq 0.$$

Proof.

$$\begin{aligned} \int_{D_i} \frac{\partial g_i}{\partial u_j} \nabla u_j \cdot \nabla\psi_i dx &= -\int_{D_i} \frac{\partial g_i}{\partial u_j} u_j \Delta\psi_i dx \\ &= -\int_{D_i} \frac{\partial g_i}{\partial u_j} u_j \left(-f_i(u) - g_i(u)|\nabla u_i|^2\right) dx. \\ &\geq 0. \end{aligned}$$

□

Theorem 2.2. *Under the hypotheses* H_1, H_2, H_3, H_4, H_5 *if* $\phi_i(x) \geq (1+\varepsilon_0)\psi_i(x)$ *in* D_i *for some small* $\varepsilon_0 > 0$ *and* $\phi_i(x) > 0$ *in* Ω *for* $i = 1, \cdots, m$, *the solution* u *of (1) must blow up in finite time.*

Proof. From standard of parabolic PDE theory, there exists a unique solution $u(x,t) \in C^{1,0}(\overline{\Omega} \times [0,T^*)) \cap C^{2,1}(\overline{\Omega} \times (0,T^*))$ to (1) where T^* is the maximum time such that u exists and is bounded (see [1]). By the maximum principle, we have $u_i \geq 0$ in D_i. Set

$$h_i^n(t) = \int_{D_i} \frac{\psi_i^{(n+2)}(x)}{u_i^n(x,t)} dx, \tag{4}$$

for $t \in [0,T^*)$ and any real number $n > 0$, the following method was used in ([5,6]).

If ∂D_i is located inside of Ω, then (4) is well defined because $u_i(x,t) > 0$ on $\overline{D}_i$. Since u_i is continuous in Ω, there exists $\bar{t} > 0$ such that $u_i(x,t) \geq \psi_i(x)$ in D_i for $0 < t \leq \bar{t}$ and $1 \leq i \leq m$.

If there exists a point $x_0 \in \partial\Omega \cap \partial D_i$ then

$$\frac{1}{1+\varepsilon_0} \geq \lim_{x \to x_0} \frac{\psi_i(x)}{\phi_i(x)} = \lim_{x \to x_0} \frac{\psi_i(x) - \psi_i(x_0)}{\phi_i(x) - \phi_i(x_0)} = \frac{\partial\psi_i(x_0)/\partial\overrightarrow{n}}{\partial\phi_i(x_0)/\partial\overrightarrow{n}}, \tag{5}$$

where $x \in \Omega$ and is located on the normal line that passes through x_0. Since $\partial u_i(x_0,t)/\partial\overrightarrow{n} < 0$ by the strong maximum principle and $\partial u_i(x,t)/\partial\overrightarrow{n}$ is continuous on Ω, we have

$$\lim_{x \to x_0} \frac{\psi_i(x)}{u_i(x,t)} = \frac{\partial\psi_i(x_0)/\partial\overrightarrow{n}}{\partial u_i(x_0,t)/\partial\overrightarrow{n}} \leq 1, \tag{6}$$

for sufficiently small $t > 0$. Hence, there exists t_1, such that (4) is well defined and $u_i(x,t) \geq \psi_i(x,t)$ in $\overline{D}_i$ for $0 < t \leq t_1 \leq \bar{t}$ and all $i = 1, \cdots, m$.

Differentiating (4), substituting in the equation (1), and integrating by parts. (notice that the boundary values are always zero) we have

$$\begin{aligned} \frac{d}{dt} h_i^n(t) = & -n \int_{D_i} \frac{\psi_i^{n+2}}{u_i^{n+1}} \left(\Delta u_i + f_i(u) + g_i(u)|\nabla u_i|^2\right) dx \\ = & -n(n+1) \int_{D_i} \frac{\psi_i^{n+2}}{u_i^{n+2}} |\nabla u_i|^2 dx + n(n+2) \int_{D_i} \frac{\psi_i^{n+1}}{u_i^{n+1}} \nabla u_i \cdot \nabla\psi_i dx \\ & -n \int_{Di} \frac{\psi_i^{n+2}}{u_i^{n+1}} f_i(u) dx - n \int_{D_i} \frac{\psi_i^{n+2}}{u_i^{n+1}} g_i(u)|\nabla u_i|^2 dx. \end{aligned} \tag{7}$$

To simply (7), we write

$$\psi_i^2 |\nabla u_i|^2 = |\psi_i \nabla u_i - u_i \nabla\psi_i|^2 + 2u_i\psi_i \nabla u_i \cdot \nabla\psi_i - u_i^2 |\nabla\psi_i|^2.$$

It follows from (7) that

$$\begin{aligned} \frac{d}{dt} h_i^n(t) = & -n \int_{D_i} \frac{\psi_i^n}{u_i^{n+2}} |\psi_i \nabla u_i - u_i \nabla\psi_i|^2 \left(n+1+\left(1-\frac{1}{n}\right) u_i g_i(u)\right) dx \\ & -n^2 \int_{D_i} \frac{\psi_i^{n+1}}{u_i^{n+1}} \nabla u_i \cdot \nabla\psi_i dx + n(n+1) \int_{D_i} \frac{\psi_i^n}{u_i^n} |\nabla\psi_i|^2 dx \\ & - \int_{D_i} \frac{\psi_i^{n+2}}{u_i^{n+1}} g_i(u)|\nabla u_i|^2 dx + (n-1) \int_{D_i} \frac{\psi_i^n}{u_i^{n-1}} g_i(u)|\nabla\psi_i|^2 dx \\ & -2(n-1) \int_{D_i} \frac{\psi_i^{n+1}}{u_i^n} g_i(u) \nabla u_i \cdot \nabla\psi_i dx - n \int_{D_i} \frac{\psi_i^{n+2}}{u_i^{n+1}} f_i(u) dx. \end{aligned} \tag{8}$$

Using the identity

$$-n^2\int_{D_i}\frac{\psi_i^{n+1}}{u_i^{n+1}}\nabla u_i\cdot\nabla\psi_i dx = n\int_{D_i}\psi_i^{n+1}\nabla\left(\frac{1}{u_i^n}\right)\cdot\nabla\psi_i dx = -n(n+1)\int_{D_i}\frac{\psi_i^n}{u_i^n}|\nabla\psi_i|^2dx - n\int_{D_i}\frac{\psi_i^{n+1}}{u_i^n}\Delta\psi_i dx,$$

and $\Delta\psi_i = -f_i(\psi) - g_i(\psi)|\nabla\psi_i|^2$. We obtain from (8) that

$$\begin{aligned}\frac{d}{dt}h_i^n(t)\leq\ & -n\int_{D_i}\frac{\psi_i^{n+1}}{u_i^n}\Delta\psi_i dx - \int_{D_i}\frac{\psi_i^{n+2}}{u_i^{n+1}}g_i(u)|\nabla u_i|^2dx\\ & -(n+3)\int_{D_i}\frac{\psi_i^n}{u_i^{n-1}}g_i(u)|\nabla\psi_i|^2dx - 2\int_{D_i}\frac{\psi_i^{n+1}}{u_i^{n-1}}\sum_{j=1}^m\frac{\partial g_i}{\partial u_j}\nabla u_j\cdot\nabla\psi_i dx\\ & -2\int_{D_i}\frac{\psi_i^{n+1}}{u_i^{n-1}}g_i(u)\Delta\psi_i dx - n\int_{D_i}\frac{\psi_i^{n+2}}{u_i^{n+1}}f_i(u)dx.\end{aligned}\tag{9}$$

Using the Lemma 2.1, we have

$$-2\int_{D_i}\frac{\psi_i^{n+1}}{u_i^{n-1}}\sum_{j=1}^m\frac{\partial g_i}{\partial u_j}\nabla u_j\cdot\nabla\psi_i dx \leq -2\int_{D_i}\frac{\psi_i^{n+1}}{u_i^{n-1}}\frac{\partial g_i}{\partial u_i}\nabla u_i\cdot\nabla\psi_i dx.$$

It follows from (9) that

$$\begin{aligned}&\frac{d}{dt}h_i^n(t)\\ &\leq -n\int_{D_i}\frac{\psi_i^{n+2}}{u_i^n}\left(\frac{f_i(u)}{u_i}-\frac{f_i(\psi)}{\psi_i}\right)dx + n\int_{D_i}\frac{\psi_i^{n+1}}{u_i^n}g_i(\psi)|\nabla\psi_i|^2dx\\ &\quad -\int_{D_i}\frac{\psi_i^n}{u_i^{n+1}}g_i(u)\left|\psi_i\nabla u_i + \frac{u_i^2\partial g_i/\partial u_i}{g_i(u)}\nabla\psi_i\right|^2dx\\ &\quad -(n+3)\int_{D_i}\frac{\psi_i^n}{u_i^{n-1}}g_i(u)|\nabla\psi_i|^2dx + 2\int_{D_i}\frac{\psi_i^{n+1}}{u_i^{n-1}}f_i(\psi)g_i(u)dx\\ &\quad +2\int_{D_i}\frac{\psi_i^{n+1}}{u_i^{n-1}}g_i(\psi)g_i(u)|\nabla\psi_i|^2dx + \int_{D_i}\frac{\psi_i^n}{u_i^{n-3}}\frac{(\partial g_i/\partial u_i)^2}{g_i(u)}|\nabla\psi_i|^2dx\\ &\leq -n\int_{D_i}\frac{\psi_i^n}{u_i^{n-1}}g_i(u)|\nabla\psi_i|^2\left(1-\frac{\psi_i g_i(\psi)}{u_i g_i(u)} - \frac{2}{n}\psi_i g_i(\psi) - \frac{u_i^2(\partial g_i/\partial u_i)^2}{n g_i^2(u)}\right)dx\\ &\quad -\frac{n}{2}\int_{D_i}\frac{\psi_i^{n+2}}{u_i^n}\left(2\frac{f_i(u)}{u_i} - 2\frac{f_i(\psi)}{\psi_i} - \frac{4}{n}u_i g_i(u)\frac{f_i(\psi)}{\psi_i}\right)dx.\end{aligned}\tag{10}$$

Using assumption H_1 and H_4, we can take n sufficiently large that for $t\in(0,t_1]$.

$$2\frac{f_i(u)}{u_i} - 2\frac{f_i(\psi)}{\psi_i} - \frac{4}{n}u_i g_i(u)\frac{f_i(\psi)}{\psi_i} \geq \frac{f_i(u)}{u_i} - \frac{f_i(\psi)}{\psi_i} > 0,\tag{11}$$

$$1 - \frac{\psi_i g_i(\psi)}{u_i g_i(u)} - \frac{2}{n}\psi_i g_i(\psi) - \frac{u_i^2(\partial g_i/\partial u_i)^2}{n^2 g_i(u)} \geq 0.\tag{12}$$

Thus we obtain $(h_i^n(t))' \leq 0$ for any $0 < t \leq t_1$. So we get $h_i^n(t) \leq h_i^n(0)$ for $t \in (0, t_1]$. Taking the nth roots and letting $n \to \infty$, we have

$$\frac{\psi_i(x)}{u_i(x,t)} \leq \max_{\overline{D_i}} \frac{\psi_i(x)}{u_i(x,t)} \leq \max_{\overline{D_i}} \frac{\psi_i(x)}{\phi_i(x)} \leq \frac{1}{1+\varepsilon_0} \text{ for } t \in [0, t_1]. \tag{13}$$

So we have $u_i(x,t_1) \geq (1+\varepsilon_0)\psi_i(x)$. Taking the $u_i(x,t_1)$ as the initial value of $u_i(x,t)$. Noticing the continuity of u_i and $\partial u_i/\partial \overrightarrow{n}$, using the similar means as above, we know that there exists $t_2 > t_1$, such that $u_i(x,t) \geq \psi_i(x)$ for $t_1 < t \leq t_2$ and $i = 1, \cdots, m$. Therefore, (10)and(13) hold for $0 < t \leq t_2$. According to this method we can extend u_i to $[0, T^*)$, and (13) holds for $0 \leq t < T^*$. We need to show that T^* is a finite time. From (13) and assumption H_2 and H_4, we have

$$1 - \frac{f_i(\psi)/\psi_i}{f_i(u)/u_i} > 0 \text{ in } D_i, \quad 1 - \frac{f_i(\psi)/\psi_i}{f_i(u)/u_i} \geq C_1 > 0 \text{ on } \overline{B}_r.$$

From (10) (11) (12) and assumption H_2 and H_5 choosing $n = \sigma - 1$ we have

$$\begin{aligned}
\frac{d}{dt} h_k^{\sigma-1}(t) &\leq -\frac{\sigma-1}{2} \int_{D_k} \frac{\psi_k^{\sigma+1}}{u_k^{\sigma-1}} \left(\frac{f_k(u)}{u_k} - \frac{f_k(\psi)}{\psi_k} \right) dx \\
&\leq -\frac{\sigma-1}{2} \int_{B_r} \frac{\psi_k^{\sigma+1}}{u_k^{\sigma-1}} \frac{f_k(u)}{u_k} \left(1 - \frac{f_k(\psi)/\psi_k}{f_k(u)/u_k} \right) dx \\
&\leq -\frac{\sigma-1}{2} C_0 C_1 \int_{B_r} \psi_k^{\sigma+1} dx = -C \int_{B_r} \psi_k^{\sigma+1} dx.
\end{aligned}$$

Hence $0 < h_k^{\sigma-1}(t) \leq h_k^{\sigma-1}(0) - Ct \int_{B_r} \psi_k^{1+\sigma} dx$, or $t \leq \dfrac{h_k^{\sigma-1}(0)}{C \int_{B_r} \psi_k^{1+\sigma} dx}$, which means that T^* can not increase to infinite. Therefore, $u_k(x,t)$ must blow up in finite time. □

Remark. At first glance, assumption H_2 may seem strong and complicated. However, it is reasonable. In fact, for a complicate domain Ω, it is not clear whether a positive solution of $\Delta\psi_i + f_i(\psi) + g_i(\psi)|\nabla\psi_i|^2 = 0$ in Ω with $\psi_i|_{\partial\Omega} = 0$ for $i = 1, \cdots, m$ exists or whether such solution, if it does exist, is unique. However, we can easily find a positive solution of $\Delta\psi_i + f_i(\psi) + g_i(\psi)|\nabla\psi_i|^2 = 0$ in D_i with $\psi_i|_{\partial D_i} = 0$ for $i = 1, \cdots, m$, where D_i is any subdomain of Ω.

Theorem 2.3. *If $0 \leq \phi_i(x) \leq \lambda\psi_i(x)$ with $\lambda < 1$ $i = 1, \cdots, m$ and the assumptions H_1, $H_{2'}$, H_3, H_4 hold, the solution u of (1) exists for all $t > 0$ and decay exponentially in t.*

Proof. Similar to the proof of Theorem 2.2, set $E_i^n(t) = \displaystyle\int_\Omega \frac{u_i^{n+2}(x,t)}{\psi_i{}^n(x)} dx$. We have

$$\begin{aligned}
\frac{d}{dt} E_i^n(t) = \; & (n+2) \int_\Omega \frac{u_i^{n+1}}{\psi_i^n} \left(\Delta u_i + f_i(u) + g_i(u) |\nabla u_i|^2 \right) dx \\
= \; & -(n+2) \int_\Omega \frac{u_i^n}{\psi_i^n} |\nabla u_i|^2 \left(n + 1 - \frac{n+3}{n+2} u_i g_i(u) \right) dx \\
& + n(n+2) \int_\Omega \frac{u_i^{n+1}}{\psi_i^{n+1}} \nabla u_i \nabla \psi_i dx \\
& - \int_\Omega \frac{u_i^{n+1}}{\psi_i^n} g_i(u) |\nabla u_i|^2 dx + (n+2) \int_\Omega \frac{u_i^{n+1}}{\psi_i^n} f_i(u) dx.
\end{aligned} \tag{14}$$

Using the Lemma 2.1. We get $-\int_\Omega \sum_{j=1}^m \frac{\partial g_i}{\partial u_j}\nabla u_j\cdot\nabla\psi_i dx \le -\int_\Omega \frac{\partial g_i}{\partial u_i}\nabla u_i\cdot\nabla\psi_i dx$. It follows from (14) that

$$\begin{aligned}
&\frac{d}{dt}E_i^n(t)\\
\le\ & (n+2)\int_\Omega \frac{u_i^{n+1}}{\psi_i^n} f_i(u)dx - \int_\Omega \frac{u_i^{n+1}}{\psi_i^n} g_i(u)|\nabla u_i|^2 dx\\
&-(n+2)\int_\Omega \frac{u_i^n}{\psi_i^{n+2}}|\psi_i\nabla u_i - u_i\nabla\psi_i|^2\left(n+1-\frac{n+3}{n+2}u_i g_i(u)\right)dx\\
&+(n+2)\int_\Omega \frac{u_i^{n+2}}{\psi_i^{n+1}}\Delta\psi_i dx - (n+3)\int_\Omega \frac{u_i^{n+3}}{\psi_i^{n+2}} g_i(u)|\nabla\psi_i|^2 dx\\
&+2(n+3)\int_\Omega \frac{u_i^{n+2}}{\psi_i^{n+1}} g_i(u)\nabla u_i\cdot\nabla\psi_i dx\\
\le\ & -(n+2)\int_\Omega \frac{u_i^{n+2}}{\psi_i^{n+1}} f_i(\psi)\left(1-\frac{f_i(u)/u_i}{f_i(\psi)/\psi_i}-\frac{2}{n+2}u_i g_i(u)\right)dx\\
&-(n+2)\int_\Omega \frac{u_i^n}{\psi_i^{n+2}}|\psi_i\nabla u_i - u_i\nabla\psi_i|^2\left(n+1-\frac{n+3}{n+2}u_i g_i(u)\right)dx\\
&-(n+2)\int_\Omega \frac{u_i^{n+2}}{\psi_i^{n+1}} g_i(\psi)|\nabla\psi_i|^2\left(1-\frac{u_i g_i(u)}{\psi_i g_i(\psi)}-\frac{2u_i g_i(u)}{n+2}-\frac{u_i^3(\partial g_i/\partial u_i)^2}{(n+2)\psi_i g_i(\psi) g_i(u)}\right)dx\\
&-\int_\Omega \frac{u_i^{n+1}}{\psi_i^{n+2}} g_i(u)\left|\psi_i\nabla u_i + \frac{u_i^2\partial g_i/\partial u_i}{g_i(u)}\nabla\psi_i\right|^2 dx - 3\int_\Omega \frac{u_i^{n+3}}{\psi_i^{n+2}} g_i(u)|\nabla\psi_i|^2 dx.
\end{aligned} \tag{15}$$

For $0 \le u_i \le \lambda\psi_i$ (by comparison theorem) and assumption H_1 and H_4, we take n sufficiently large such that for sufficiently small $\varepsilon_n > 0$

$$1-\frac{f_i(u)/u_i}{f_i(\psi)/\psi_i}-\frac{2}{n+2}u_i g_i(u) \ge \varepsilon_n, \quad n+1-\frac{n+3}{n+2}u_i g_i(u) \ge 0,$$

$$1-\frac{u_i g_i(u)}{\psi_i g_i(\psi)}-\frac{2}{n+2}u_i g_i(u)-\frac{u_i^3(\partial g_i/\partial u_i)^2}{(n+2)\psi_i g_i(\psi) g_i(u)} \ge 0.$$

For any $\varepsilon > 0$. Let $\Omega_\varepsilon := \{x\in\Omega | \mathrm{dist}(x,\partial\Omega) < \varepsilon\}$. Using (15), we obtain

$$\begin{aligned}
\frac{d}{dt}E_i^n(t) \le\ & -(n+2)\varepsilon_n \int_\Omega \frac{u_i^{n+2}}{\psi_i^{n+1}} f_i(\psi)dx\\
\le\ & -(n+2)\varepsilon_n\varepsilon_1 \int_{\Omega-\Omega_\varepsilon} \frac{u_i^{n+2}}{\psi_i^n}dx\\
\equiv\ & -(n+2)\varepsilon_n\varepsilon_1 E_{i\varepsilon}^n(t),
\end{aligned}$$

where $\varepsilon_1 = \min_{\{\Omega-\Omega_\varepsilon\}} \frac{f_i(\psi)}{\psi_i}$. Then $E_{i\varepsilon}^n(t) < E_i^n(t) \le E_i^n(0) - (n+2)\varepsilon_n\varepsilon_1\int_0^t E_{i\varepsilon}^n(\tau)d\tau$, which implies for fixed n

$$\frac{1}{\max_{\{\Omega-\Omega_\varepsilon\}}\psi_i^n}\int_{\Omega-\Omega_\varepsilon} u_i^{n+2}(x,t) < E_{i\varepsilon}^n(t) \to 0, \ as\, t\to+\infty.$$

Since $u_i = 0$ on the boundary and $u_i(x,t) \le \psi_i(x)$. we have $\int_\Omega u_i^{n+2}dx \to 0, \ as\, t\to+\infty$, for $i = 1,\cdots,m$.

If $\lim_{\psi_i \to 0} \frac{f_i(\psi)}{\psi_i} = 0$ by [12], $u_i(x,t)$ decays exponentially and exists globally.

If $\frac{f_i(\psi)}{\psi_i} \geq C_2 > 0$, for any $x \in \Omega$ and n sufficiently large, we get from (15) and H_4 that

$$\begin{aligned}
\frac{d}{dt}E_i^n(t) \leq & -(n+2)\int_\Omega \frac{u_i^{n+2}}{\psi_i^{n+1}} f_i(\psi)\left(1-\frac{f_i(u)/u_i}{f_i(\psi)/\psi_i}-\frac{2}{n+2}u_i g_i(u)\right)dx \\
= & -\frac{n+2}{2}\int_\Omega \frac{u_i^{n+2}}{\psi_i^n}\left(2\frac{f_i(\psi)}{\psi_i}-2\frac{f_i(u)}{u_i}-\frac{4}{n+2}u_i g_i(u)\frac{f_i(\psi)}{\psi_i}\right)dx \\
\leq & -\frac{n+2}{2}\int_\Omega \frac{u_i^{n+2}}{\psi_i^n}\left(\frac{f_i(\psi)}{\psi_i}-\frac{f_i(u)}{u_i}\right)dx \\
= & -\frac{n+2}{2}\int_\Omega \frac{u_i^{n+2}}{\psi_i^n}\frac{f_i(\psi)}{\psi_i}\left(1-\frac{f_i(u)/u_i}{f_i(\psi)/\psi_i}\right)dx \\
\leq & -\frac{n+2}{2}C_2\varepsilon_n\int_\Omega \frac{u_i^{n+2}}{\psi_i^n}dx
\end{aligned}$$

We obtain $\frac{d}{dt}E_i^n(t) \leq -CE_i^n(t)$, which implies that $u_i(x,t)$ decays exponentially and exists globally. □

Remark. As a result of Theorem 2.3, the solution of (1) is unstable.

References

[1] H. Amann, Nonlinear analysis: a collection of papers in honor of Erich Rothe, *Academic Press,New York.* 1978, 1-29.

[2] J. Bebernes and A. Lace, Finite time blowup for a particular parabolic system, *SIAM J.Math.Anal.* **21**(1990), 1415-1425.

[3] J. Bebernes and A. Lace, Finite time blowup for semilinear reactive-diffusion system, *J.Differential Equation* **95**(1992) 105-129.

[4] H.Chen, Global existence and blowup for a nonlinear reaction diffusion system, *J.Math.Anal.Appl.* **212**(1997) 481-492.

[5] S. Chen and W. R. Derrick, Global Existence and blow up of solutions for semilinear parabolic system, *Rocky Mountain. J.Math*, **29** (1999), 449-457.

[6] S. Chen, A sufficient condition for blowup solutions of nonlinear heat equations, *J. Math. Anal.Appl.* **293** (2004), 227-236.

[7] K. Deng, Stabilization of solution of a nonlinear parabolic equation with a gradient term, *Math.Z.* **216** (1994), 147-155.

[8] Lawrence C. Evances, Partial Differential Equations, *Graduate Studies in Mathematics, American Mathematical Society Providence, Rhode Island* Vol. 19(1998), 194-195.

[9] M. Fila, Remarks on blow up for a nonlinear parabolic equation with a gradient term, *Proc. Amer. Math. Soc,* **111** (1991), 795-801.

[10] V. A. Galaktionov, On new exact blow-up solutions for nonlinear heat conduction equations with source and applications, *Diff. and Int. Eqs* **3** (1990), 863-874.

[11] L. Gang, B. D. Sleeman, Non-existence of global solution to system of semi-linear parabolic equations, *J.Differential Equations* **104**(1993), 147-168.

[12] D. Henry, Geometric theory of semilinear parabolic equation, *Springer-Verlag,Berlin* Vol.840 (1981).

[13] B. Kawohl and L. A. Peletier, observations on blow up and dead cores for nonlinear parabolic equations, *Math.Z.* **202** (1989), 207-217.

[14] H. A. Levine, Some nonexistence and instability theorems for solutions of the form $pu_t = -Au + F(u)$, *Arch.Rational Mech.Anal.* **51**(1973), 371-385.

[15] A. Lacy, Mathematical analysis of thermal runaway for spatially inhomogeneous reactions, *SIAM.J.Appl.Math.* **43**(1983), 1350-1366.

[16] C. V. Pao, Nonlinear Parabolic and Elliptic Equations, *Plenum, New York*, 1992.

[17] P. Souplet, *Résultats d*'explosion en temps fini pour une *équation* de la chaleur non *linéaire*, *C.R.Acad. Sc. Paris* **321** (1995), *SérieI* 721-726 .

[18] Z. Tan, The reaction-diffusion equation with Lewis function and critical Sobolev exponent, *J.Math.Anal.Appl.* **272**(2002) 480-495.

In: Progress in Evolution Equations
Editor: Gaston M. N'Guerekata, pp. 135-147
ISBN: 978-1-60456-328-3

Chapter 8

THE REPRESENTATION OF EXACT SOLUTION OF THE COMPOUND BURGERS-KORTEWEG-DE VRIES EQUATION

***Fazhan Geng*[*] *and Minggen Cui*[†]**
Dept. of Math., Harbin Institute of Technology,
Weihai, ShanDong, P. R. China. 264209

Abstract

In this paper, the compound Burgers-Korteweg-de Vries (cBKdV) equation with initial-boundary conditions is investigated in a reproducing kernel space. Its exact solution is represented in the form of series. In the mean time, the approximate solution $u_n(t,x)$ is obtained by the n-term intercept of series and is proved to converge to the exact solution. Moreover, the error between the approximate solution $u_n(t,x)$ and the exact solution $u(t,x)$ is monotone decreasing. Some numerical examples are studied to demonstrate the accuracy of the present method. Results obtained by the method are compared with the exact solution of each example and are found to be in good agreement with each other.

2000 Mathematics Subject Classification: Primary 35C10;47B32

Key words: Compound Burgers-Korteweg-de Vries equation; reproducing kernel.
Foundation item: Supported by National Natural Science Foundation of China(60572125).

1. Introduction

In this paper, we consider the following cBKdV equation with initial-boundary conditions in a reproducing kernel space

$$\begin{cases} u_t + b_1(x,t)uu_x + b_2(x,t)u^2u_x + b_3(x,t)u_{xx} + b_4(x,t)u_{xxx} = f(x,t), \\ u(0,x) = 0, \;\; 0 \le t \le 1, 0 \le x < \infty, \\ u(t,0) = 0, u_x(t,0) = 0, u_{xx}(t,0) = 0, \end{cases} \tag{1.1}$$

[*]E-mail address: gengfazhan@sina.com. The first author was supported in part by NSF Grant #000000.
[†]E-mail address: cmgyfs@263.net

where $b_1(x,t)$, $b_2(x,t)$, $b_3(x,t)$, $b_4(x,t)$ and $f(x,t)$ are continuous.

The KdV equation is the champion of model of nonlinear wave. It was studied by Korteweg and de Vries late in the 19th century as wave equation, and after a long period of sleep, revived as one of most fundamental equations of soliton phenomena. The equations arise in a variety of physical context. The general KdV equations have been studied by many authors (See [1-3]). However, the detailed studies for the cBKdV equation only began several years ago (see [4-8]). The cBKdV equation has attracted much attention. In particular, the travelling wave solution to the cBKdV equation has been widely investigated. Wang treated the cBKdV equation using the homogeneous balance method and presented an exact solution (see [4]). However, his approach is quite complicated. Feng studied the cBKdV equation by applying the first integral method and the method of variation of parametres repectively (see [5-6]). Zhang and his co-workers considered the equation by means of proper transformation and the method of undetermined coefficients (see [7]). The authors studied the equation by the first integral method (see [8]). These methods and results for the cBKdV equation mentioned above have contributed to our understanding of nonlinear pysical phenomena and propagation.

In this paper, we will give the representation of exact solution to $Eq.(1.1)$ in a reproducing kernel space under the assumption that the solution to $Eq.(1.1)$ is unique. The approach is simple and effective.

Put $Lu \equiv u_t + b_3(x,t)u_{xx} + b_4(x,t)u_{xxx}$, then $Eq.(1.1)$ can be converted into following form

$$\begin{cases} Lu = f(x,t) - b_1(x,t)uu_x - b_2(x,t)u^2u_x, 0 \le t \le 1, 0 \le x < \infty, \\ u(0,x) = 0, \\ u(t,0) = 0, u_x(t,0) = 0, u_{xx}(t,0) = 0, \end{cases} \tag{1.2}$$

where $f \in W_{(1,1)}(D), u \in W_{(2,4)}(D), D = [0,1] \times [0,\infty)$. $W_{(1,1)}(D)$ and $W_{(2,4)}(D)$ are defined in the following section.

2. Several Reproducing Kernel Spaces

1. The reproducing kernel space $W_2[0,1]$

Inner product space $W_2[0,1]$ is defined as $W_2[0,1] = \{u(t) \mid u, u'$ are absolutely continuous real value functions, $u, u', u'' \in L^2[0,1], u(0) = 0\}$. The inner product in $W_2[0,1]$ is given by

$$(u(t), v(t))_{W_2} = \int_0^1 (4uv + 5u'v' + u''v'')dt, \tag{2.1}$$

$u, v \in W_2[0,1]$ and the norm $\| u \|_{W_2}$ is denoted by $\| u \|_{W_2} = \sqrt{(u,u)_{W_2}}$.

Theorem 2.1. *The space $W_2[0,1]$ defined as above is a reproducing kernel space. That is, for any $u(y) \in W_2[0,1]$ and each fixed $t \in [0,1]$, there exists $R_t^{\{2\}}(y) \in W_2[0,1]$, $y \in [0,1]$, such that $(u(y), R_t^{\{2\}}(y))_{W_2} = u(t)$. The reproducing kernel $R_t^{\{2\}}(y)$ can be denoted by*

$$R_t^{\{2\}}(y) = \begin{cases} c_{11}e^y + c_{12}e^{-y} + c_{13}e^{2y} + c_{14}e^{-2y}, & y \le t \\ d_{11}e^y + d_{12}e^{-y} + d_{13}e^{2y} + d_{14}e^{-2y}, & y > t \end{cases} \tag{2.2}$$

where

$c_{11} = \frac{e^{-2t}(4e^3 - 7e^{3t} - 9e^{2+t} + 7e^{6+t} + 9e^{4+3t} - 4e^{3+4t})}{6(-7 - 9e^2 + 9e^4 + 7e^6)}$,

$$c_{12} = \frac{e^{-2t}(-4e^3+7e^{3t}+9e^{2+t}-7e^{6+t}-9e^{4+3t}+4e^{3+4t})}{6(-7-9e^2+9e^4+7e^6)},$$
$$c_{13} = \frac{e^{-2t}(-9e^4-7e^6+7e^{4t}+8e^{3+t}-8e^{3+3t}+9e^{2+4t})}{12(-7-9e^2+9e^4+7e^6)},$$
$$c_{14} = \frac{e^{-2t}(9e^4+7e^6-7e^{4t}-8e^{3+t}+8e^{3+3t}-9e^{2+4t})}{12(-7-9e^2+9e^4+7e^6)},$$
$$d_{11} = \frac{e^{-2t}(4e^3-7e^{3t}-9e^{2+t}+7e^{6+t}+9e^{4+3t}-4e^{3+4t})}{6(-7-9e^2+9e^4+7e^6)},$$
$$d_{12} = \frac{e^{2-2t}(-4e+9e^t-9e^{3t}-7e^{4+t}+7e^{4+3t}+4e^{1+4t})}{6(-7-9e^2+9e^4+7e^6)},$$
$$d_{13} = \frac{e^{-2t}(-1+e^{2t})(7+9e^2+7e^{2t}-8e^{3+t}+9e^{2+2t})}{12(-7-9e^2+9e^4+7e^6)},$$
$$d_{14} = \frac{-e^{3-2t}(-1+e^{2t})(9e+7e^3-8e^t+9e^{1+2t}+7e^{3+2t})}{12(-7-9e^2+9e^4+7e^6)}.$$

The proof of Theorem (2.1) is given in Appendix A.

2. The reproducing kernel space $W_4[0,\infty)$

Inner product space $W_4[0,\infty)$ is defined as $W_4[0,\infty) = \{u(x) \mid u, u', u''$ *and* $u^{(3)}$ are absolutely continuous real value functions, $u, u', u'', u^{(3)}, u^{(4)} \in L^2[0,\infty), u(0)=0, u'(0)=0, u''(0)=0\}$. The inner product in $W_4[0,\infty)$ is given by

$$(u(y), v(y))_{W_4} = \int_0^\infty (576uv + 820u'v' + 273u''v'' + 30u^{(3)}v^{(3)} + u^{(4)}v^{(4)})dy, \tag{2.3}$$

and the norm $\| u \|_{W_4}$ is denoted by $\| u \|_{W_4} = \sqrt{(u,u)_{W_4}}$, where $u, v \in W_4[0,\infty)$.

In the same way as the proof of Theorem (2.1), we can prove that $W_4[0,\infty)$ is a reproducing kernel space and its reproducing kernel can be denoted by

$$R_x^{\{4\}}(y) = \begin{cases} c_1e^y + c_2e^{-y} + c_3e^{2y} + c_4e^{-2y} + c_5e^{3y} + c_6e^{-3y} + c_7e^{4y} + c_8e^{-4y}, \\ y \le x, \\ d_1e^y + d_2e^{-y} + d_3e^{2y} + d_4e^{-2y} + d_5e^{3y} + d_6e^{-3y} + d_7e^{4y} + d_8e^{-4y}, \\ y > x. \end{cases} \tag{2.4}$$

where

$$c_1 = \frac{e^{-x}}{720}, c_2 = \frac{e^{-4x}(2-9e^x+14e^{2x}-8e^{3x})}{720}, c_3 = \frac{-e^{-2x}}{720}, c_4 = \frac{e^{-4x}(-4+18e^x-27e^{2x}+14e^{3x})}{720},$$
$$c_5 = \frac{e^{-3x}}{1680}, c_6 = \frac{e^{-4x}(6-28e^x+42e^{2x}-21e^{3x})}{1680}, c_7 = \frac{-e^{-4x}}{10080}, c_8 = \frac{e^{-4x}(-7+36e^x-56e^{2x}+28e^{3x})}{10080},$$
$$d_2 = \frac{e^{-4x}(-1+e^x)^3(-2+3e^x+e^{2x})}{720}, d_4 = \frac{-e^{-4x}(-1+e^x)^3(-4+6e^x+3e^{2x}+e^{3x})}{720},$$
$$d_6 = \frac{e^{-4x}(6-28e^x+42e^{2x}-21e^{3x}+e^{7x})}{1680}, d_8 = \frac{e^{-4x}(-7+36e^x-56e^{2x}+28e^{3x}-e^{8x})}{10080},$$
$$d_1 = 0, d_3 = 0, d_5 = 0, d_7 = 0.$$

3. The reproducing kernel space $W_1[0,1]$

Inner product space $W_1[0,1]$ is defined by $W_1[0,1] = \{u(x) \mid u$ is absolutely continuous real value function, $u, u' \in L^2[0,1]\}$. The inner product and norm in $W_1[0,1]$ are given respectively by

$$(u(x), v(x))_{W_1} = \int_0^1 (uv + u'v')dx, \; \| u \|_{W_1} = \sqrt{(u,u)_{W_1}},$$

where $u(x), v(x) \in W_1[0,1]$. In [9], the authors proved that $W_1[0,1]$ is a reproducing kernel space and its reproducing kernel is

$$R_x^{\{1\}}(y) = \frac{1}{2\sinh(1)}[\cosh(x+y-1) + \cosh(|x-y|-1)].$$

4. The reproducing kernel space $W_{(2,4)}(D)$
Put $D=[0,1]\times[0,\infty)$. Assume that $\{p_i(t)\}_{i=1}^{\infty}$ is the complete normal orthogonal system of $W_2[0,1]$ and $\{q_i(x)\}_{i=1}^{\infty}$ is the complete normal orthogonal system of $W_4[0,\infty)$. Now we define $W_{(2,4)}(D)$ by

$$W_{(2,4)}(D)=\{u(t,x)|u(t,x)=\sum_{i,j=1}^{\infty}c_{ij}p_i(t)q_j(x),\{c_{i,j}\}_{i,j=1}^{\infty}\in l^2\}.$$

The inner product of $W_{(2,4)}(D)$ is defined dy

$$(u_1,u_2)_{W_{(2,4)}}=\sum_{i,j=1}^{\infty}c_{ij}d_{ij},$$

where $u_1=\sum_{i,j=1}^{\infty}c_{ij}p_i(t)q_j(x)$ and $u_2=\sum_{k,l=1}^{\infty}d_{kl}p_k(t)q_l(x)$. The norm is denoted by $\|u\|_{W_{(2,4)}}^2=(u,u)_{W_{(2,4)}}$.
From [10], we can know that $W_{(2,4)}(D)$ is a Hilbert space and obtain the following Proposition 2.1, Proposition 2.2.

Proposition 2.1. *If $u(t,x)=u_1(t)u_2(x)$, $v(t,x)=v_1(t)v_2(x)\in W_{(2,4)}(D)$, then*

$$(u(t,x),v(t,x))_{W_{(2,4)}}=(u_1(t),v_1(t))_{W_2}(u_2(x),v_2(x))_{W_4} \tag{2.5}$$

Proposition 2.2. *$W_{(2,4)}(D)$ is a reproducing kernel space and the reproducing kernel is*

$$K_{(\xi,\eta)}(t,x)=R_{\xi}^{\{2\}}(t)R_{\eta}^{\{4\}}(x), \tag{2.6}$$

where $R_{\xi}(t)$, $R_{\eta}^{\{4\}}(x)$ are given by (2.2)and (2,4) respectively.

Similar to the definition of $W_{(2,4)}(D)$, we can define $W_{(1,1)}(D)$. $W_{(1,1)}(D)$ is also a reproducing kernel space and its reproducing kernel is $\overline{K}_{(\xi,\eta)}(t,x)=R_{\xi}^{\{1\}}(t)R_{\eta}^{\{1\}}(x)$.

3. The Solution of Eq.(1.2)

In this section, the solution of $Eq.(1.2)$ is given in the reproducing kernel space $W_{(2,4)}(D)$.

Note that $Lu=u_t+b_3(x,t)u_{xx}+b_4(x,t)u_{xxx}$ in $Eq.(1.2)$. It is clear that $L:W_{(2,4)}(D)\to W_{(1,1)}(D)$ is a bounded linear operator. Put $M=(t,x),M_i=(t_i,x_i)$, $\varphi_i(M)=\overline{K}_{M_i}(M)$ and $\psi_i(M)=L^*\varphi_i(M)$, where $\overline{K}$ is the reproducing kernel of $W_{(1,1)}(D)$ and L^* is the adjoint operator of L . The orthonormal system $\{\overline{\psi}_i(M)\}_{i=1}^{\infty}$ of $W_{(2,4)}(D)$ can be derived from Gram-Schmidt orthogonalization process of $\{\psi_i(M)\}_{i=1}^{\infty}$,

$$\overline{\psi}_i(M)=\sum_{k=1}^{i}\beta_{ik}\psi_k(M),(\beta_{ii}>0,i=1,2,...). \tag{3.1}$$

Theorem 3.1. *For $Eq.(1.2)$, if $\{M_i\}_{i=1}^{\infty}$ is dense on D, then $\{\psi_i(M)\}_{i=1}^{\infty}$ is the complete system of $W_{(2,4)}(D)$ and $\psi_i(M)=L_{M'}K_M(M')|_{M'=M_i}$.*

Proof. Notice that

$$\begin{aligned}\psi_i(M) &= (L^*\varphi_i)(M) = ((L^*\varphi_i)(M'), K_M(M')) \\ &= (\varphi_i(M'), L_{M'}K_M(M')) = L_{M'}K_M(M')|_{M'=M_i}.\end{aligned}$$

The subscript M' by the operator L indicates that the operator L applies to the function of M'.

Clearly, $\psi_i(M) \in W_{(2,4)}(D)$.

For each fixed $u(M) \in W_{(2,4)}(D)$, let $(u(M), \psi_i(M)) = 0, (i = 1, 2, ...)$, which means that,

$$(u(M), (L^*\varphi_i)(M)) = (Lu(\cdot), \varphi_i(\cdot)) = (Lu)(M_i) = 0. \tag{3.2}$$

Since $\{M_i\}_{i=1}^{\infty}$ is dense on D, we must have, $(Lu)(M) = 0$. It follows that $u \equiv 0$ from the existence of L^{-1} and the proof of the Theorem 3.1 is complete. □

Theorem 3.2. *If $\{M_i\}_{i=1}^{\infty}$ is dense on D and the solution of $Eq.(1.2)$ is unique, then the solution of $Eq.(1.2)$ satisfies the form*

$$u(M) = \sum_{i=1}^{\infty}\sum_{k=1}^{i} \beta_{ik} F(M_k, u(M_k), u_x(M_k))\overline{\psi}_i(M), \tag{3.3}$$

where $F(M, u(M), u_x(M)) = f(M) - b_1(M)uu_x(M) - b_2(M)u^2u_x(M)$.

Proof. Applying Theorem 3.1, it is easy to know that $\{\overline{\psi}_i(M)\}_{i=1}^{\infty}$ is the complete normal orthogonal basis of $W_{(2,4)}(D)$.

Note that $(v(M), \varphi_i(M)) = v(M_i)$ for each $v(M) \in W_{(1,1)}(D)$. Hence

$$\begin{aligned}u(M) &= \sum_{i=1}^{\infty} (u(M), \overline{\psi}_i(M))\overline{\psi}_i(M) \\ &= \sum_{i=1}^{\infty}\sum_{k=1}^{i} \beta_{ik}(u(M), L^*\varphi_k(M))\overline{\psi}_i(M) \\ &= \sum_{i=1}^{\infty}\sum_{k=1}^{i} \beta_{ik}(Lu(M), \varphi_k(M))\overline{\psi}_i(M) \\ &= \sum_{i=1}^{\infty}\sum_{k=1}^{i} \beta_{ik}(F(M, u(M), u_x(M)), \varphi_k(M))\overline{\psi}_i(M) \\ &= \sum_{i=1}^{\infty}\sum_{k=1}^{i} \beta_{ik}F(M_k, u(M_k), u_x(M_k))\overline{\psi}_i(M)\end{aligned} \tag{3.4}$$

and the proof of the theorem is complete. □

4. The Implementation Method

In this section, a new method of obtaining the solution of $Eq.(1.2)$ shall be presented.

(3.3) can be denoted by

$$u(M) = \sum_{i=1}^{\infty} A_i \overline{\psi}_i(M), \tag{4.1}$$

where $A_i = \sum_{k=1}^{i} \beta_{ik} F(M_k, u(M_k), u_x(M_k))$. Let $M_1 = (0,0)$, it follows that $u(M_1)$, $u_x(M_1)$ is known from the initial and boundary conditions of $Eq.(1.2)$.
So $F(M_1, u(M_1), u_x(M_1))$ is known. Considering the numerical computation, we put $u_0(M_1) = u(M_1)$ and define the n-term approximation to $u(M)$ by

$$u_n(M) = \sum_{i=1}^{n} B_i \overline{\psi}_i(M), \tag{4.2}$$

where

$$\begin{aligned}
&B_1 = \beta_{11} F((M_1, u_0(M_1), (u_0)_x(M_1)),\\
&u_1(M) = B_1 \overline{\psi}_1(M),\\
&B_2 = \sum_{k=1}^{2} \beta_{2k} F((M_k, u_{k-1}(M_k), (u_{k-1})_x(M_k)),\\
&u_2(M) = \sum_{i=1}^{2} B_i \overline{\psi}_i(M),\\
&\cdots\cdots\\
&u_{n-1}(M) = \sum_{i=1}^{n-1} B_i \overline{\psi}_i(M),\\
&B_n = \sum_{k=1}^{n} \beta_{nk} F((M_k, u_{k-1}(M_k), (u_{k-1})_x(M_k)).
\end{aligned} \tag{4.3}$$

Next, the convergence of $u_n(M)$ will be proved. Now, two Lemmas are given first.

Lemma 4.1. *If $u(t,x) \in W_{(2,4)}(D)$, then there exists a constant $C > 0$ such that $|u(t,x)| \le C \| u(t,x) \|_{W_{(2,4)}(D)}$, $|u_t(t,x)| \le C \| u(t,x) \|_{W_{(2,4)}(D)}$, $|u_x(t,x)| \le C \| u(M) \|_{W_{(2,4)}(D)}$, $|u_{xt}(t,x)| \le C \| u(t,x) \|_{W_{(2,4)}(D)}$, and $|u_{xx}(t,x)| \le C \| u(t,x) \|_{W_{(2,4)}(D)}$.*

Proof. For any $(t,x) \in D$,

$$\begin{aligned}
|u(t,x)| &= |(u(\xi,\eta), R_t^{\{2\}}(\xi) R_x^{\{4\}}(\eta))_{W_{(2,4)}}| \\
&\le \| u(\xi,\eta) \|_{W_{(2,4)}} \| R_t^{\{2\}}(\xi) \|_{W_2} \| R_x^{\{4\}}(\eta) \|_{W_4},
\end{aligned}$$

then there exists $C_1 > 0$ such that

$$|u(t,x)| \le C_1 \| u(\xi,\eta) \|_{W_{(2,4)}}.$$

Note that

$$\begin{aligned}
|u_x(t,x)| &= |(u(\xi,\eta), R_t^{\{2\}}(\xi) (R_x^{\{4\}}(\eta))_x)_{W_{(2,4)}}| \\
&\le \| u(\xi,\eta) \|_{W_{(2,4)}} \| R_t^{\{2\}}(\xi) \|_{W_2} \| (R_x^{\{4\}}(\eta))_x \|_{W_4}.
\end{aligned}$$

Hence there exists $C_2 > 0$ such that

$$|u_x(t,x)| \le C_2 \| u(\xi,\eta) \|_{W_{(2,4)}}.$$

In like manner, there exist $C_3 > 0, C_4 > 0, C_5 > 0$ such that

$$|u_t(t,x)| \le C_3 \| u(t,x) \|_{W_{(2,4)}(D)}, \ |u_{xt}(t,x)| \le C_4 \| u(t,x) \|_{W_{(2,4)}(D)}$$

and

$$|u_{xx}(t,x)| \leq C_5 \parallel u(t,x) \parallel_{W_{(2,4)}(D)} .$$

Let $C = \max\{C_1, C_2, C_3, C_4, C_5\}$, then the proof is complete. □

Lemma 4.2. *If* $u_n \longrightarrow \overline{u}(n \to \infty)$ *in the sense of* $\parallel \cdot \parallel_{W_{(2,4)}(D)}$, $M_n = (t_n, x_n) \to M = (t,x)(n \to \infty)$ *and* $\parallel u_n \parallel_{W_{(2,4)}(D)}$ *is bounded, then*

$$F(M_n, u_{n-1}(M_n), (u_{n-1})_x(M_n)) \to F(M, \overline{u}(M), \overline{u}_x(M)) \ \ as\ n \to \infty.$$

Proof. Observing that

$$\begin{aligned} & |u_{n-1}(M_n) - \overline{u}(M)| \\ = \ & |u_{n-1}(M_n) - u_{n-1}(M) + u_{n-1}(M) - \overline{u}(M)| \\ \leq \ & |(u_{n-1})_t(\xi_1)|\,|t_n - t| + |(u_{n-1})_x(\eta_1)|\,|x_n - x| + |u_{n-1}(M) - \overline{u}(M)| \end{aligned}$$

and

$$\begin{aligned} & |(u_{n-1})_x(M_n) - \overline{u}_x(M)| \\ = \ & |(u_{n-1})_x(M_n) - (u_{n-1})_x(M) + (u_{n-1})_x(M) - \overline{u}_x(M)| \\ \leq \ & |(u_{n-1})_{xt}(\xi_2)|\,|t_n - t| + |(u_{n-1})_{xx}(\eta_2)|\,|x_n - x| + |(u_{n-1})_x(M) - \overline{u}_x(M)|. \end{aligned}$$

From the given condition and Lemma 4.1, it follows that, for any $M \in D$

$$|u_{n-1}(M) - \overline{u}(M)| \to 0(n \to \infty)\ ,\ |(u_{n-1})_x(M) - \overline{u}_x(M)| \to 0(n \to \infty)(uniformly),$$

$$|(u_{n-1})_t(\xi_1)| \leq C \parallel u_{n-1} \parallel_{W_{(2,4)}(D)}, \qquad |(u_{n-1})_x(\eta_1)| \leq C \parallel u_{n-1} \parallel_{W_{(2,4)}(D)},$$

$$|(u_{n-1})_{xt}(\xi_2)| \leq C \parallel u_{n-1} \parallel_{W_{(2,4)}(D)}\ \ and\ \ |(u_{n-1})_{xx}(\eta_2)| \leq C \parallel u_{n-1} \parallel_{W_{(2,4)}(D)} .$$

By the boundedness of $\parallel u_n \parallel_{W_{(2,4)}(D)}$, we get

$$|u_{n-1}(M_n) - \overline{u}(M)| \to 0,\ and\ |(u_{n-1})_x(M_n) - \overline{u}_x(M)| \to 0\ \ as\ n \to \infty.$$

The continuation of $F(M, u(M), u_x(M))$ implies that

$$F(M_n, u_{n-1}(M_n), (u_{n-1})_x(M_n)) \to F(M, \overline{u}(M), \overline{u}_x(M)) \ \ as\ n \to \infty.$$

The proof is complete. □

Theorem 4.3. *Suppose that* $\parallel u_n \parallel$ *is bounded in (4.2) and Eq.(1.2) has a unique solution. If* $\{M_i\}_{i=1}^{\infty}$ *is dense on D, then the n-term approximate solution* $u_n(M)$ *derived from the above method converges to the exact solution* $u(M)$ *of Eq.(1.2) and*

$$u(M) = \sum_{i=1}^{\infty} B_i \overline{\psi}_i(M), \tag{4.4}$$

where B_i *is given by (4.3).*

Proof. First, we will prove the convergence of u_n.
From (4.2), we infer that

$$u_{n+1}(M) = u_n(M) + B_{n+1}\overline{\psi}_{n+1}(M). \tag{4.5}$$

The orthonormality of $\{\overline{\psi}_i\}_{i=1}^{\infty}$ yields that

$$\| u_{n+1} \|^2 = \| u_n \|^2 + (B_n)^2 = \cdots\cdots = \sum_{i=1}^{n+1} (B_i)^2 \tag{4.6}$$

In terms of (4.6), it holds that $\| u_{n+1} \| \geq \| u_n \|$. Due to the condition that $\| u_n \|$ is bounded, $\| u_n \|$ is convergent and there exists a constant c such that

$$\sum_{i=1}^{\infty} (B_i)^2 = c.$$

This implies that

$$B_i \in l^2, i = 1, 2, \cdots.$$

If $m > n$, then

$$\| u_m - u_n \|^2 = \| u_m - u_{m-1} + u_{m-1} - u_{m-2} + \cdots + u_{n+1} - u_n \|^2 .$$

In view of $(u_m - u_{m-1}) \perp (u_{m-1} - u_{m-2}) \perp \cdots \perp (u_{n+1} - u_n)$, it follows that

$$\| u_m - u_n \|^2 = \| u_m - u_{m-1} \|^2 + \cdots + \| u_{n+1} - u_n \|^2 .$$

Furthermore

$$\| u_m - u_{m-1} \|^2 = (B_m)^2.$$

Consequently,

$$\| u_m - u_n \|^2 = \sum_{l=n+1}^{m} (B_l)^2 \to 0 \quad as\ n \to \infty.$$

The completeness of $W_{(2,4)}(D)$ shows that $u_m \to \overline{u}\ \ as\ m \to \infty$ in the sense of $\| \cdot \|_{W_{(2,4)}(D)}$.
Second, we will prove that $\overline{u}$ is the solution of $Eq.(1.2)$.
It follows that, on taking limits in (4.2)

$$\overline{u}(M) = \sum_{i=1}^{\infty} B_i \overline{\psi}_i(x).$$

Since

$$\begin{aligned} (L\overline{u})(M_n) &= \sum_{i=1}^{\infty} B_i (L\overline{\psi}_i, \varphi_n) \\ &= \sum_{i=1}^{\infty} B_i (\overline{\psi}_i, L^*\varphi_n) \\ &= \sum_{i=1}^{\infty} B_i (\overline{\psi}_i, \psi_n), \end{aligned}$$

we must have

$$\begin{aligned}\sum_{j=1}^{n} \beta_{nj}(L\overline{u})(M_j) &= \sum_{i=1}^{\infty} B_i(\overline{\psi}_i, \sum_{j=1}^{n} \beta_{nj}\psi_j) \\ &= \sum_{i=1}^{\infty} B_i(\overline{\psi}_i, \overline{\psi}_n) \\ &= B_n.\end{aligned}$$

If $n = 1$, then

$$(L\overline{u})(M_1) = F(M_1, u_0(M_1), (u_0)_x(M_1)).$$

If $n = 2$, then $\beta_{21}(L\overline{u})(M_1) + \beta_{22}(L\overline{u})(M_2) = \beta_{21}F(M_1, u_0(M_1), (u_0)_x(M_1)) + \beta_{22}F(M_2, u_1(M_2), (u_1)_x(M_2))$.
It is clear that

$$(L\overline{u})(M_2) = F(M_2, u_1(M_2), (u_1)_x(M_2)).$$

Moreover, it is easy to see by induction that

$$(L\overline{u})(M_j) = F(M_j, u_{j-1}(M_j), (u_{j-1})_x(M_j)), j = 1, 2, \cdots. \tag{4.7}$$

Since $\{M_i\}_{i=1}^{\infty}$ is dense on D, for $\forall\, Y \in D$, there exists a subsequence $\{M_{n_j}\}_{j=1}^{\infty}$ such that

$$M_{n_j} \to Y \quad as\ j \to \infty.$$

Let $j \to \infty$ in (4.7). Using the convergence of u_n and Lemma 4.2, one can then derive that

$$(L\overline{u})(Y) = F(Y, \overline{u}(Y), \overline{u}_x(Y)).$$

Using Theorem (3.1), one can find that $\overline{\psi}_i(x) \in W_{(2,4)}(D)$. Clearly, $\overline{u}(Y)$ satisfies the initial-boundary conditions of $Eq.(1.2)$.
Therefore, $\overline{u}(M)$ is the solution of $Eq.(1.2)$. The application of the uniqueness of solution to $Eq.(1.2)$ then yields that

$$u(M) = \sum_{i=1}^{\infty} B_i \overline{\psi}_i(M). \tag{4.8}$$

The proof is complete. □

Theorem 4.4. *Assume $u(M)$ is the solution of $Eq.(1.2)$ and $r_n(x)$ is the error between the approximate $u_n(M)$ and the exact solution $u(M)$, where $u_n(M)$ is given by* (4.2). *Then the error $r_n(x)$ is monotone decreasing in the sense of $\|\cdot\|_{W_{(2,4)}(D)}$.*

Proof. From (4.2), (4.8), it follows that

$$\begin{aligned}\| r_n \|^2_{W_{(2,4)}} &= \| \sum_{i=n+1}^{\infty} B_i \overline{\psi}_i(x) \|^2_{W_{(2,4)}} \\ &= \sum_{i=n+1}^{\infty} (B_i)^2.\end{aligned} \tag{4.9}$$

(4.9) shows that the error r_n is monotone decreasing in the sense of $\|\cdot\|_{W_{(2,4)}(D)}$.
The proof is complete. □

5. Numerical Example

In this section, some numerical examples are studied to demonstrate the accuracy of the present method. In the process of computation, all the symbolic and numerical computations performed by using Mathematica 4.2. Results obtained by the method are compared with the exact solution of each example and are found to be in good agreement with each other.

Example 1

Consider equation

$$\begin{cases} u_t + xuu_x + t^2u^2u_x + u_{xx} + u_{xxx} = f(x,t), \\ u(0,x) = 0, \\ u(t,0) = 0, u_x(t,0) = 0, u_{xx}(t,0) = 0, \end{cases}$$

where $0 \le t \le 1, 0 \le x \le 1$ and $f(t,x) = e^{-3x^2}(-x^5e^{x^2}t^2(2x^2-3)+x^8t^3(3-2x^2)+e^{2x^2}(x^3(1-14t)+6t+6xt-54x^2t+48x^4t+4x^5t-8x^6t))$. The true solution is $x^3e^{-x^2}t$. Using our method, we choose 168 points on $D_1 = [0,1]\times[0,1]$ and obtain approximate solution $u_{168}(t,x)$ on D_1. The numerical results are given in the following Table 1,2.

Example 2

Consider equation

$$\begin{cases} u_t + \sin(x)uu_x + t^2u^2u_x + u_{xx} + u_{xxx} = f(x,t), \\ u(0,x) = 0, \\ u(t,0) = 0, u_x(t,0) = 0, u_{xx}(t,0) = 0, \end{cases}$$

where $0 \le t \le 1$, $0 \le x \le 1$ and $f(t,x) = e^{-3x^2}(-\sin(xt)(e^{2x^2}(-2-4x^4+8x^5+6x(4+t^2)-6x^3(6+t^2)+x^2(10+t^2))+2x^3e^{x^2}(x^2-1)\sin(x)\sin(xt)+2x^5(x^2-1)t^2(\sin(xt))^2)+\cos(xt)(e^{2x^2}(x^3(1-4t)+6t+4xt+12x^4t-x^2t(30+t^2))+x^4te^{x^2}\sin(x)\sin(xt)+x^6t^3(\sin(xt))^2))$. The true solution is $x^2e^{-x^2}\sin(xt)$. Using our method, we choose 168 points on $D_1 = [0,1]\times[0,1]$ and obtain approximate solution $u_{168}(t,x)$ on D_1. The numerical results are given in the following Table 3,4.

6. Appendix

A The Proof of Theorem(2.1)

Proof. Through several integrations by part, one can derive that

$$\begin{aligned}(u(y),R_t^{\{2\}}(y))_{W_2} = & \int_0^1 4R_t^{\{2\}}(y) - 5\tfrac{d^2}{dy^2}R_t^{\{2\}}(y) + \tfrac{d^4}{dy^4}R_t^{\{2\}}(y)dy + u(y) \\ & (5\tfrac{d}{dy}R_t^{\{2\}}(y) - \tfrac{d^3}{dy^3}R_t^{\{2\}}(y))|_0^1 + u'(y)\tfrac{d^2}{dy^2}R_t^{\{2\}}(y)|_0^1.\end{aligned} \tag{1.1}$$

Since $R_t^{\{2\}}(y) \in W_2[0,1]$, we must have

$$R_t^{\{2\}}(0) = 0 \tag{1.2}$$

Table 1. Comparison of results at t=0.48.

x	True solution u(x,t)	Approximate solution u_{168}	Absolute error
0.08	0.000244	0.000240	$3.8E^{-6}$
0.16	0.001916	0.001885	0.00003
0.24	0.006264	0.006164	0.00009
0.32	0.014197	0.013972	0.00022
0.40	0.026177	0.025763	0.00041
0.48	0.042160	0.041494	0.00066
0.56	0.061604	0.060629	0.00097
0.64	0.083539	0.082215	0.00132
0.72	0.106684	0.104986	0.00169
0.80	0.129587	0.127513	0.00207
0.88	0.15079	0.148361	0.00242
0.96	0.16897	0.166227	0.00274

Table 2. Comparison of results

(t,x)	True solution u(x,t)	Approximate solution u_{168}	Absolute error
(0.08,0.08)	0.000040	0.000040	$4.9E^{-7}$
(0.16.0.16)	0.000638	0.000629	$8.8E^{-6}$
(0.24.0.24)	0.003132	0.003086	0.00004
(0.32,0.32)	0.009465	0.009322	0.00014
(0.40,0.40)	0.021814	0.021478	0.00033
(0.48,0.48)	0.042160	0.041494	0.00066
(0.56,0.56)	0.071871	0.070705	0.00116
(0.64,0.64)	0.111387	0.109540	0.00184
(0.72,0.72)	0.160026	0.157302	0.00272
(0.80,0.80)	0.215979	0.212228	0.00375
(0.88,0.88)	0.276448	0.271596	0.00485
(0.96,0.96)	0.337940	0.331129	0.00681

Table 3. Comparison of results at t=0.48.

x	True solution u(x,t)	Approximate solution u_{168}	Absolute error
0.08	0.000244	0.000240	$3.1E^{-6}$
0.16	0.001914	0.001889	0.00002
0.24	0.006250	0.006167	0.00008
0.32	0.014142	0.013956	0.00018
0.40	0.026017	0.025677	0.00034
0.48	0.041788	0.041245	0.00054
0.56	0.060865	0.060079	0.00078
0.64	0.082232	0.081180	0.00105
0.72	0.104573	0.103250	0.00132
0.80	0.126426	0.124847	0.00157
0.88	0.146345	0.144548	0.00179
0.96	0.163053	0.161093	0.00195

Table 4. Comparison of results

(t,x)	True solution u(x,t)	Approximate solution u_{168}	Absolute error
(0.08,0.08)	0.000040	0.000040	$4.7E^{-7}$
(0.16.0.16)	0.000638	0.000630	$8.3E^{-6}$
(0.24.0.24)	0.003130	0.003088	0.00004
(0.32,0.32)	0.009448	0.009320	0.00012
(0.40,0.40)	0.021721	0.021432	0.00028
(0.48,0.48)	0.041788	0.041245	0.00054
(0.56,0.56)	0.070699	0.069821	0.00087
(0.64,0.64)	0.108298	0.107041	0.00125
(0.72,0.72)	0.152955	0.151317	0.00163
(0.80,0.80)	0.201534	0.199740	0.00179
(0.88,0.88)	0.249634	0.247976	0.00165
(0.96,0.96)	0.292093	0.291066	0.00102

If

$$\left.\begin{array}{l} 5\frac{d}{dy}R_t^{\{2\}}(1)-\frac{d^3}{dy^3}R_t^{\{2\}}(1)=0,\ \frac{d^2}{dy^2}R_t^{\{2\}}(0)=0 \\ \frac{d^2}{dy^2}R_t^{\{2\}}(1)=0 \end{array}\right\} \tag{1.3}$$

then (A.1) yields that

$$(u(y),R_t^{\{2\}}(y))_{W_2}=\int_0^1 4R_t^{\{2\}}(y)-5\frac{d^2}{dy^2}R_t^{\{2\}}(y)+\frac{d^4}{dy^4}R_t^{\{2\}}(y)dy.$$

If $R_t^{\{2\}}(y)$ also satisfies

$$4R_t^{\{2\}}(y)-5\frac{d^2}{dy^2}R_t^{\{2\}}(y)+\frac{d^4}{dy^4}R_t^{\{2\}}(y)=\delta(y-t), \tag{1.4}$$

then $(u(y),R_t^{\{2\}}(y))_{W_2}=u(t)$.
Characteristic equation of (A.4) is given by $4-5\lambda^2+\lambda^4=0$,
then we can find characteristic values $\lambda_1=1,\lambda_2=-1,\lambda_3=2$ and $\lambda_4=-2$. Let

$$R_t^{\{2\}}(y)=\begin{cases} c_{11}e^y+c_{12}e^{-y}+c_{13}e^{2y}+c_{14}e^{-2y}, & y\le t \\ d_{11}e^y+d_{12}e^{-y}+d_{13}e^{2y}+d_{14}e^{-2y}, & y>t \end{cases}$$

On the other hand, for (A.4), let $R_t^{\{2\}}(y)$ satisfies

$$\frac{d^k}{dy^k}R_t^{\{2\}}(t+0)=\frac{d^k}{dy^k}R_t^{\{2\}}(t-0), k=0,1,2 \tag{1.5}$$

Integrating (A.4) from $t-\varepsilon$ to $t+\varepsilon$ with respect to y and let $\varepsilon\to 0$, one obtains the jump degree of $R_t^{\{2\}}(y)$ at y=t

$$\frac{d^3}{dy^3}R_t^{\{2\}}(t+0)-\frac{d^3}{dy^3}R_t^{\{2\}}(t-0)=1 \tag{1.6}$$

Finally, the required unknown coefficient of (2.2) follows from (A.2),(A.3), (A.5), (A.6).
The proof is complete. □

References

[1] Bona JL, Schonbek ME. Travelling wave solutions to the Korteweg-de Vries-Burgers equation. *Proc R Soc Edin burgn Sect A* **101**(1985), 207-226.

[2] Coffey MW. On series expansins giving closed-form of Korteweg-de Vries-like equations. *SIAM J. Appl. Math.* **50**(1990), 1580-1592.

[3] Shaher Momani. An explicit and numerical solutions of the fractional KdV equation. *Mathematics and Computers in simulation* **70**(2005), 110-118.

[4] M.L. Wang. Exact solutions for a compound KdV-Burgers equation. *Phys. Lett.A* **213**(1996), 279-287.

[5] Z. Feng. On explicit exact solutions to the compound Burgers-KdV equation. *Phys. Lett.A* **293**(2002), 57-66.

[6] Z. Feng. A note on "Explicit exact solutions to the compound Burgers-Korteweg-de Vries equation". *Phys. Lett.A* **312**(2003), 65-70.

[7] W.G. Zhang, Q.S. Chang,B.G. Jiang. Explicit exact solitary-wave solutions for compound KdV-type and compound KdV-Burgers-type equations with nonlinear terms of any order. Chaos,*Soliton and Fractals* **13**(2002), 311-319.

[8] Z.S. Zhang, Goony Chen. Solitary wave solutions of the compound Burgers-Korteweg-de Vries equation. *Physica A* **352**(2005), 419-435.

[9] Chunli Li, Minggen Cui. The exact solution for solving a class nonlinear operator equations in the reproducing kernel space. *Appl.Math.Compu.* **143**(2003), 393-399.

[10] N.Aronszajn. Theory of Reproducing kernel. *Trans,A.M.S.* **168**(1950), 1-50.

In: Progress in Evolution Equations
Editor: Gaston M. N'Guerekata, pp. 149-157
ISBN: 978-1-60456-328-3

Chapter 9

CRITICAL CURVES FOR A DEGENERATE PARABOLIC SYSTEM WITH NONLINEAR BOUNDARY CONDITIONS

Zhongping Li*
Department of Mathematics, China West Normal University,
Nanchong 637002, P. R. China

Abstract

We establish the critical global existence curve and critical Fujita curve for a degenerate parabolic system with nonlinear boundary conditions in multi-dimension.

2000 Mathematics Subject Classification: Primary 35K57, 35K65

Key words: Critical curve, Nonlinear degenerate parabolic system, Nonlinear boundary conditions

1. Introduction

In this paper, we consider the following degenerate system with nonlinear boundary conditions

$$\begin{cases} u_t = \triangle u^m, \quad v_t = \triangle v^n, & x \in \mathbb{R}_+^N, \quad t > 0, \\ -\frac{\partial u^m}{\partial x_1} = u^\alpha v^p, \quad -\frac{\partial v^n}{\partial x_1} = u^q v^\beta, & x_1 = 0, \quad t > 0, \\ u(x,0) = u_0(x), \quad v(x,0) = v_0(x), & x \in \mathbb{R}_+^N, \end{cases} \tag{1.1}$$

where $\mathbb{R}_+^N = \{(x_1, x') | x' \in \mathbb{R}^{N-1}, x_1 > 0\}$, $m, n > 1$, $\alpha, \beta \geq 0$, $p, q > 0$ and $u_0(x), v_0(x)$ are continuous, nonnegative bounded functions and satisfying the compatibility conditions

$$-\frac{\partial u_0^m}{\partial x_1} = u_0^\alpha v_0^p; \quad -\frac{\partial v_0^n}{\partial x_1} = u_0^q v_0^\beta, \quad x_1 = 0$$

and are locally supported near some point, namely, for some $x_0,\ y_0 \in \mathbb{R}_+^N$, $supp u_0 \subset B_R(x_0) \cap \mathbb{R}_+^N$, $supp v_0 \subset B_R(y_0) \cap \mathbb{R}_+^N$ and $u_0, v_0 \neq 0$. However, the last assumption is not a

*E-mail address: zplimath@163.com.

real requirement for deriving our results. What we want to show is that even for the initial datums $u_0(x), v_0(x)$ vanishing except for a small ball, the solutions may still blow up in finite time.

Nonlinear parabolic equations such as (1.1) appear in population dynamics, chemical reactions, heat transfer, and so on, where u and v represent the densities of two biological population during a migration, the thickness of two kinds of chemical reactants in chemical reaction, or the temperatures of two kinds of porous materials during a propagation. The problem of determining critical curves is very interesting for various nonlinear parabolic equations of mathematical physics. See the surveys [1, 5, 9], where a full list of references can be found.

In [4], Galaktionov and Levine studied the following degenerate problem

$$\begin{cases} u_t = (u^m)_{xx}, & x > 0,\ 0 < t < T, \\ -(u^m)_x(0,t) = u^p(0,t), & 0 < t < T, \\ u(x,0) = u_0(x), & x > 0, \end{cases} \tag{1.2}$$

and obtained the following results on the critical Fujita exponent p_c:
(1) If $\frac{m+1}{2} < p \le p_c = m+1$, then the solutions of (1.2) blow up in finite time for all nontrivial u_0.
(2) If $p > p_c = m+1$, then the solutions of (1.2) are global for small u_0 and blow up in finite time for large u_0.

In [7], Huang, Yin and Wang studied the following problem

$$\begin{cases} u_t = \Delta u^m, & x \in \mathbb{R}^N_+,\ 0 < t < T, \\ -\frac{\partial u^m}{\partial x_1} = u^p(x,t), & x_1 = 0, 0 < t < T, \\ u(x,0) = u_0(x), & x \in \mathbb{R}^N_+. \end{cases} \tag{1.3}$$

They obtained that (i) if $p > m + \frac{1}{N}$, then any nontrivial nonnegative solution of (1.3) blows up in finite time for large u_0; (ii) any nontrivial nonnegative solution of (1.3) blows up in finite time if $\frac{m+1}{2} < p < m + \frac{1}{N}$; (iii) if $p > m + \frac{1}{N}$, then any nontrivial nonnegative solution of (1.3) is global for small u_0; (iv) any nontrivial nonnegative solution of (1.3) is global if $0 < p < \frac{m+1}{2}$.

Recently, Quirós and Rossi [8] considered the degenerate equations coupled via variational nonlinear boundary flux

$$\begin{cases} u_t = (u^m)_{xx}, \quad v_t = (v^n)_{xx}, & x \in \mathbb{R}_+, \quad t > 0, \\ -(u^m)_x(0,t) = v^p(0,t), & t > 0, \\ -(v^n)_x(0,t) = u^q(0,t), & t > 0, \\ u(x,0) = u_0(x), \quad v(x,0) = v_0(x), & x \in \mathbb{R}_+, \end{cases} \tag{1.4}$$

with notations

$$\alpha_1 = \frac{2p+n+1}{(m+1)(n+1)-4pq}, \qquad \alpha_2 = \frac{2q+m+1}{(m+1)(n+1)-4pq},$$

$$\beta_1 = \frac{p(m-1-2q)+(n+1)m}{(m+1)(n+1)-4pq}, \qquad \beta_2 = \frac{q(n-1-2p)+(m+1)n}{(m+1)(n+1)-4pq}.$$

They proved that the solutions of (1.4) are global if $pq \le \frac{(m+1)(n+1)}{4}$ and may blow up in finite time if $pq > \frac{(m+1)(n+1)}{4}$. In the case of $pq > \frac{(m+1)(n+1)}{4}$, if $\alpha_1 + \beta_1 \le 0$ or $\alpha_2 + \beta_2 \le 0$, then every nonnegative, nontrivial solution of (1.4) blows up in finite time; if $\alpha_1+\beta_1>0$ and $\alpha_2+\beta_2>0$, then there exist blow-up solutions for large initial data and global solutions for small initial data. The critical Fujita curve to (1.4) are described by $\alpha_i+\beta_i=0, i=1,2$.

In [9], Zheng, Song and Jing considered the degenerate parabolic equations

$$\begin{cases} u_t=(u^m)_{xx},\ v_t=(v^n)_{xx}, & x\in\mathbb{R}_+,\ t>0,\\ -(u^m)_x(0,t)=u^{\alpha}(0,t)v^{p}(0,t), & t>0,\\ -(v^n)_x(0,t)=u^{q}(0,t)v^{\beta}(0,t), & t>0,\\ u(x,0)=u_0(x),\ v(x,0)=v_0(x), & x\in\mathbb{R}_+, \end{cases} \tag{1.5}$$

with notations

$$k_1=\frac{2p+n+1-2\beta}{4pq-(m+1-2\alpha)(n+1-2\beta)},\quad k_2=\frac{2q+m+1-2\alpha}{4pq-(m+1-2\alpha)(n+1-2\beta)}, \tag{1.6}$$

$$l_1=\frac{1-k_1(m-1)}{2},\quad l_2=\frac{1-k_2(n-1)}{2}. \tag{1.7}$$

They proved that (i) the solutions of (1.5) may blow up in finite time if $\alpha>\frac{m+1}{2}$ or $\beta>\frac{n+1}{2}$; (ii) every solution of (1.5) exists globally if $\alpha<\frac{m+1}{2}$, $\beta<\frac{n+1}{2}$ and $pq\le(\frac{m+1}{2}-\alpha)(\frac{n+1}{2}-\beta)$; (iii) if $\alpha\le\frac{m+1}{2}$, $\beta\le\frac{n+1}{2}$ and $pq>(\frac{m+1}{2}-\alpha)(\frac{n+1}{2}-\beta)$, then every nonnegative, nontrivial solution of (1.5) blows up in finite time for $l_1<k_1$ or $l_2<k_2$ or $l_1=k_1$ and $l_2=k_2$, and the solutions of (1.5) are global for small initial data and blow up in finite time with large initial data for $l_1>k_1$ and $l_2>k_2$.

The purpose of this paper is to investigate the effect of space dimension on critical curves of the problem (1.1). This paper can be thought of as a nature continuation of [9] to muti-dimensional case.

We also define k_1, k_2, l_1,l_2 as in (1.6) and (1.7). Now we state our results as follows.

Theorem 1.1. *Assume $\alpha>\frac{m+1}{2}$ or $\beta>\frac{n+1}{2}$. If the solution (u,v) of (1.1) is nondecreasing in time, then the (u,v) blows up in finite time with any nontrivial nonnegative large initial data.*

Remark 1.2. If Δu_0^m, $\Delta v_0^n\ge 0$, applying the comparison principle, we prove that the solution (u,v) of (1.1) is nondecreasing in time (see [9] for details).

Theorem 1.3. *If $\alpha<\frac{m+1}{2}$, $\beta<\frac{n+1}{2}$ and $pq\le(\frac{m+1}{2}-\alpha)(\frac{n+1}{2}-\beta)$, then every solution of (1.1) exists globally.*

Remark 1.4. Theorems 1.1 and 1.3 show that the global existence curve of (1.1) is $\max\{\alpha-\frac{m+1}{2},\beta-\frac{n+1}{2}\}=0$ if $pq\le(\frac{m+1}{2}-\alpha)(\frac{n+1}{2}-\beta)$.

Theorem 1.5. *Assume $\alpha\le\frac{m+1}{2}$, $\beta\le\frac{n+1}{2}$ and $pq>(\frac{m+1}{2}-\alpha)(\frac{n+1}{2}-\beta)$.*

1. *If $Nl_1>k_1$ and $Nl_2>k_2$, then the solutions of (1.1) are global with small initial data for α, $\beta\ge 1$ and blow up in finite time with any nontrivial nonnegative large initial data.*

2. *If $Nl_1<k_1$ or $Nl_2<k_2$, then every nontrivial nonnegative solution of (1.1) blows up in finite time.*

Remark 1.6. Theorem 1.5 shows that the critical Fujita curve is given by $\min\{Nl_1-k_1, Nl_2-k_2\}=0$ if $1<\alpha\le\frac{m+1}{2}$, $1<\beta\le\frac{n+1}{2}$ and $pq>(\frac{m+1}{2}-\alpha)(\frac{n+1}{2}-\beta)$.

Remark 1.7. From Remarks 1.4, 1.6, we see that the critical global existence curve of (1.1) does not depend on space dimension N, while the critical Fujita curve depends on N.

2. Proof of Theorems

Proof of Theorem 1.1. Without loss of generality, we assume that $\alpha > \frac{m+1}{2}$. Since $u_t \geq 0$, $v_t \geq 0$, we have $u^\alpha(x,t)v^p(x,t)|_{x_1=0} \geq u^\alpha(x,t)v_0^p(x)$. Consider the following single equation

$$\begin{cases} w_t = \Delta w^m, & x \in \mathbb{R}_+^N,\ t>0, \\ -\frac{\partial w^m}{\partial x_1}|_{x_1=0} = w^\alpha v_0^p, & t>0, \\ w(x,0) = u_0(x), & x \in \mathbb{R}_+^N. \end{cases}$$

Clearly, (w, v_0) is a subsolution of (1.1). By the result of [7], we know that the solution w blows up in finite time with large initial data, and so does the solution of (1.1). □

Proof of Theorem 1.3. A similar proof can be found in [9]. We arrange the proof here for the convenience of the reader.

Set

$$\overline{u}(x,t) = e^{kt}(M + e^{-L_1 x_1 e^{\frac{k(1-m)t}{2}}})^{\frac{1}{m}},$$
$$\overline{v}(x,t) = e^{\frac{k(m+1-2\alpha)t}{2p}}(M + e^{-L_2 x_1 e^{\frac{k(m+1-2\alpha)(1-n)t}{4p}}})^{\frac{1}{n}}$$

with

$$M = \max\{||u_0||_\infty^m,\ ||v_0||_\infty^n,\ 1\},\ L_1 = (M+1)^{\frac{\alpha}{m}+\frac{p}{n}},$$
$$L_2 = (M+1)^{\frac{p}{n}+\frac{q}{m}},\quad k = \max\{L_1^2 M^{-\frac{1}{m}},\ \frac{2p}{m+1-2\alpha}L_2^2 M^{-\frac{1}{n}}\}.$$

It is easy to see that $\overline{u}(x,0) \geq u_0(x)$, $\overline{v}(x,0) \geq v_0(x)$ for $x \in \mathbb{R}_+^N$. After a computation we have

$$\begin{aligned} \overline{u}_t &= ke^{kt}(M + e^{-L_1 x_1 e^{\frac{k(1-m)t}{2}}})^{\frac{1}{m}} \\ &\quad + \frac{kL_1(m-1)x_1}{2m}e^{kt}(M + e^{-L_1 x_1 e^{\frac{k(1-m)t}{2}}})^{\frac{1}{m}-1} e^{\frac{k(1-m)t}{2}} e^{-L_1 x_1 e^{\frac{k(1-m)t}{2}}} \\ &\geq ke^{kt}(M + e^{-L_1 x_1 e^{\frac{k(1-m)t}{2}}})^{\frac{1}{m}}, \\ \Delta \overline{u}^m &= L_1^2 e^{kt} e^{-L_1 x_1 e^{\frac{k(1-m)t}{2}}} \leq L_1^2 e^{kt} \end{aligned}$$

and

$$\overline{v}_t \geq \frac{k(m+1-2\alpha)}{2p} e^{\frac{k(m+1-2\alpha)t}{2p}}(M + e^{-L_2 x_1 e^{\frac{k(m+1-2\alpha)(1-n)t}{4p}}})^{\frac{1}{n}},$$
$$\Delta \overline{v}^n = L_2^2 e^{\frac{k(m+1-2\alpha)t}{2p}} e^{-L_2 x_1 e^{\frac{k(m+1-2\alpha)(1-n)t}{4p}}} \leq L_2^2 e^{\frac{k(m+1-2\alpha)t}{2p}}$$

in $\mathbb{R}_+^N \times \mathbb{R}_+$. We have on the boundary that

$$-\frac{\partial \overline{u}^m}{\partial x_1}|_{x_1=0} = L_1 e^{\frac{k(m+1)t}{2}},\quad -\frac{\partial \overline{v}^n}{\partial x_1}|_{x_1=0} = L_2 e^{\frac{k(m+1-2\alpha)(n+1)}{4p}t},$$
$$\overline{u}^\alpha|_{x_1=0} = e^{k\alpha t}(M+1)^{\frac{\alpha}{m}},\quad \overline{v}^p|_{x_1=0} = e^{\frac{k(m+1-2\alpha)t}{2}}(M+1)^{\frac{p}{n}},$$
$$\overline{u}^q|_{x_1=0} = e^{kqt}(M+1)^{\frac{q}{m}},\quad \overline{v}^\beta|_{x_1=0} = e^{\frac{k(m+1-2\alpha)\beta t}{2p}}(M+1)^{\frac{\beta}{n}}.$$

By the definitions of k, M, L_1, L_2 and the assumption $pq \leq (\frac{m+1}{2}-\alpha)(\frac{n+1}{2}-\beta)$, we know that $\overline{u}_t \geq \Delta\overline{u}^m$, $\overline{v}_t \geq \Delta\overline{v}^n$ in $\mathbb{R}_+^N \times \mathbb{R}_+$ and

$$-\frac{\partial \overline{u}^m}{\partial x_1}|_{x_1=0} \geq \overline{u}^\alpha \overline{v}^p,\quad -\frac{\partial \overline{v}^n}{\partial x_1}|_{x_1=0} \geq \overline{u}^q \overline{v}^\beta \quad \text{for } t>0.$$

We have shown that $(\overline{u},\ \overline{v})$ is a supersolution of (1.1), which implies that every solution of (1.1) is global provided that $pq \leq (\frac{m+1}{2}-\alpha)(\frac{n+1}{2}-\beta)$. □

The proof of Theorem 1.5 consists of three following lemmas.

Lemma 2.1. *Assume* $\alpha \le \frac{m+1}{2}$, $\beta \le \frac{n+1}{2}$ *and* $pq > (\frac{m+1}{2}-\alpha)(\frac{n+1}{2}-\beta)$. *If* $Nl_1 > k_1$, $Nl_2 > k_2$ *and* α, $\beta \ge 1$, *then the solutions of (1.1) are global for small initial data.*

Proof. Set

$$\overline{u}(x,t) = (T+t)^{-k_1}A_1 f(\eta), \quad \overline{v}(x,t) = (T+t)^{-k_2}A_2 g(\xi)$$

where $\eta = |\varepsilon|$, $\varepsilon_1 = x_1(T+t)^{-l_1}+h_1$, $\varepsilon_i = x_i(T+t)^{-l_1}$, $(i=2,...,N)$, and $\xi = |\zeta|$, $\zeta_1 = x_1(T+t)^{-l_2}+h_2$, $\zeta_i = x_i(T+t)^{-l_2}$, $(i=2,...,N)$, and $T>0$ is a given positive constant. It is easy to know from (1.6) and (1.7) that

$$\begin{aligned} k_1+1 &= k_1 m + 2l_1, \quad k_2+1 = k_2 n + 2l_2,\\ k_1 m + l_1 &= k_1\alpha + k_2 p, \quad k_2 n + l_2 = k_1 q + k_2\beta. \end{aligned}$$

We see that $(\overline{u}(x,t),\overline{v}(x,t))$ is a supersolution of (1.1) if $f(\eta), g(\xi) \ge 0$ satisfy

$$\begin{aligned} &\frac{A_1^{m-1}}{\eta^{N-1}}(\eta^{N-1}(f^m)')' + l_1\eta f' + k_1 f - \frac{l_1\varepsilon_1 h_1}{\eta}f' \le 0,\\ &\frac{A_2^{n-2}}{\xi^{N-1}}(\xi^{N-1}(g^n)')' + l_2\xi g' + k_2 g - \frac{l_2\zeta_1 h_2}{\xi}g' \le 0 \end{aligned} \tag{2.1}$$

and

$$-A_1^m\frac{\partial f^m}{\partial\varepsilon_1}|_{\varepsilon_1=h_1} \ge A_1^\alpha A_2^p f^\alpha g^p, \quad -A_2^n\frac{\partial g^n}{\partial\zeta_1}|_{\zeta_1=h_2} \ge A_1^q A_2^\beta f^q g^\beta, \tag{2.2}$$

where $\varepsilon \in \{\eta > 0 | f(\eta) \ge 0\}, A_1 > 0$ and $\zeta \in \{\xi > 0 | g(\xi \ge 0\}, A_2 > 0$.
Take

$$f(\eta) = B_1(c_1^2-\eta^2)_+^{\frac{1}{m-1}}, \qquad g(\xi) = B_2(c_2^2-\xi^2)_+^{\frac{1}{n-1}},$$

where

$$B_1 = \{\frac{m-1}{2m[N(m-1)+2]}\}^{\frac{1}{m-1}}, \qquad B_2 = \{\frac{n-1}{2n[N(n-1)+2]}\}^{\frac{1}{n-1}}$$

and $0 < h_1 < \sqrt{2} < c_1$, $0 < h_2 < \sqrt{2} < c_2$.
Now we show that f, g satisfy (2.1). Using

$$\begin{aligned} \frac{1}{\eta^{N-1}}(\eta^{N-1}(f^m)')' &= -\frac{1}{N(m-1)+2}\eta f' - \frac{N}{N(m-1)+2}f,\\ \frac{1}{\xi^{N-1}}(\xi^{N-1}(g^n)')' &= -\frac{1}{N(n-1)+2}\xi g' - \frac{N}{N(n-1)+2}g, \end{aligned}$$

we see that (2.1) is valid if f, g satisfy

$$\begin{aligned} &A_1^{m-1}\{-\frac{1}{N(m-1)+2}\eta f' - \frac{N}{N(m-1)+2}f\} + l_1\eta f' + k_1 f - l_1 h_1 f' \le 0,\\ &A_2^{n-1}\{-\frac{1}{N(n-1)+2}\xi g' - \frac{N}{N(n-1)+2}g\} + l_2\xi g' + k_2 g - l_2 h_2 g' \le 0, \end{aligned}$$

namely

$$\begin{aligned} &\{\frac{2}{m-1}[\frac{A_1^{m-1}}{N(m-1)+2}-l_1]-[k_1-\frac{NA_1^{m-1}}{N(m-1)+2}]\}\eta^2+[k_1-\frac{NA_1^{m-1}}{N(m-1)+2}]c_1^2+\frac{2l_1h_1\eta}{m-1}\le 0,\\ &\{\frac{2}{n-1}[\frac{A_2^{n-1}}{N(n-1)+2}-l_2]-[k_2-\frac{NA_2^{n-1}}{N(n-1)+2}]\}\xi^2+[k_2-\frac{NA_2^{n-1}}{N(n-1)+2}]c_2^2+\frac{2l_2h_2\xi}{n-1}\le 0. \end{aligned} \tag{2.3}$$

Since $k_1 < Nl_1$, $k_2 < Nl_2$, we choose suitable constants A_1, A_2 such that

$$\begin{aligned} &\frac{A_1^{m-1}}{N(m-1)+2} < l_1, \ k_1 < \frac{NA_1^{m-1}}{N(m-1)+2}, \qquad \frac{A_2^{n-1}}{N(n-1)+2} < l_2, \ k_2 < \frac{NA_2^{n-1}}{N(n-1)+2},\\ &h_1 \le \frac{1}{l_1}\min\{[l_1 - \frac{A_1^{m-1}}{N(m-1)+2}], \ \frac{m-1}{2}[\frac{NA_1^{m-1}}{N(m-1)+2}-k_1]\},\\ &h_2 \le \frac{1}{l_2}\min\{[l_2 - \frac{A_2^{n-1}}{N(n-1)+2}], \ \frac{n-1}{2}[\frac{NA_2^{n-1}}{N(n-1)+2}-k_2]\}. \end{aligned}$$

Thus (2.3) holds. (2.2) is equivalent to

$$\begin{aligned} \tfrac{2mh_1}{m-1} &\geq (B_1A_1)^{(\alpha-m)}(B_2A_2)^p(c_1^2-\eta^2)^{\frac{\alpha-1}{m-1}}(c_2^2-\xi^2)^{\frac{p}{n-1}}, \\ \tfrac{2nh_2}{n-1} &\geq (B_2A_2)^{(\beta-n)}(B_1A_1)^q(c_1^2-\eta^2)^{\frac{q}{m-1}}(c_2^2-\xi^2)^{\frac{\beta-1}{n-1}}, \end{aligned} \tag{2.4}$$

where $h_1 < \eta < c_1$ and $h_2 < \xi < c_2$.
For $h_1 < \eta < c_1$ and $h_2 < \xi < c_2$, if the following two inequalities

$$\begin{aligned} \tfrac{2mh_1}{m-1} &\geq (B_1A_1)^{(\alpha-m)}(B_2A_2)^p(c_1^2-h_1^2)^{\frac{\alpha-1}{m-1}}(c_2^2-h_2^2)^{\frac{p}{n-1}}, \\ \tfrac{2nh_2}{n-1} &\geq (B_2A_2)^{(\beta-n)}(B_1A_1)^q(c_1^2-h_1^2)^{\frac{q}{m-1}}(c_2^2-h_2^2)^{\frac{\beta-1}{n-1}} \end{aligned} \tag{2.5}$$

hold, (2.4) is true. The assumption $pq > (\frac{m+1}{2}-\alpha)(\frac{n+1}{2}-\beta)$ implies that $\frac{m+1-2\alpha}{2p} < \frac{2q}{n+1-2\beta}$. Therefore, for any positive constants $\mu_1,\ \mu_2$, there exist positive constants $h_1,\ h_2$ small enough such that $\mu_1 h_1^{\frac{m+1-2\alpha}{2p}} > h_2^{\frac{m-1}{n-1}} > \mu_2 h_1^{\frac{2q}{n+1-2\beta}}$. Thus, setting $c_1 = \rho_1 h_1,\ \rho_1 > 1$ and $c_2 = \rho_2 h_2,\ \rho_2 > 1$, we have

$$\begin{aligned} \tfrac{2m}{m-1} &\geq (B_1A_1)^{(\alpha-m)}(B_2A_2)^p(\rho_1^2-1)^{\frac{\alpha-1}{m-1}}h_1^{\frac{2\alpha-(m+1)}{m-1}}(\rho_2^2-1)^{\frac{p}{n-1}}h_2^{\frac{2p}{n-1}}, \\ \tfrac{2n}{n-1} &\geq (B_2A_2)^{(\beta-n)}(B_1A_1)^q(\rho_1^2-1)^{\frac{q}{m-1}}h_1^{\frac{2q}{m-1}}(\rho_2^2-1)^{\frac{\beta-1}{n-1}}h_2^{\frac{2\beta-(n+1)}{n-1}} \end{aligned}$$

for small positive constants $h_1,\ h_2$, which imply that (2.4) is valid.
Thus, $(\overline{u}(x,t), \overline{v}(x,t))$ is a global supersolution of (1.1), which implies the global existence of solution to (1.1) with small initial data. □

The following lemma deals with the case of large initial data.

Lemma 2.2. *Assume* $\alpha \leq \frac{m+1}{2}$, $\beta \leq \frac{n+1}{2}$ *and* $pq > (\frac{m+1}{2}-\alpha)(\frac{n+1}{2}-\beta)$. *If* $Nl_1 > k_1$ *and* $Nl_2 > k_2$, *then the solutions of (1.1) blow up in finite time with large initial data.*

Proof. We begin with the construction of a nonglobal subsolution of the self-similar form

$$\underline{u}(x,t) = (T-t)^{-k_1} f_1(\eta), \quad \underline{v}(x,t) = (T-t)^{-k_2} f_2(\xi) \tag{2.6}$$

where $T > 0$ is a given positive constant and $\eta = |\varepsilon|,\ \varepsilon = x(T-t)^{-l_1}$ and $\xi = |\zeta|,\ \zeta = x(T-t)^{-l_2}$.
We see that $(\underline{u}(x,t), \underline{v}(x,t))$ is a subsolution of (1.1) if the functions $f_1,\ f_2$ satisfy

$$\tfrac{1}{\eta^{N-1}}(\eta^{N-1}(f_1^m)')' - l_1\eta f_1' - k_1 f_1 \geq 0, \tag{2.7}$$

$$\tfrac{1}{\xi^{N-1}}(\xi^{N-1}(f_2^n)')' - l_2\xi f_2' - k_2 f_2 \geq 0 \tag{2.8}$$

and

$$-\tfrac{\partial f_1^m}{\partial \varepsilon_1}|_{\varepsilon_1=0} \leq f_1^\alpha f_2^p, \quad -\tfrac{\partial f_2^n}{\partial \zeta_1}|_{\zeta_1=0} \leq f_1^q f_2^\beta, \tag{2.9}$$

where $\varepsilon \in \{\eta > 0 | f(\eta) \geq 0\}$ and $\zeta \in \{\xi > 0 | g(\xi \geq 0\}$.
Take

$$f_1(\eta) = A(a-\eta)_+^{\frac{1}{m-1}}(\eta-b)_+^{\frac{1}{m-1}}, \qquad f_2(\xi) = B(a-\xi)_+^{\frac{1}{n-1}}(\xi-b)_+^{\frac{1}{n-1}},$$

where $\eta,\ \xi \in (b,a)$ and A, B, a, b are positive constants.
First, such $f_1,\ f_2$ satisfy (2.9), since $x_1 = 0$ implies $\varepsilon_1,\ \zeta_1 = 0$ and

$$-\frac{\partial f_1^m}{\partial \varepsilon_1} = -\frac{\partial f_1^m}{\partial \eta}\frac{\partial \eta}{\partial \varepsilon_1} = -\frac{\partial f_1^m}{\partial \eta}\frac{\varepsilon_1}{\eta}, \quad -\frac{\partial f_2^m}{\partial \zeta_1} = -\frac{\partial f_2^m}{\partial \xi}\frac{\partial \xi}{\partial \zeta_1} = -\frac{\partial f_2^m}{\partial \xi}\frac{\zeta_1}{\xi},$$

which imply

$$-\frac{\partial f_1^m}{\partial \varepsilon_1}|_{\varepsilon_1=0} = 0 \le f_1^\alpha f_2^p, \quad -\frac{\partial f_2^m}{\partial \zeta_1}|_{\zeta_1=0} = 0 \le f_1^q f_2^\beta.$$

We first prove (2.7). By a simple calculation we get

$$f_1' = -A\frac{1}{m-1}(a-\eta)_+^{\frac{1}{m-1}-1}(\eta-b)_+^{\frac{1}{m-1}} + A\frac{1}{m-1}(a-\eta)_+^{\frac{1}{m-1}}(\eta-b)_+^{\frac{1}{m-1}-1},$$

$$\begin{aligned}(f_1^m)'' = A^m\frac{m}{m-1}\Big[&\frac{1}{m-1}(a-\eta)_+^{\frac{1}{m-1}-1}(\eta-b)_+^{\frac{m}{m-1}} - \frac{m}{m-1}(a-\eta)_+^{\frac{1}{m-1}}(\eta-b)_+^{\frac{1}{m-1}}\\ &-\frac{m}{m-1}(a-\eta)_+^{\frac{1}{m-1}}(\eta-b)_+^{\frac{1}{m-1}} + \frac{1}{m-1}(a-\eta)_+^{\frac{m}{m-1}}(\eta-b)_+^{\frac{m}{m-1}-1}\Big].\end{aligned}$$

Substitution those into (2.7) and multiplying (2.7) by $(a-\eta)_+^{1-\frac{1}{m-1}}(\eta-b)_+^{1-\frac{1}{m-1}}$, we have

$$\begin{aligned}&\frac{A^m m}{(m-1)^2}(\eta-b)_+^2 - \frac{2A^m m^2}{(m-1)^2}(a-\eta)_+(\eta-b)_+ + \frac{A^m m}{(m-1)^2}(a-\eta)_+^2\\ &-\frac{N-1}{\eta}\frac{A^m m}{m-1}(a-\eta)_+(\eta-b)_+^2 + \frac{N-1}{\eta}\frac{A^m m}{m-1}(a-\eta)_+^2(\eta-b)_+\\ &-l_1\eta[-\frac{A}{m-1}(\eta-b)_+ + \frac{A}{m-1}(a-\eta)_+]\\ &-Ak_1(a-\eta)_+(\eta-b)_+ \ge 0.\end{aligned}$$

Define

$$\lambda_1 = \frac{2A^m m}{(m-1)^2} + \frac{2A^m m^2}{(m-1)^2} + (N-1)\frac{2A^m m}{(m-1)} + \frac{A}{m-1},$$

$$\begin{aligned}\lambda_2 \ = &-\frac{2A^m m}{(m-1)^2}(a+b) - \frac{2A^m m^2}{(m-1)^2}(a+b) - (N-1)\frac{A^m m}{(m-1)}3(a+b)\\ &-l_1\frac{A}{m-1}(a+b) - Ak_1(a+b),\end{aligned}$$

$$\lambda_3 = \frac{A^m m}{(m-1)^2}(a^2+b^2) + \frac{2A^m m^2}{(m-1)^2}(ab) + Ak_1 ab + (N-1)\frac{A^m m}{(m-1)}ab(a^2+b^2+4ab),$$

$$\lambda_4 = -(N-1)\frac{A^m m}{m-1}ab(a+b).$$

We observe that (2.7) holds if $\lambda_1\eta^3+\lambda_2\eta^2+\lambda_3\eta+\lambda_4 \ge 0,\ 0<b<\eta<a$. If we choose A large enough, it reduces to show

$$\overline{\lambda}_1\eta^3+\overline{\lambda}_2\eta^2+\overline{\lambda}_3\eta+\overline{\lambda}_4 \ge 0, \quad 0<b<\eta<a, \tag{2.10}$$

where

$$\overline{\lambda}_1 = \frac{2+2m}{m-1} + 2(N-1),\ \overline{\lambda}_2 = -\frac{2+2m}{m-1}(a+b) - 3(N-1)(a+b),$$

$$\overline{\lambda}_3 = \frac{1}{m-1}(a^2+b^2) + \frac{2m}{m-1}ab + (N-1)(a^2+b^2+4ab),\ \overline{\lambda}_4 = -(N-1)ab(a+b).$$

To do this, set $Y(\eta) = \overline{\lambda}_1\eta^3+\overline{\lambda}_2\eta^2+\overline{\lambda}_3\eta+\overline{\lambda}_4$. Notice that $\overline{\lambda}_1 > 0$ implies $\lim_{\eta\to\infty} Y(\eta) = +\infty$. Now, we consider the the cubic equation $x^3+px+q=0$. There exist three roots $x_1 = M+N$, $x_{2,3} = -\frac{M+N}{2} \pm \frac{(M-N)\sqrt{3}}{2}i$, where $M = \sqrt[3]{-\frac{q}{2}+\sqrt{Q}}$, $N = \sqrt[3]{-\frac{q}{2}-\sqrt{Q}}$, and $Q = (\frac{p}{3})^3 + (\frac{q}{2})^2$. We see that the cubic equation has one real root and two conjugate complex roots if $Q>0$.
Letting $a = \rho b,\ \rho > 1$, we want to show that if $\rho \to 1^+$, (2.10) holds. We know that when

$\rho \to 1^+$, the equation $Y(\eta) = 0$ has one real root and two conjugate complex roots. We only need to show that if $\rho \to 1^+$, $Y(b) \geq 0$, namely,

$$\begin{aligned} Y(b) &= [\tfrac{2+2m}{m-1} + 2(N-1)]b^3 + [-\tfrac{(2+2m)(1+\rho)}{m-1} - 3(N-1)(1+\rho)]b^3 \\ &+ [\tfrac{1}{m-1}(1+\rho^2) + \tfrac{2m}{m-1}\rho + (N-1)(1+\rho^2+4\rho)]b^3 - [(N-1)\rho(\rho+1)]b^3. \end{aligned}$$

Let

$$\begin{aligned} y(\rho) &= [\tfrac{2+2m}{m-1} + 2(N-1)] + [-\tfrac{(2+2m)(1+\rho)}{m-1} - 3(N-1)(1+\rho)] \\ &+ [\tfrac{1}{m-1}(1+\rho^2) + \tfrac{2m}{m-1}\rho + (N-1)(1+4\rho+\rho^2)] - [(N-1)\rho(\rho+1)], \end{aligned}$$

then $Y(b) = y(\rho)b^3$, where $y(\rho) = \frac{(1-\rho)^2}{m-1} \geq 0$, So $Y(b) \geq 0$. It imply that (2.7) holds. By similar arguments we confirm that (2.8) holds.
If for any given T, A, B, a, b and ρ satisfying (2.7)-(2.9), the initial data u_0, v_0 are large enough such that $u_0(x) \geq \underline{u}(x,0)$, $v_0(x) \geq \underline{v}(x,0)$, $x \in \mathbb{R}^N_+$, thus we have verified that $(\underline{u}(x,t), \underline{v}(x,t))$ is a weakly subsolution of (1.1), which blows up in finite time. From the comparison principle, $(u(x,t), v(x,t))$ blows up in finite time which is not larger than T. □

Lemma 2.3. *Assume $\alpha \leq \frac{m+1}{2}$, $\beta \leq \frac{n+1}{2}$ and $pq > (\frac{m+1}{2} - \alpha)(\frac{n+1}{2} - \beta)$. If $Nl_1 < k_1$ or $Nl_2 < k_2$, then every nonnegative, nontrivial solution of (1.1) blows up in finite time.*

Proof. Set

$$u_s = (\tau+t)^{-\frac{N}{N(m-1)+2}}\phi(\eta),\ v_s = (\tau+t)^{-\frac{N}{N(n-1)+2}}\psi(\xi)$$

where $\tau > 0$ is an arbitrary constant and $\eta = |\varepsilon|$, $\varepsilon = \frac{x}{(T-t)^{\frac{1}{N(m-1)+2}}}$ and $\xi = |\zeta|$, $\zeta = \frac{x}{(T-t)^{\frac{1}{N(n-1)+2}}}$.
We see that $(u_s(x,t), v_s(x,t))$ is a subsolution of (1.1) if the functions ϕ, ψ satisfy

$$\frac{1}{\eta^{N-1}}(\eta^{N-1}(\phi^m)')' + \frac{1}{N(m-1)+2}\eta\phi' + \frac{N}{N(m-1)+2}\phi \geq 0, \tag{2.11}$$

$$\frac{1}{\xi^{N-1}}(\xi^{N-1}(\psi^n)')' + \frac{1}{N(n-1)+2}\xi\psi' + \frac{N}{N(n-1)+2}\psi \geq 0 \tag{2.12}$$

and

$$-\frac{\partial\phi^m}{\partial\varepsilon_1}|_{\varepsilon_1=0} = 0, \quad -\frac{\partial\psi^n}{\partial\zeta_1}|_{\zeta_1=0} = 0, \tag{2.13}$$

here $\varepsilon \in \{\eta > 0 | f(\eta) \geq 0\}$ and $\zeta \in \{\xi > 0 | g(\xi \geq 0\}$.
Take

$$\phi(\eta) = E(c^2 - \eta^2)_+^{\frac{1}{m-1}}, \quad \psi(\xi) = F(c^2 - \xi^2)_+^{\frac{1}{n-1}},$$

where

$$E = \{\frac{m-1}{2m[N(m-1)+2]}\}^{\frac{1}{m-1}}, \qquad F = \{\frac{n-1}{2n[N(n-1)+2]}\}^{\frac{1}{n-1}}$$

and η, $\xi \in (0,c)$.
By a simple calculation, we see that ϕ, ψ satisfy (2.11)-(2.13). By using the properties of the weak solution of the problem (1.1), we deduce that there exist $t_0 \geq 0$ such that $u(0,t_0), v(0,t_0) > 0$. Since $u(x,t_0), v(x,t_0)$ are continuous functions, there exist $\tau > 0$ large enough and small $c > 0$ such that $u(x,t_0) \geq u_s(x,t_0)$, $v(x,t_0) \geq v_s(x,t_0), x \in \mathbb{R}^N_+$. Then by the comparison principle we deduce that

$$u(x,t) \geq u_s(x,t),\ v(x,t) \geq v_s(x,t), \quad x \in \mathbb{R}^N_+,\ t \geq t_0. \tag{2.14}$$

Without loss of generality, assume $k_1 > Nl_1$. We now prove that there exist $t' \geq t_0$ and T large enough so that

$$u_s(x,t') \geq \underline{u}(x,0), \quad x \in \mathbb{R}^N_+, \tag{2.15}$$

where $\underline{u}(x,t)$ is given by (2.6). By using the space-time structure of both u_s and $\underline{u}$, we choose suitable constants a,b such that $0 < a-b < 1$. If

$$(\tau+t')^{-\frac{N}{N(m-1)+2}} \gg T^{-k_1},\ (\tau+t')^{\frac{1}{N(m-1)+2}} \gg T^{l_1} \tag{2.16}$$

is satisfied, (2.15) is valid. We see from (2.16) that such t' and T exist if $T^{k_1} \gg T^{Nl_1}$ for arbitrarily large T. This implies that $k_1 > Nl_1$. Thus, from (2.14) and (2.15), using the comparison principle we confirm that $(u(x,t),v(x,t))$ blows up in finite time under the conditions of the lemma. □

By Lemmas 2.1 , 2.2 and 2.3, we see that Theorem 1.5 holds.

References

[1] K. Deng and H. A. Levine, The role of critical exponents in blow-up theorems: the sequel. *J. Math. Anal. Appl.,* **243** (2000), 85-126.

[2] H. Fujita, On the blowing up of solutions of the Cauchy problem for $u_t = \Delta u + u^{1+\alpha}$. *J. Fac. Sci. Univ. Tokyo Sect. IA Math.,* **13** (1966), 105-113.

[3] V. A. Galaktionov, Blow-up for quasilinear heat equations with critical fujita's exponents. *Proceedings of the Royal Society of Edinburgh*, **124** (1994), 517-525.

[4] V. A. Galaktionov and H. A. Levine, On critical Fujita exponents for heat equations with nonlinear flux boundary condition. *Israel J. Math.*, **94** (1996), 125-146.

[5] H. A. Levine, The role of critical exponents in blow-up theorems. *SIAM Rev.*, **32** (1990), 262-288.

[6] B. Hu and H. M. Yin, On critical Fujita exponents for heat equations with nonlinear flux boundary condition. *Ann, Inst. H. Poincaré,* **13** (1996), 707-732.

[7] W. M. Huang, J. X. Yin, and Y. F. Wang, On critical Fujita exponents for the porous equation with nonlinear boundary condition. *J. Math. Anal. Appl.,* **286** (2003), 369-377.

[8] F. Quirós and J. D. Rossi, Blow-up sets and Fujita type curves for a parabolic system with nonlinear boundary conditions. *Indiana. Univ. Math. J.*, **50** (2001), 629-654.

[9] S. N. Zheng, X. F. Song and Z. X. Jiang, Critical Fujita exponents for degenerate parabolic equations couple via nonlinear boundary flux. *J. Math. Anal. Appl.,* **298** (2004), 308-324.

In: Progress in Evolution Equations
Editor: Gaston M. N'Guerekata, pp. 159-181
ISBN: 978-1-60456-328-3

Chapter 10

POSITIVE SOLUTIONS FOR P-LAPLACIAN DYNAMIC DELAY DIFFERENTIAL EQUATIONS ON TIME SCALES*

Hua Su†
School of Mathematics and System Sciences,
Shandong University, Jinan Shandong, 250100, China

Abstract

Let **T** be a time scale. We study the existence of positive solutions for the nonlinear four-point singular boundary value problem with p-Laplacian dynamic delay differential equations on time scales, subject to some boundary conditions. By using the fixed-point index theory, the existence of positive solution and many positive solutions for nonlinear four-point singular boundary value problem with p-Laplacian operator are obtained.

2000 Mathematics Subject Classification: Primary 34B 16, 39A 10

Key words: Time scale, p-Laplacian dynamic delay differential equations, positive solutions, fixed-point index theory.

1. Introduction

The study of dynamic equations on time scales goes back to its founder Stefan Hilger [22], and is a new area of still fairly theoretical exploration in mathematics. Boundary value problems for delay differential equations arise in a variety of areas of applied mathematics, physics and variational problems of control theory (see [10,11]). In recent years, many authors have begun to pay attention to the study of boundary-value problems or with p-Laplacian equations or with p-Laplacian dynamic equations on time scales (see [4-7, 13-15, 17, 19-20, 24, 27-28] and the references therein).

*Research supported by the NNSF of China (10471075), Shandong University(306001).
†E-mail address: jnsuhua@163.com

In [7], Sun and Li considered the existence of positive solution of the following dynamic equations on time scales:

$$u^{\Delta\nabla}(t)+a(t)f(t,u(t))=0, t\in(0,T), \tag{1.1}$$

$$\beta u(0)-\gamma u^{\Delta}(0)=0,\ \alpha u(\eta)=u(T), \tag{1.2}$$

where $\beta,\ \gamma\geq 0,\ \beta+\gamma>0,\ \eta\in(0,\rho(T)),\ 0<\alpha<T/\eta$. They obtained the existence of single and multiple positive solutions of the problem (1.1) and (1.2) by using fixed point theorem and Leggett-Williams fixed point theorem (see [25]), respectively.

In [4], Anderson discussed the following dynamic equation on time scales:

$$\begin{cases} u^{\Delta\nabla}(t)+a(t)f(u(t))=0,\ t\in(0,T), \\ u(0)=0,\ \alpha u(\eta)=u(T). \end{cases} \tag{1.3}$$

He obtained some results for the existence of one positive solution of the problem (1.3) based on the limits $f_0=\lim\limits_{u\to 0^+}\dfrac{f(u)}{u}$ and $f_\infty=\lim\limits_{u\to 0}\dfrac{f(u)}{u}$.

In [5], Kaufmann studied the problem (1.3) and obtained the existence results of at least two positives solutions.

In [24], Wang discussed the following dynamic equation on time scales by using Avery-Peterson fixed theorem (see [24]):

$$(\phi_p(u'))'+q(t)f(t,u(t),u(t-1),u'(t))=0,\ t\in(0,1), \tag{1.4}$$

$$u(t)=\xi(t),\ -1\leq t\leq 0,\ u(1)=0, \tag{1.5}$$

and

$$u(t)=\xi(t),\ -1\leq t\leq 0,\ u'(1)=0, \tag{1.5'}$$

They obtained some results for the existence three positive solutions of the problem (1.4), (1.5) and (1.4), (1.5′), respectively.

In [25], Lee and Sim discussed the following equation

$$\begin{cases} (\phi_p(u'))'+\lambda h(t)f(u(t))=0,\ \text{a. e. } t\in(0,1), \\ u(0)=u(1)=0, \end{cases} \tag{1.6}$$

By applying the global bifurcation theorem and figuring the shape of unbounded subcontinua of solutions, they obtain many different types of global existence results of positive solutions.

However, there are not many concerning the p-Laplacian problems on time scales. Especially, for the singular multi-point boundary value problems for p-Laplacian dynamic delay differential equations on time scales, with the author's acknowledge, no one has studied the existence of positive solutions in this case.

Recently, in [8], we study the existence of positive solutions for the following nonlinear two-point singular boundary value problem with p-Laplacian operator

$$\begin{cases} (\phi_p(u'))'+a(t)f(u(t))=0,\quad 0<t<1, \\ \alpha\phi_p(u(0))-\beta\phi_p(u'(0))=0,\ \gamma\phi_p(u(1))+\delta\phi_p(u'(1))=0, \end{cases} \tag{1.7}$$

by using the fixed-point theorem of cone expansion and compression of norm type, the existence of positive solution and infinitely many positive solutions for nonlinear singular boundary value problem (1.7) with p-Laplacian operator are obtained.

Now, motivated by the results mentioned above, in this paper, we study the existence of positive solutions for the following nonlinear four-point singular boundary value problem with higher-order p-Laplacian dynamic delay differential equations operator on time scales (SBVP):

$$\left(\phi_p\left(u^{\Delta}(t)\right)\right)^{\nabla}+g(t)f(u(t-\tau),u(t))=0,\ 0<t<T,\ \tau>0 \tag{1.8}$$

$$\begin{cases} u(t)=\zeta(t), & -\tau\le t\le 0,\\ \alpha\phi_p\left(u(0)\right)-\beta\phi_p\left(u^{\Delta}(\xi)\right)=0, \\ \gamma\phi_p\left(u(T)\right)+\delta\phi_p\left(u^{\Delta}(\eta)\right)=0, \end{cases} \tag{1.9}$$

or

$$\begin{cases} u(t)=\zeta(t), & -\tau\le t\le 0,\\ u(0)-B_0\left(u^{\Delta}(\xi)\right)=0, \\ u(T)+B_1\left(u^{\Delta}(\eta)\right)=0, \end{cases} \tag{1.10}$$

where $\phi_p(s)$ is p-Laplacian operator, i.e., $\phi_p(s)=|s|^{p-2}s,\ p>1,\ \phi_q=\phi_p^{-1},\ \frac{1}{p}+\frac{1}{q}=1.$ $\xi,\ \eta\in(0,T),\tau\in[0,T]$ is prescribed and $\xi<\eta, g:(0,T)\to[0,\infty),\alpha>0,\beta\ge 0,\gamma>0,\delta\ge 0$ and $B_0,\ B_1$ are both nondecreasing continuous odd functions defined on $(-\infty,+\infty)$.

In this paper, by constructing one integral equation which is equivalent to the problem (1.8), (1.9) and (1.8), (1.10), we research the existence of positive solutions for nonlinear singular boundary value problem (1.8), (1.9) and (1.8), (1.10) when g and f satisfy some suitable conditions.

Our main tool of this paper is the following fixed point index theory.

Theorem 1.1 (2,3). *Suppose E is a real Banach space, $K\subset E$ is a cone, let $\Omega_r=\{u\in K: \|u\|\le r\}$. Let operator $T:\Omega_r\longrightarrow K$ be completely continuous and satisfy $Tx\ne x,\ \forall\, x\in\partial\Omega_r$. Then*

(i) If $\|Tx\|\le\|x\|,\ \forall\, x\in\partial\Omega_r$, then $i(T,\Omega_r,K)=1$;

(ii) If $\|Tx\|\ge\|x\|,\ \forall\, x\in\partial\Omega_r$, then $i(T,\Omega_r,K)=0$.

This paper is organized as follows. In section 2, we present some preliminaries and lemmas that will be used to prove our main results. In section 3, we discuss the existence of single solution of the systems (1.8), (1.9) and (1.8), (1.10). In section 4, we study the existence of at least two solutions of the systems (1.8), (1.9) and (1.8), (1.10). In section 5, we give two examples as the application.

2. Preliminaries and Lemmas

For convenience, we list here the following definitions which are needed later. We begin by presenting some basic definitions which can be found in [22,23,12].

A time scale $\mathbf{T}$ is an arbitrary nonempty closed subset of real numbers R. The operators σ and ρ from $\mathbf{T}$ to $\mathbf{T}$,

$$\sigma(t) = \inf\{\tau \in \mathbf{T} \mid \tau > t\} \in \mathbf{T},\ \rho(t) = \sup\{\tau \in \mathbf{T} \mid \tau < t\} \in \mathbf{T},$$

are called the forward jump operator and the backward jump operator, respectively.

The point $t \in \mathbf{T}$ is left-dense, left-scattered, right-dense, right-scattered if $\rho(t) = t$, $\rho(t) < t$, $\sigma(t) = t$, $\sigma(t) > t$, respectively. If $\mathbf{T}$ has a right scattered minimum m, define $\mathbf{T}_k = \mathbf{T} - \{m\}$; otherwise set $\mathbf{T}_k = \mathbf{T}$. If $\mathbf{T}$ has a left scattered maximum M, define $\mathbf{T}^k = \mathbf{T} - \{M\}$; otherwise set $\mathbf{T}^k = \mathbf{T}$.

Let $f : \mathbf{T} \to R$ and $t \in \mathbf{T}^k$ (assume t is not left-scattered if $t = \sup \mathbf{T}$), then the delta derivative of f at the point t is defined to be the number $f^{\Delta}(t)$ (provided it exists) with the property that for each $\varepsilon > 0$ there is a neighborhood U of t such that

$$|f(\sigma(t)) - f(s) - f^{\Delta}(t)(\sigma(t) - s)| \leq |\sigma(t) - s|, \quad \text{for all } s \in U.$$

Similarly, for $t \in \mathbf{T}$ (assume t is not right-scattered if $t = \inf \mathbf{T}$), the nabla derivative of f at the point t is defined in [22] to be the number $f^{\nabla}(t)$ (provided it exists) with the property that for each $\varepsilon > 0$ there is a neighborhood U of t such that

$$|f(\rho(t)) - f(s) - f^{\nabla}(t)(\rho(t) - s)| \leq |\rho(t) - s|, \quad \text{for all } s \in U.$$

If $\mathbf{T} = R$, then $x^{\Delta}(t) = x^{\nabla}(t) = x'(t)$. If $\mathbf{T} = Z$, then $x^{\Delta}(t) = x(t+1) - x(t)$ is the forward difference operator while $x^{\nabla}(t) = x(t) - x(t-1)$ is the backward difference operator.

A function f is left-dense continuous (i.e., ld-continuous), if f is continuous at each left-dense point in $\mathbf{T}$ and its right-sided limit exists at each right-dense point in $\mathbf{T}$. It is well-known that if f is ld-continuous, then there is a function $F(t)$ such that $F^{\nabla}(t) = f(t)$.

If $F^{\nabla}(t) = f(t)$, then we define the nabla integral by

$$\int_a^b f(t)\nabla t = F(b) - F(a).$$

If $F^{\Delta}(t) = f(t)$, then we define the delta integral by

$$\int_a^b f(t)\Delta t = F(b) - F(a).$$

In the rest of this article, $\mathbf{T}$ is closed subset of R with $0 \in \mathbf{T}_k, T \in \mathbf{T}^k$. And let $B = \{u \in C_{ld}[-\tau, T]\}$, then B is a Banach space with the norm $\|u\| = \max_{t \in [-\tau, T]} |u(t)|$. And let

$$K = \{u \in B : \ u(t) \geq 0,\ u(t) \text{ is concave function}, t \in [0, T]\}.$$

Obviously, K is a cone in B. Set $K_r = \{u \in K : \|u\| \leq r\}$.

Definition 2.1. $u(t)$ is called a solution of SBVP (1.8) and (1.9) if it satisfies the following

(1) $u \in C[-\tau, 0] \cap C_{ld}(0, T)$;

(2) $u(t) > 0$ for all $t \in (0, T)$ and satisfies conditions (1.9);

(3) $\left(\phi_p\left(u^{\Delta}(t)\right)\right)^{\nabla} = -g(t) f(u(t-\tau), u(t))$ holds for $t \in (0, T)$.

In the rest of the paper, we also make the following assumptions:
(H_1) $f \in C_{ld}([0,+\infty)^2,[0,+\infty))$;
(H_2) $g(t) \in C_{ld}((0,T),[0,+\infty))$ and there exists $t_0 \in (0,T)$, such that

$$g(t_0) > 0,\ 0 < \int_0^T g(s)\,\nabla s < +\infty;$$

(H_3) $\zeta(t) \in C([-\tau,0],\ \zeta(t) > 0$ on $[-\tau,0)$ and $\zeta(0) = 0$.
(H_4) $B_0,\ B_1$ are both increasing, continuous, odd functions defined on $(-\infty,+\infty)$ and at least one of them satisfies the condition that there exists one $b > 0$ such that

$$0 < B_i(v) \le bv,\ \forall\, v \ge 0,\ i = 0 \text{ or } 1.$$

It is easy to check that condition (H_2) implies that

$$0 < \int_0^T \phi_q\left(\int_0^s g(s_1)\nabla s_1\right)\Delta s < +\infty.$$

We can easily get the following Lemmas.

Lemma 2.2. *Suppose condition (H_2) holds. Then there exists a constant $\theta \in (0,\frac{1}{2})$ satisfies*

$$0 < \int_\theta^{T-\theta} g(t)\nabla t < \infty.$$

Furthermore, the function

$$A(t) = \int_\theta^t \phi_q\left(\int_s^t g(s_1)\nabla s_1\right)\Delta s + \int_t^{T-\theta} \phi_q\left(\int_t^s g(s_1)\nabla s_1\right)\nabla s,\ t \in [\theta, T-\theta]$$

is positive continuous functions on $[\theta, T-\theta]$, therefore $A(t)$ has minimum on $[\theta, T-\theta]$. Hence we suppose that there exists $L > 0$ such that $A(t) \ge L,\ t \in [\theta, T-\theta]$.

Proof. At first, it is easily seen that $A(t)$ is continuous on $[\theta, T-\theta]$. Next, let

$$A_1(t) = \int_\theta^t \phi_q\left(\int_s^t g(s_1)\nabla s_1\right)\Delta s,\ A_2(t) = \int_t^{T-\theta} \phi_q\left(\int_t^s g(s_1)\nabla s_1\right)\Delta s.$$

Then, from condition (H_2), we have the function $A_1(t)$ is strictly monotone nondecreasing on $[\theta, T-\theta]$ and $A_1(\theta) = 0$, the function $A_2(t)$ is strictly monotone nonincreasing on $[\theta, T-\theta]$ and $A_2(T-\theta) = 0$, which implies $L = \min_{t \in [\theta, T-\theta]} A(t) > 0$. The proof is complete. □

Lemma 2.3 (8). *Let $u \in K$ and θ of Lemma 2.2, then*

$$u(t) \ge \theta\|u\|,\ t \in [\theta, T-\theta].$$

Lemma 2.4. *Suppose that conditions (H_1), (H_2), (H_3), (H_4) hold, $u(t) \in B \cap C_{ld}(0,1)$ is a solution of the following boundary value problems*

$$\left(\phi_p\left(u^\Delta(t)\right)\right)^\nabla + g(t)f(u(t-\tau)+h(t-\tau),u(t)) = 0,\ 0<t<T, \tag{2.1}$$

$$\begin{cases} u(t)=0, \quad -\tau \le t \le 0, \\ \alpha\phi_p\left(u(0)\right) - \beta\phi_p\left(u^\Delta(\xi)\right) = 0, \\ \gamma\phi_p\left(u(T)\right) + \delta\phi_p\left(u^\Delta(\eta)\right) = 0, \end{cases} \tag{2.2}$$

or

$$\begin{cases} u(t)=0, \quad -\tau \le t \le 0, \\ u(0) - B_0\left(u^\Delta(\xi)\right) = 0, \\ u(T) + B_1\left(u^\Delta(\eta)\right) = 0, \end{cases} \tag{2.2$'$}$$

where

$$h(t) = \begin{cases} \zeta(t), & -\tau \le t \le 0, \\ 0, & 0 \le t \le T. \end{cases}$$

Then, $\overline{u}(t) = u(t) + h(t)$, $-\tau \le t \le T$ is a positive solution to the SBVP (1.8), (1.9) or (1.8), (1.10).

Proof. It is easy to check that $\overline{u}(t)$ satisfies (1.8), (1.9) or (1.8), (1.10). □

So in the rest section of this paper, we focus on SBVP (2.1) (2.2) and (2.1) (2.2').

Lemma 2.5. *Suppose that conditions (H_1), (H_2), (H_3) or (H_1), (H_2), (H_3), (H_4) hold, $u(t) \in B \cap C_{ld}(0,1)$ is a solution of boundary value problems (2.1), (2.2) or (2.1), (2.2') respectively if and only if $u(t) \in B$ is a solution of the following integral equation, respectively*

$$u(t) = \begin{cases} \zeta(t), & -\tau \le t \le 0, \\ \int_0^t w(s)\Delta s, & 0 \le t \le T, \end{cases}$$

$$u(t) = \begin{cases} \zeta(t), & -\tau \le t \le 0, \\ \int_0^t \overline{w}(s)\Delta s, & 0 \le t \le T, \end{cases}$$

where

$$w(t) = \begin{cases} \phi_q\left(\frac{\beta}{\alpha}\int_\xi^\sigma g(s)f(u(s-\tau)+h(s-\tau),u(s))\nabla s\right) \\ \quad + \int_0^t \phi_q\left(\int_s^\sigma g(r)f(u(r-\tau)+h(r-\tau),u(r))\nabla r\right)\Delta s, & 0 \le t \le \sigma, \\ \phi_q\left(\frac{\delta}{\gamma}\int_\sigma^\eta g(s)f(u(s-\tau)+h(s-\tau),u(s))\nabla s\right) \\ \quad + \int_t^T \phi_q\left(\int_\sigma^s g(r)f(u(r-\tau)+h(r-\tau),u(r))\nabla r\right)\Delta s, & \sigma \le t \le T. \end{cases} \tag{2.3}$$

$$\overline{w}(t)=\begin{cases} B_0\circ\phi_q\left(\int_\xi^\rho g(s)f(u(s-\tau)+h(s-\tau),u(s))\nabla s\right) \\ +\int_0^t \phi_q\left(\int_s^\rho g(r)f(u(r-\tau)+h(r-\tau),u(r))\nabla r\right)\Delta s, & 0\le t\le\rho, \\ B_1\circ\phi_q\left(\int_\rho^\eta g(s)f(u(s-\tau)+h(s-\tau),u(s))\nabla s\right) \\ +\int_t^T \phi_q\left(\int_\rho^s g(r)f(u(r-\tau)+h(r-\tau),u(r))\nabla r\right)\Delta s, & \rho\le t\le T. \end{cases} \tag{2.3'}$$

Here σ, ρ is unique solution of the equation, respectively

$$g_1(t)=g_2(t),\ \overline{g}_1(t)=\overline{g}_2(t)$$

where

$$\begin{aligned} g_1(t)= &\ \phi_q\left(\frac{\beta}{\alpha}\int_\xi^\sigma g(s)f(u(s-\tau)+h(s-\tau),u(s))\nabla s\right) \\ &+\int_0^t \phi_q\left(\int_s^\sigma g(r)f(u(r-\tau)+h(r-\tau),u(r))\nabla r\right)\Delta s, \\ g_2(t)= &\ \phi_q\left(\frac{\delta}{\gamma}\int_\sigma^\eta g(s)f(u(s-\tau)+h(s-\tau),u(s))\nabla s\right) \\ &+\int_t^T \phi_q\left(\int_\sigma^s g(r)f(u(r-\tau)+h(r-\tau),u(r))\nabla r\right)\Delta s. \\ \overline{g}_1(t)= &\ B_0\circ\phi_q\left(\int_\xi^\rho g(s)f(u(s-\tau)+h(s-\tau),u(s))\nabla s\right) \\ &+\int_0^t \phi_q\left(\int_s^\rho g(r)f(u(r-\tau)+h(r-\tau),u(r))\nabla r\right)\Delta s, \\ \overline{g}_2(t)= &\ B_1\circ\phi_q\left(\int_\rho^\eta g(s)f(u(s-\tau)+h(s-\tau),u(s))\nabla s\right) \\ &+\int_t^T \phi_q\left(\int_\rho^s g(r)f(u(r-\tau)+h(r-\tau),u(r))\nabla r\right)\Delta s. \end{aligned}$$

Equation $g_1(t)=g_2(t)$, $\overline{g}_1(t)=\overline{g}_2(t)$ has unique solution in $(0,T)$. Because $g_1(t)$, $\overline{g}_1(t)$ is strictly monotone increasing on $[0,T)$, and $g_1(0)=0$, $\overline{g}_1(0)=0$, $g_2(t)$, $\overline{g}_2(t)$ is strictly monotone decreasing on $(0,T]$, and $g_2(T)=0$, $\overline{g}_2(T)=0$.

Proof. We only proof the first section of the results.

Necessity. Obviously, for $t\in(-\tau,0)$, we have $u(t)=\zeta(t)$.

If $t\in(0,T)$, by the equation of the boundary condition, we have $u^\Delta(\xi)\ge 0$, $u^\Delta(\eta)\le 0$, then there exist a constant $\sigma\in[\xi,\eta]\subset(0,T)$ such that $u^\Delta(\sigma)=0$.

Firstly, by integrating the equation of the problems (2.1) on (σ,t), we have

$$\begin{aligned} \phi_p(u^\Delta(t))= &\ \phi_p(u^\Delta(\sigma)) \\ &-\int_\sigma^t g(s)f(u(s-\tau)+h(s-\tau),u(s))\nabla s, \end{aligned} \tag{2.4}$$

then

$$u^{\Delta}(t) = -\phi_q\left(\int_{\sigma}^{t} g(s)f(u(s-\tau)+h(s-\tau),u(s))\nabla s\right),$$

thus

$$u(t) = u(\sigma) - \int_{\sigma}^{t} \phi_q\left(\int_{\sigma}^{s} g(r)f(u(r-\tau)+h(r-\tau),u(r))\nabla r\right)\Delta s. \tag{2.5}$$

By $u^{\Delta}(\sigma) = 0$ and condition (2.4), let $t = \eta$ on (2.4), we have

$$\phi_p(u^{\Delta}(\eta)) = -\int_{\sigma}^{\eta} g(s)f(u(s-\tau)+h(s-\tau),u(s))\nabla s.$$

By the equation of the boundary condition (2.2), we have

$$\phi_p(u(T)) = -\frac{\delta}{\gamma}\phi_p(u^{\Delta}(\eta)),$$

then

$$u(T) = \phi_q\left(\frac{\delta}{\gamma}\int_{\sigma}^{\eta} g(s)f(u(s-\tau)+h(s-\tau),u(s))\nabla s\right).$$

Then, by (2.5) and let let $t = T$ on (2.5), we have

$$\begin{aligned} u(\sigma) = {} & \phi_q\left(\frac{\delta}{\gamma}\int_{\sigma}^{\eta} g(s)f(u(s-\tau)+h(s-\tau),u(s))\nabla s\right) \\ & + \int_{\sigma}^{T} \phi_q\left(\int_{\sigma}^{s} g(r)f(u(r-\tau)+h(r-\tau),u(r))\nabla r\right)\Delta s. \end{aligned} \tag{2.6}$$

Then

$$\begin{aligned} u(t) = {} & \phi_q\left(\frac{\delta}{\gamma}\int_{\sigma}^{\eta} g(s)f(u(s-\tau)+h(s-\tau),u(s))\nabla s\right) \\ & + \int_{t}^{T} \phi_q\left(\int_{\sigma}^{s} g(r)f(u(r-\tau)+h(r-\tau),u(r))\nabla r\right)\Delta s. \end{aligned} \tag{2.7}$$

Similarly, for $t \in (0,\sigma)$, by integrating the equation of problems (2.1) on $(0,\sigma)$, we have

$$\begin{aligned} u(t) = {} & \phi_q\left(\frac{\beta}{\alpha}\int_{\xi}^{\sigma} g(s)f(u(s-\tau)+h(s-\tau),u(s))\nabla s\right)\Delta s \\ & + \int_{0}^{t} \phi_q\left(\int_{s}^{\sigma} g(r)f(u(r-\tau)+h(r-\tau),u(r))\nabla r\right)\Delta s. \end{aligned}$$

Therefore, for any $t \in [0,T]$, $u(t)$ can be expressed as equation

$$u(t) = \begin{cases} \zeta(t), & -\tau \le t \le 0, \\ \int_{0}^{t} w(s)\Delta s, & 0 \le t \le T, \end{cases}$$

where $w(t)$ is expressed as (2.3).

Sufficiency. Suppose that $u(t)=\int_0^t w(s)\Delta s_{n-2}\Delta s,\ 0\le t\le T$. Then by (2.3), we have

$$u^{\Delta}(t)=\begin{cases}\phi_q\Big(\int_t^{\sigma} g(s)f(u(s-\tau)+h(s-\tau),u(s))\nabla s\Big)\ge 0, & 0\le t\le\sigma,\\ -\phi_q\Big(\int_{\sigma}^{t} g(s)f(u(s-\tau)+h(s-\tau),u(s))\nabla s\Big)\le 0, & \sigma\le t\le T,\end{cases}\tag{2.8}$$

So, $(\phi_p(u^{\Delta})^{\nabla}+g(t)f(u(t-\tau)+h(t-\tau),u(t))=0,\ 0<t<T$. These imply that the equation (2.1) holds. Furthermore, by letting $t=0$ and $t=T$ on (2.3) and (2.8), we can obtain the boundary value equations of (2.2). The proof is complete. □

Now, we define a operetorequation T given by

$$(Tu)(t)=\begin{cases}\zeta(t), & -\tau\le t\le 0,\\ \int_0^t w(s)\Delta s, & 0\le t\le T,\end{cases}$$

$$(\overline{T}u)(t)=\begin{cases}\zeta(t), & -\tau\le t\le 0,\\ \int_0^t \overline{w}(s)\Delta s, & 0\le t\le T,\end{cases}$$

where $w(t)$, $\overline{w}(t)$ is given by (2.3) (2.3').

From the definition of T, $\overline{T}$ and above discussion, we deduce that for each $u\in K$, Tu, $\overline{T}u\in K$. Moreover, we have the following Lemma.

Lemma 2.6. *T, $\overline{T}:K\to K$ is completely continuous.*

Proof. We only proof the completely continuous of T.

Because

$$(Tu)^{\Delta}(t)=w^{\Delta}(t)$$
$$=\begin{cases}\phi_q\left(\int_t^{\sigma} g(s)f(u(s-\tau)+h(s-\tau),u(s))\nabla s\right)\ge 0, & 0\le t\le\sigma,\\ -\phi_q\left(\int_{\sigma}^{t} g(s)f(u(s-\tau)+h(s-\tau),u(s))\nabla s\right)\le 0, & \sigma\le t\le T,\end{cases}$$

is continuous, decreasing on $[0,T]$ and satisfies $(Tu)^{\Delta}(\sigma)=0$. Then, $Tu\in K$ for each $u\in K$ and $(Tu)(\sigma)=\max_{t\in[0,T]}(Tu)(t)$. This shows that $TK\subset K$. Furthermore, it is easy to check by Arzela-ascoli Theorem that $T:K\to K$ is completely continuous. □

Lemma 2.7. *Suppose that conditions (H_1), (H_2), (H_3) hold, the solution $u(t)\in K$ of problem (2.1), (2.2) satisfy:*

$$\max_{0\le t\le T}|u(t-\tau)+h(t-\tau)|\le\max_{-\tau\le t\le 0}|\zeta(t)|.$$

Proof. Firstly, we can have

$$\begin{aligned}\max_{0\le t\le T}|u(t-\tau)+h(t-\tau)| &\le \max_{0\le t\le T}|u(t-\tau)|+\max_{0\le t\le T}|h(t-\tau)|\\ &= \max_{-\tau\le t\le T-\tau}|u(t)|+\max_{-\tau\le t\le T-\tau}|h(t)|\\ &= \max_{-\tau\le t\le 0}|\zeta(t)|.\end{aligned}$$

The proof is complete. □

For convenience, we set

$$H=\max_{-\tau\le t\le 0}|\zeta(t)|,\ \theta^*=\frac{2}{L},$$

$$\theta_*=\frac{1}{\left(T+\phi_q\left(\frac{\beta}{\alpha}\right)\right)\phi_q\left(\int_0^T g(r)\nabla r\right)},\quad \theta_{**}=\frac{1}{(b+1)\phi_q\left(\int_0^T g(r)\nabla r\right)}.$$

where L is the constant from Lemma 2.1. By Lemma 2.5, we can also set

$$f^0=\lim_{u_2\to 0}\max_{0\le u_1\le H}\frac{f(u_1,u_2)}{u_2^{p-1}},\ f^\infty=\lim_{u_2\to\infty}\max_{0\le u_1\le H}\frac{f(u_1,u_2)}{u_n^{p-1}},$$

$$f_0=\lim_{u_2\to 0}\min_{0\le u_1\le H}\frac{f(u_1,u_2)}{u_2^{p-1}},\ f_\infty=\lim_{u_2\to\infty}\min_{0\le u_1\le H}\frac{f(u_1,u_2)}{u_n^{p-1}}.$$

3. The Existence of Single Positive Solution to (1.8) (1.9)

In this section, we present our main results.

Theorem 3.1. *Suppose that condition (H_1), (H_2), (H_3) hold. Assume that f also satisfy*
(A_1): $f(u_1,u_2)\ge (mr)^{p-1}$, for $\theta r\le u_2\le r$, $0\le u_1\le H$;
(A_2): $f(u_1,u_2)\le (MR)^{p-1}$, for $0\le u_2\le R$, $0\le u_1\le H$,
where $m\in(\theta^*,\infty)$, $M\in(0,\theta_*)$.
Then, the (SBVP) (2.1), (2.2) has a solution u such that $\|u\|$ lies between r and R.

Proof. Without loss of generality, we suppose that $r<R$. For any $u\in K$, by Lemma 2.3, we have

$$u(t)\ge\theta\|u\|,\quad t\in[\theta,T-\theta]. \tag{3.1}$$

We define two open subset Ω_1 and Ω_2 of E:

$$\Omega_1=\{u\in K:\|u\|<r\},\ \Omega_2=\{u\in K:\|u\|<R\}.$$

For any $u\in\partial\Omega_1$, by (3.1) we have

$$r=\|u\|\ge u(t)\ge\theta\|u\|=\theta r,\quad t\in[\theta,T-\theta].$$

For $t \in [\theta, T-\theta]$ and $u \in \partial\Omega_1$, we shall discuss it from three perspectives.
(i) If $\sigma \in [\theta, T-\theta]$, thus for $u \in \partial\Omega_1$, by (A_1) and Lemma 2.4, we have

$$\begin{aligned} 2\|Tu\| &= 2(Tu)(\sigma) \\ &\geq \int_0^\sigma \phi_q\left(\int_s^\sigma g(r)f(u(r-\tau)+h(r-\tau),u(r))\nabla r\right)\Delta s \\ &\quad + \int_\sigma^T \phi_q\left(\int_\sigma^s g(r)f(u(r-\tau)+h(r-\tau),u(r))\nabla r\right)\Delta s \\ &\geq \int_\theta^\sigma \phi_q\left(\int_s^\sigma g(r)f(u(r-\tau)+h(r-\tau),u(r))\nabla r\right)\Delta s \\ &\quad + \int_\sigma^{T-\theta} \phi_q\left(\int_\sigma^s g(r)f(u(r-\tau)+h(r-\tau),u(r))\nabla r\right)\Delta s \\ &\geq mrA(\sigma) \geq mrL \geq 2r = 2\|u\|. \end{aligned}$$

(ii) If $\sigma \in (T-\theta, T]$, thus for $u \in \partial\Omega_1$, by (A_1) and Lemma 2.4, we have

$$\begin{aligned} \|Tu\| &= (Tu)(\sigma) \\ &\geq \phi_q\left(\frac{\beta}{\alpha}\int_\xi^\sigma g(s)f(u(s-\tau)+h(s-\tau),u(s))\nabla s\right) \\ &\quad + \int_0^\sigma \phi_q\left(\int_s^\sigma g(r)f(u(r-\tau)+h(r-\tau),u(r))\nabla r\right)\Delta s \\ &\geq \int_\theta^{T-\theta} \phi_q\left(\int_s^{T-\theta} g(r)f(u(r-\tau)+h(r-\tau),u(r))\nabla r\right)\Delta s \\ &\geq mrA(T-\theta) \geq mrL \geq 2r > r = \|u\|. \end{aligned}$$

(iii) If $\sigma \in (0, \theta)$, thus for $u \in \partial\Omega_1$, by (A_1) and Lemma 2.4, we have

$$\begin{aligned} \|Tu\| &= (Tu)(\sigma) \\ &\geq \phi_q\left(\frac{\delta}{\gamma}\int_\sigma^\eta g(s)f(u(s-\tau)+h(s-\tau),u(s))\nabla s\right) \\ &\quad + \int_\sigma^T \phi_q\left(\int_\sigma^s g(r)f(u(r-\tau)+h(r-\tau),u(r))\nabla r\right)\Delta s \\ &\geq \int_\theta^{T-\theta} \phi_q\left(\int_\theta^s g(r)f(u(r-\tau)+h(r-\tau),u(r))\nabla r\right)\Delta s \\ &\geq mrA(\theta) \geq mrL \geq 2r > r = \|u\|. \end{aligned}$$

Therefore, no matter under which condition, we all have

$$\|Tu\| > \|u\|, \ \forall\, u \in \partial\Omega_1.$$

Then by Theorem 1.1, we have

$$i(T, \Omega_1, K) = 0. \tag{3.2}$$

On the other hand, for $u \in \partial\Omega_2$, we have $u(t) \le \|u\| = R$, by (A_2) we know

$$\begin{aligned} \|Tu\| &= (Tu)(\sigma) \\ &\le \phi_q\left(\frac{\beta}{\alpha}\int_{\xi}^{\sigma} g(s)f(u(s-\tau)+h(s-\tau),u(s))\nabla s\right) \\ &\quad + \int_0^T \phi_q\left(\int_s^{\sigma} g(r)f(u(r-\tau)+h(r-\tau),u(r))\nabla r\right)\Delta s \\ &\le \left(T+\phi_q\left(\frac{\beta}{\alpha}\right)\right) MR\phi_q\left(\int_0^T g(r)\nabla r\right) \le R = \|u\|. \end{aligned}$$

thus

$$\|Tu\| \le \|u\|, \forall\, u \in \partial\Omega_2.$$

Then, by Theorem 1.1, we have

$$i(T,\Omega_2,K) = 1. \tag{3.3}$$

Therefore, by (3.2), (3.3), $r < R$, we have

$$i(T,\Omega_2\setminus\overline{\Omega}_1,K) = 1.$$

Then operator T has a fixed point $u \in (\Omega_1\setminus\overline{\Omega}_2)$, and $r \le \|u\| \le R$. This completes the proof of Theorem 3.1. □

Theorem 3.2. *Suppose that condition (H_1), (H_2), (H_3) hold. Assume that f also satisfy*

(A_3): $f^0 = \varphi \in \left[0, \left(\frac{\theta_*}{4}\right)^{p-1}\right)$;

(A_4): $f_\infty = \lambda \in \left(\left(\frac{2\theta^*}{\theta}\right)^{p-1}, \infty\right)$.

Then, the (SBVP) (2.1), (2.2) has a solution u which is bounded in $\|\cdot\|$.

Proof. First, by $f^0 = \varphi \in \left[0, \left(\frac{\theta_*}{4}\right)^{p-1}\right)$, for $\varepsilon = \left(\frac{\theta_*}{4}\right)^{p-1} - \varphi$, there exists an adequately small positive number ρ, as $0 \le u_2 \le \rho,\ u_2 \ne 0,\ u_1 \le H$, we have

$$f(u_1,u_2) \le (\varphi+\varepsilon)(u_2)^{p-1} \le \left(\frac{\theta_*}{4}\right)^{p-1}\rho^{p-1} = \left(\frac{\theta_*}{4}\rho\right)^{p-1}. \tag{3.4}$$

Then let $R = \rho,\ M = \dfrac{\theta_*}{4} \in (0,\theta_*)$, thus by (3.4)

$$f(u_1,u_2) \le (MR)^{p-1},\ 0 \le u_2 \le R.$$

So condition (A_2) holds.

Next, by condition (A_4), $f_\infty = \lambda \in \left(\left(\frac{2\theta^*}{\theta}\right)^{p-1}, \infty\right)$, then for $\varepsilon = \lambda - \left(\frac{2\theta^*}{\theta}\right)^{p-1}$, there exists an appropriately big positive number $r \neq R$, as $u_2 \geq \theta r$, $u_1 \leq H$, we have

$$f(u_1,u_2) \geq (\lambda-\varepsilon)(u_2)^{p-1} \geq \left(\frac{2\theta^*}{\theta}\right)^{p-1} (\theta r)^{p-1} = (2\theta^* r)^{p-1}, \tag{3.5}$$

Let $m = 2\theta^* > \theta^*$, thus by (3.5), condition (A_1) holds. Therefor by Theorem 3.1 we know that the results of Theorem 3.2 holds. The proof of Theorem 3.2 is complete. □

Theorem 3.3. *Suppose that condition* (H_1), (H_2), (H_3) *hold. Assume that* f *also satisfy*

(A_5): $f^\infty = \lambda \in \left[0, \left(\frac{\theta_*}{4}\right)^{p-1}\right)$;

(A_6): $f_0 = \varphi \in \left(\left(\frac{2\theta^*}{\theta}\right)^{p-1}, \infty\right)$.

Then, the (SBVP) (2.1), (2.2) has a solution u *which is bounded in* $\|\cdot\|$.

Proof. First, by condition (A_6), $f_0 = \varphi \in \left(\left(\frac{2\theta^*}{\theta}\right)^{p-1}, \infty\right)$, then for $\varepsilon = \varphi - \left(\frac{2\theta^*}{\theta}\right)^{p-1}$, there exists an adequately small positive number r, as $0 \leq u_2 \leq r$, $u_2 \neq 0$, $u_1 \leq H$, we have

$$f(u_1,u_2) \geq (\varphi-\varepsilon)(u_2)^{p-1} = \left(\frac{2\theta^*}{\theta}\right)^{p-1} (u_2)^{p-1},$$

thus when $\theta r \leq u_2 \leq r$, $u_1 \leq H$, we have

$$f(u_1,u_2) \geq \left(\frac{2\theta^*}{\theta}\right)^{p-1} (\theta r)^{p-1} = (2\theta^* r)^{p-1}. \tag{3.6}$$

Let $m = 2\theta^* > \theta^*$, so by (3.6), condition (A_1) holds.

Next, by condition (A_5): $f^\infty = \lambda \in \left[0, \left(\frac{\theta_*}{4}\right)^{p-1}\right)$, then for $\varepsilon = \left(\frac{\theta_*}{4}\right)^{p-1} - \lambda$, there exists an suitably big positive number $\rho \neq r$, as $u_2 \geq \rho$, $u_1 \leq H$, we have

$$f(u_1,u_2) \leq (\lambda+\varepsilon)(u_2)^{p-1} \leq \left(\frac{\theta_*}{4}\right)^{p-1} (u_2)^{p-1}. \tag{3.7}$$

If f is unbounded, by the continuity of f on $[0,\infty)^2$, then exists constant $R\ (\neq r\ ) \geq \rho$, and a point $(u_{01},u_{02}) \in [0,\infty)^2$ such that

$$\rho \leq u_{02} \leq R$$

and

$$f(u_1,u_2) \leq f(u_{01},u_{02}), \quad 0 \leq u_2 \leq R,\ u_1 \leq H.$$

Thus, by $\rho \leq u_{02} \leq R$, $u_1 \leq H$, we know

$$f(u_1,u_2) \leq f(u_{01},u_{02}) \leq \left(\frac{\theta_*}{4}\right)^{p-1} (u_{02})^{p-1} \leq \left(\frac{\theta_*}{4}R\right)^{p-1}.$$

Choose $M = \frac{\theta_*}{4} \in (0,\theta_*)$. Then, we have

$$f(u_1,u_2) \leq (MR)^{p-1},\ 0 \leq u_2 \leq R,\ u_1 \leq H.$$

If f is bounded, we suppose $f(u_1,u_2) \leq \overline{M}^{p-1}$, $u_2 \in [0,\infty)$, $\overline{M} \in R_+$, there exists an appropriately big positive number $R > \frac{4}{\theta_*}\overline{M}$, then choose $M = \frac{\theta_*}{4} \in (0,\theta_*)$, we have

$$f(u_1,u_2) \leq \overline{M}^{p-1} \leq \left(\frac{\theta_*}{4}R\right)^{p-1} = (MR)^{p-1},\ 0 \leq u_2 \leq R,\ u_1 \leq H.$$

Therefore, condition (A_2) holds. Therefore, by Theorem 3.1, we know that the results of Theorem 3.3 holds. The proof of Theorem 3.3 is complete. □

4. The Existence of Many Positive Solutions to (1.8) (1.9)

Next, we will discuss the existence of many positive solutions.

Theorem 4.1. *Suppose that conditions (H_1), (H_2), (H_3) and (A_2) in Theorem 3.1 hold. Assume that f also satisfy*

(A_7): $f_0 = +\infty$;

(A_8): $f_\infty = +\infty$.

Then, the (SBVP) (2.1), (2.2) have at last two solutions u_1, u_2 such that

$$0 < \|u_1\| < R < \|u_2\|.$$

Proof. First, by condition (A_7), for any $N > \frac{2}{\theta L}$, there exists a constant $\rho_* \in (0,R)$ such that

$$f(u_1,u_2) \geq (Nu_2)^{p-1},\ 0 < u_2 \leq \rho_*,\ u_1 \leq H. \tag{4.1}$$

Set $\Omega_{\rho_*} = \{u \in K : \|u\| < \rho_*\}$, for any $u \in \partial\Omega_{\rho_*}$, by (4.1) and Lemma 2.3, similar to the previous proof of Theorem 3.1, we can have from three perspectives

$$\|Tu\| \geq \|u\|,\ \forall\, u \in \partial\Omega_{\rho_*}.$$

Then by Theorem 1.1, we have

$$i(T,\Omega_{\rho_*},K) = 0. \tag{4.2}$$

Next, by condition (A_8), for any $\overline{N} > \frac{2}{\theta L}$, there exists a constant $\rho_0 > 0$ such that

$$f(u_1,u_2) \geq (\overline{N}u_2)^{p-1},\ u_2 > \rho_0,\ u_1 \leq H. \tag{4.3}$$

We choose a constant $\rho^* > \max\left\{R, \frac{\rho_0}{\theta}\right\}$, obviously $\rho_* < R < \rho^*$. Set $\Omega_{\rho^*} = \{u \in K : \|u\| < \rho^*\}$. For any $u \in \partial\Omega_{\rho^*}$, by Lemma 2.3, we have

$$u(t) \geq \theta\|u\| = \theta\rho^* > \rho_0,\ t \in [\theta, T-\theta].$$

Then by (4.3) and also similar to the previous proof of Theorem 3.1, we can also have from three perspectives

$$\|Tu\| \geq \|u\|,\ \forall\, u \in \partial\Omega_{\rho^*}.$$

Then by Theorem 1.1, we have

$$i(T, \Omega_{\rho^*}, K) = 0. \tag{4.4}$$

Finally, set $\Omega_R = \{u \in K : \|u\| < R\}$, For any $u \in \partial\Omega_R$, by (A_2), Lemma 2.3 and also similar to the latter proof of Theorem 3.1, we can also have

$$\|Tu\| \leq \|u\|,\ \forall\, u \in \partial\Omega_R.$$

Then by Theorem 1.1, we have

$$i(T, \Omega_R, K) = 1. \tag{4.5}$$

Therefore, by (4.2), (4.4), (4.5), $\rho_* < R < \rho^*$ we have

$$i(T, \Omega_R \setminus \overline{\Omega}_{\rho_*}, K) = 1,\ i(T, \Omega_{\rho^*} \setminus \overline{\Omega}_R, K) = -1.$$

Then T have fixed point $u_1 \in \Omega_R \setminus \overline{\Omega}_{\rho_*}$, and fixed point $u_2 \in \Omega_{\rho^*} \setminus \overline{\Omega}_R$. Obviously, u_1, u_2 are all positive solutions of problem (2.1),(2.2) and $\rho_* < \|u_1\| < R < \|u_2\| < \rho^*$. The proof of Theorem 4.1 is complete. □

Theorem 4.2. *Suppose that conditions (H_1), (H_2), (H_3) and (A_1) in Theorem 3.1 hold. Assume that f also satisfy*

(A_9): $f^0 = 0$;

(A_{10}): $f^\infty = 0$.

Then, the (SBVP) (2.1), (2.2) have at last two solutions u_1, u_2 such that $0 < \|u_1\| < r < \|u_2\|$.

Proof. First, by $f^0 = 0$, for $\varepsilon_1 \in (0, \theta_*)$, there exists a constant $\rho_* \in (0, r)$ such that

$$f(u_1, u_2) \leq (\varepsilon_1 u_2)^{p-1},\ 0 < u_2 \leq \rho_*,\ u_1 \leq H. \tag{4.6}$$

Set $\Omega_{\rho_*} = \{u \in K : \|u\| < \rho_*\}$, for any $u \in \partial\Omega_{\rho_*}$, by (4.6), we have

$$\begin{aligned}
\|Tu\| &= (Tu)(\sigma) \\
&\leq \phi_q\left(\frac{\beta}{\alpha}\int_\xi^\sigma g(s)f(u(s-\tau)+h(s-\tau),u(s))\nabla s\right) \\
&\quad +\int_0^T \phi_q\left(\int_s^\sigma g(r)f(u(r-\tau)+h(r-\tau),u(r))\nabla r\right)\Delta s \\
&\leq \phi_q\left(\frac{\beta}{\alpha}\int_\xi^\sigma g(s)f(u(s-\tau)+h(s-\tau),u(s))\nabla s\right) \\
&\quad +T\phi_q\left(\int_0^T g(r)f(u(r-\tau)+h(r-\tau),u(r))\nabla r\right) \\
&\leq \left(T+\phi_q\left(\frac{\beta}{\alpha}\right)\right)\varepsilon_1\rho_*\phi_q\left(\int_0^T g(r)\nabla r\right) \leq \rho_* = \|u\|.
\end{aligned}$$

i.e.,

$$\|Tu\| \leq \|u\|,\ \forall\, u \in \partial\Omega_{\rho_*}.$$

Then by Theorem 1.1, we have

$$i(T,\Omega_{\rho_*},K) = 1. \tag{4.7}$$

Next, let $f^*(x) = \max\limits_{0\leq u_2\leq x, u_1\leq H} f(u_1,u_2)$, note that $f^*(x)$ is monotone increasing with respect to $x \geq 0$. Then from $f^\infty = 0$, it is easy to see that

$$\lim_{x\to\infty}\frac{f^*(x)}{x^{p-1}} = 0.$$

Therefore, for any $\varepsilon_2 \in (0,\theta_*)$, there exists a constant $\rho^* > r$ such that

$$f^*(x) \leq (\varepsilon_2 x)^{p-1},\ x \geq \rho^*. \tag{4.8}$$

Set $\Omega_{\rho^*} = \{u \in K : \|u\| < \rho^*\}$, for any $u \in \partial\Omega_{\rho^*}$, by (4.8), we have

$$\begin{aligned}
\|Tu\| &= (Tu)(\sigma) \\
&\leq \phi_q\left(\frac{\beta}{\alpha}\int_\xi^\sigma g(s)f(u(s-\tau)+h(s-\tau),u(s))\nabla s\right) \\
&\quad +\int_0^T \phi_q\left(\int_s^\sigma g(r)f(u(r-\tau)+h(r-\tau),u(r))\nabla r\right)\Delta s \\
&\leq \phi_q\left(\frac{\beta}{\alpha}\int_\xi^\sigma g(s)f(u(s-\tau)+h(s-\tau),u(s))\nabla s\right) \\
&\quad +T\phi_q\left(\int_0^T g(r)f(u(r-\tau)+h(r-\tau),u(r))\nabla r\right) \\
&\leq \left(T+\phi_q\left(\frac{\beta}{\alpha}\right)\right)\phi_q\left(\int_0^T g(r)f^*(\rho^*)\nabla r\right) \\
&\leq \left(T+\phi_q\left(\frac{\beta}{\alpha}\right)\right)\varepsilon_2\rho^*\phi_q\left(\int_0^T g(r)\nabla r\right) \leq r^* = \|u\|.
\end{aligned}$$

i.e.,

$$\|Tu\| \leq \|u\|, \ \forall\, u \in \partial\Omega_{\rho^*}.$$

Then by Theorem 1.1, we have

$$i(T, \Omega_{\rho^*}, K) = 1. \tag{4.9}$$

Finally, set $\Omega_r = \{u \in K : \|u\| < r\}$, For any $u \in \partial\Omega_r$, by (A_1), Lemma 2.3 and also similar to the previous proof of Theorem 3.1, we can also have

$$\|Tu\| \geq \|u\|, \ \forall\, u \in \partial\Omega_r.$$

Then by Theorem 1.1, we have

$$i(T, \Omega_r, K) = 0. \tag{4.10}$$

Therefore, by (4.7), (4.9), (4.10), $\rho_* < r < \rho^*$, we have

$$i(T, \Omega_r \setminus \overline{\Omega}_{\rho_*}, K) = -1, \ i(T, \Omega_{\rho^*} \setminus \overline{\Omega}_r, K) = 1.$$

Then T have fixed point $u_1 \in \Omega_r \setminus \overline{\Omega}_{\rho_*}$, and fixed point $u_2 \in \Omega_{\rho_*} \setminus \overline{\Omega}_r$. Obviously, u_1, u_2 are all positive solutions of problem (1.8),(1.9) and $\rho_* < \|u_1\| < r < \|u_2\| < \rho^*$. The proof of Theorem 4.2 is complete. □

Similar to Theorem 3.1, we also obtain the following Theorems.

Theorem 4.3. *Suppose that conditions (H_1), (H_2), (H_3) and (A_2) in Theorem 3.1, (A_4) in Theorem 3.2 and (A_6) in Theorem 3.3 hold. Then, the (SBVP) (2.1), (2.2) have at last two solutions u_1, u_2 such that $0 < \|u_1\| < R < \|u_2\|$.*

Theorem 4.4. *Suppose that conditions (H_1), (H_2), (H_3) and (A_1) in Theorem 3.1, (A_3) in Theorem 3.2 and (A_5) in Theorem 3.3 hold. Then, the (SBVP) (2.1), (2.2) have at last two solutions u_1, u_2 such that $0 < \|u_1\| < r < \|u_2\|$.*

5. The Existence of Many Positive Solutions to (1.8) (1.10)

In that follows, we will deal with problem (1.8), (1.10), the method is similar to that in Section 3 and Section 4, so we omit many proof in this section.

Theorem 5.1. *Suppose that condition (H_1), (H_2), (H_3), (H_4) hold. Assume that f also satisfy*

(A_1'): $f(u_1, u_2) \geq (mr)^{p-1}$, for $\theta r \leq u_2 \leq r$, $0 \leq u_1 \leq H$;

(A_2'): $f(u_1, u_2) \leq (MR)^{p-1}$, for $0 \leq u_2 \leq R$, $0 \leq u_1 \leq H$,

where $m \in (\theta^*, \infty)$, $M \in (0, \theta_{**})$. Then, the (SBVP) (1.8), (1.10) has a solution u such that $\|u\|$ lies between r and R.

Theorem 5.2. *Suppose that condition (H_1), (H_2), (H_3), (H_4) hold. Assume that f also satisfy*

(A_3'): $f^0 = \varphi \in \left[0, \left(\frac{\theta_{**}}{4}\right)^{p-1}\right)$;

(A_4'): $f_\infty = \lambda \in \left(\left(\frac{2\theta^*}{\theta}\right)^{p-1}, \infty\right)$.

Then, the (SBVP) (1.8), (1.10) has a solution u which is bounded in $\|\cdot\|$.

Theorem 5.3. *Suppose that condition (H_1), (H_2), (H_3), (H_4) hold. Assume that f also satisfy*

(A_5'): $f^\infty = \lambda \in \left[0, \left(\frac{\theta_{**}}{4}\right)^{p-1}\right)$;

(A_6'): $f_0 = \varphi \in \left(\left(\frac{2\theta^*}{\theta}\right)^{p-1}, \infty\right)$.

Then, the (SBVP) (1.8), (1.10) has a solution u which is bounded in $\|\cdot\|$.

Theorem 5.4. *Suppose that conditions (H_1), (H_2), (H_3), (H_4) and (A_2') in Theorem 5.1 hold. Assume that f also satisfy*

(A_7'): $f_0 = +\infty$;

(A_8'): $f_\infty = +\infty$.

Then, the (SBVP) (1.8), (1.10) have at least two solutions u_1, u_2 such that

$$0 < \|u_1\| < R < \|u_2\|.$$

Theorem 5.5. *Suppose that conditions (H_1), (H_2), (H_3), (H_4) and (A_1') in Theorem 5.1 hold. Assume that f also satisfy*

(A_9'): $f^0 = 0$;

(A_{10}'): $f^\infty = 0$.

Then, the (SBVP) (1.8), (1.10) have at least two solutions u_1, u_2 such that $0 < \|u_1\| < r < \|u_2\|$.

Theorem 5.6. *Suppose that conditions (H_1), (H_2), (H_3), (H_4) and (A_2) in Theorem 5.1, (A_4) in Theorem 5.2 and (A_6') in Theorem 3.3 hold. Then, the (SBVP) (1.8), (1.10) have at least two solutions u_1, u_2 such that $0 < \|u_1\| < R < \|u_2\|$.*

Theorem 5.7. *Suppose that conditions (H_1), (H_2), (H_3), (H_4) and (A_1') in Theorem 5.1, (A_3') in Theorem 5.2 and (A_5') in Theorem 3.3 hold. Then, the (SBVP) (1.8), (1.10) have at least two solutions u_1, u_2 such that $0 < \|u_1\| < r < \|u_2\|$.*

6. Application

EXAMPLE 6.1. Consider the following 3-order singular boundary value problem (SBVP) with p-Laplacian

$$\begin{cases} (\phi_p(u^\Delta))^\nabla(t) + \dfrac{1}{20}t^{-\frac{1}{2}}u^{\frac{1}{2}}(t)\cdot \\ \qquad \left[\dfrac{1}{5} + \dfrac{\frac{94}{5}e^{2u(t)}}{120u(t-1)+7e^{u(t)}+e^{2u(t)}}\right] = 0, \quad 0<t<1, \\ u(t) = t^2-1, \ -1 \le t \le 0, \\ \phi_p(u(0)) - \phi_p(u^\Delta(\frac{1}{4})) = 0, \ \phi_p(u(1)) + \delta\phi_p(u^\Delta(\frac{1}{2})) = 0, \end{cases} \tag{6.1}$$

where

$$\alpha = \gamma = 1, \ \beta = 1, \ p = \frac{3}{2}, \ \ \delta \ge 0, \ \xi = \frac{1}{4}, \ \ \eta = \frac{1}{2}, \ \theta = \frac{1}{4}, \ \tau = T = 1.$$

So, by Lemma 2.3, we discuss the following SBVP:

$$\begin{cases} (\phi_p(u^\Delta))^\nabla(t) + \dfrac{1}{20}t^{-\frac{1}{2}}u^{\frac{1}{2}}(t)\cdot \\ \left[\dfrac{1}{5} + \dfrac{\frac{94}{5}e^{2u(t)}}{120[u(t-1)+h(t-1)]+7e^{u(t)}+e^{2u(t)}}\right] = 0, \quad 0<t<1, \\ u(t) = 0, \ -1 \le t \le 0, \\ \phi_p(u(0)) - \phi_p(u^\Delta(\frac{1}{4})) = 0, \ \phi_p(u(1)) + \delta\phi_p(u^\Delta(\frac{1}{2})) = 0, \end{cases} \tag{6.2}$$

where

$$h(t) = \begin{cases} t^2-1, & -1 \le t \le 0, \\ 0, & 0 \le t \le 1, \end{cases}, \quad g(t) = \frac{1}{20}t^{-\frac{1}{2}}, \ \zeta(t) = t^2-1,$$

$$f(u_1, u_2) = (u_2)^{\frac{1}{2}}\left[\frac{1}{5} + \frac{\frac{94}{5}e^{2u_2}}{120u_1+7e^{u_2}+e^{2u_2}}\right].$$

Then obviously,

$$q = 3, \ H = \max_{-1\le t\le 0}|\zeta(t)| = 1, \ f^0 = \varphi = \lim_{u_2\to 0^+}\max_{0\le u_1\le 1}\frac{f(u_1, u_2)}{u_2^{p-1}} = \frac{51}{20},$$

$$f_\infty = \lambda = \lim_{u_2\to\infty}\min_{0\le u_1\le 1}\frac{f(u_1, u_2)}{u_2^{p-1}} = \frac{95}{5}, \ \int_0^T g(t)\nabla t = \frac{1}{10},$$

so conditions (H_1), (H_2), (H_3) hold.

Next,

$$\theta_* = \frac{1}{\left(T+\phi_q\left(\frac{\beta}{\alpha}\right)\right)\phi_q\left(\int_0^T g(r)\nabla r\right)} = 50,$$

then $\left(\frac{\theta_*}{4}\right)^{p-1} = \frac{5\sqrt{2}}{2} > \frac{51}{20}$, i.e. $\varphi \in \left[0, \left(\frac{\theta_*}{4}\right)^{p-1}\right)$, so conditions (A_3) holds.

For $\theta = \frac{1}{4}$, it is easy see by calculating that

$$L = \min_{t\in[\theta,T-\theta]} A(t) = \frac{1}{16}\left(\frac{7}{36}+\frac{\sqrt{3}}{3}\right).$$

Because of

$$\left(\frac{2\theta^*}{\theta}\right)^{p-1} = 96 \times \left(\frac{1}{7+12\sqrt{3}}\right)^{\frac{1}{2}} < \frac{95}{5},$$

then

$$\lambda \in \left(\left(\frac{2\theta^*}{\theta}\right)^{p-1}, \infty\right),$$

so conditions (A_4) holds. Then by Theorem 3.2, SBVP (6.2) has at least a positive solution $u(t)$. So, $\overline{u}(t) = u(t)+h(t)$, $-1 < t < 1$ is the positive solution of SBVP (6.1).

EXAMPLE 6.2. Consider the following 3-order singular boundary value problem (SBVP) with p-Laplacian

$$\begin{cases} (\phi_p(u^\Delta))^\nabla(t) + \frac{1}{64\pi^4}t^{-\frac{1}{2}}(1-t)\left[u(t-1)+u^2(t)+u^4(t)\right] = 0,\ 0<t<1, \\ u(t) = -te^t,\ -1 \le t \le 0, \\ 2\phi_p(u(0)) - \phi_p(u^\Delta(\frac{1}{4})) = 0,\ \phi_p(u(1)) + \delta\phi_p(u^\Delta(\frac{1}{2})) = 0, \end{cases} \tag{6.3}$$

where

$$\beta = \gamma = 1,\ \alpha = 2,\ p = 4,\ \delta \ge 0,\ p = 4,\ \xi = \frac{1}{4},\ \eta = \frac{1}{3},\ \theta = \frac{1}{4},\ \tau = T = 1,$$

So, by Lemma 2.3, we discuss the following SBVP:

$$\begin{cases} (\phi_p(u^\Delta))^\nabla(t) + \frac{1}{64\pi^4}t^{-\frac{1}{2}}(1-t)\Big[\ [u(t-1)+h(t-1)]+ \\ \qquad\qquad u^2(t)+u^4(t)\Big] = 0,\ \ 0<t<1, \\ u(t) = 0,\ -1 \le t \le 0, \\ 2\phi_p(u(0)) - \phi_p(u^\Delta(\frac{1}{4})) = 0,\ \phi_p(u(1)) + \delta\phi_p(u^\Delta(\frac{1}{2})) = 0, \end{cases} \tag{6.4}$$

where

$$h(t) = \begin{cases} \zeta(t), & -1 \le t \le 0, \\ 0, & 0 \le t \le 1, \end{cases} \quad ; \quad \zeta(t) = -te^t.$$

$$g(t) = \frac{1}{64\pi^4} t^{-\frac{1}{2}}(1-t), \quad f(u_1, u_2) = u_1 + u_2^{\frac{2}{3}} + u_2^{\frac{4}{3}}.$$

Then obviously,

$$q = \frac{4}{3}, \quad \int_0^1 g(t)\nabla t = \frac{1}{64\pi^3}, \quad H = \max_{-1\le t\le 0} |\zeta(t)| = e, \quad f_\infty = +\infty, \quad f_0 = +\infty,$$

so conditions (H_1), (H_2), (H_3), (A_7), (A_8) hold.

Next,

$$\phi_q\left(\int_0^1 a(t)\nabla t\right) = \frac{1}{4\pi}, \quad \theta_* = \frac{4\pi}{1+\sqrt[3]{4}},$$

we choose $R = 3$, $M = 2$ and for $\theta = \frac{1}{4}$, because of the monotone increasing of $f(u_1, u_2, u_3)$ on $[0, \infty)^3$, then

$$f(u_1, u_2) \le f(e, 3) = e + 90, \; 0 \le u_2 \le 3, \; 0 \le u_1 \le e.$$

Therefore, by

$$M \in (0, \theta_*), \quad (MR)^{p-1} = (6)^3 = 216,$$

we know

$$f(u_1, u_2, u_3) \le (MR)^{p-1}, \; 0 \le u_2 \le 3, \; 0 \le u_1 \le e,$$

so conditions (A_2) holds. Then by Theorem 4.1, SBVP (6.4) has at least two positive solutions v_1, v_2 and $0 < \|v_1\| < 3 < \|v_2\|$. Then, by Lemma 2.3, $\overline{v}_1(t) = v_1(t) + h(t)$, $\overline{v}_2(t) = v_2(t) + h(t)$, $t \in (-1, 1)$ are the positive solutions of the SBVP (6.3).

References

[1] C. H. Leggett, L. R. Lee, Multiple positive fixed points of nonlinear operators on ordered Banach Spaces, *Indiana University Mathematics Journal*, **28** (1979), 637-688.

[2] D. Guo, V. Lakshmikantham, Nonlinear Problems in Abstract Cone. Academic Press, Sandiego, 1988.

[3] D. Guo, V. Lakshmikantham and X. Liu, *Nonlinear Integral Equations in Abstract Spaces.* Kluwer Academic Publishers, 1996.

[4] D. R. Anderson, Solutions to second-order three-point problems on time scales, *J. Differ. Equ. Appl.*, **8** (2002), 673-688.

[5] E. R. Kaufmann, Positive solutions of a three-point boundary value problem on a time scale, Eletron. *J. Diff. Equ.* , **82** (2003), 1-11.

[6] F. M. Atici, G. Sh. Guseinov; On Green's functions and positive solutions for boundary value problems on time scales, *J. Comput. Appl. Math.*, **141** (2002), 75-99.

[7] H. R. Sun, W. T.Li, Positive solutions for nonlinear three-point boundary value problems on time scales, *J. Math. Anal. Appl.*, **299** (2004), 508-524.

[8] H. Su, Z. Wei, F. Xu, The Existence of Positive Solutions for Nonlinear Singular Boundary Value System with p-Laplacian, *J. Appl. Math. Comp.*, **181**(2006), 826-836.

[9] H. Su, Z. Wei, F. Xu, The Existence of Countably Many Positive Solutions For A System of Nonlinear Singular Boundary Value Problems With The p-Laplacian Operator, *J. Math. Anal. Appl.*, **325** (2007), 319-332.

[10] G. B. Gustaf, K. Schmitt, Nonzero solutions of boundary value problems for second-order ordinary and delay differential equations, *J. Diff. Equ.*, **12**(1972), 129-147.

[11] L. H. Erbe, Q. K. Kong, boundary value Problems for singular second-order functional differential equations, *J. Comput. Appl. Math.*, **53** (1994), 377-388.

[12] M. Bohner, A. Peterson, Advances in Dynamic Equations on time scales, Birkh¨user Boston, Cambridge, MA, 2003.

[13] J. Henderson, H. B. Thompson, Multiple symmetric positive solutions for a second-order boundary value problem, *Proc. Amer. Math. Soc.*, **128** (2000), 2373-2379.

[14] J. W. Lee, D. O'Regan, Existence results for differential delay equations-I, *J. Diff. Equ.*, **102** (1993), 342-359.

[15] J. W. Lee, D. O'Regan, Existence results for differential delay equations-II,*Nonlinear Analysis (TMA)*, **102** (1993), 342-359.

[16] J. P. Sun, Existence of solution and positive solution of BVP for nonlinear third-order dynamic equation, *Nonlinear Analysis (TMA)*, **64** (2006) 629-636.

[17] J. Liang, T. J. Xiao, Z. C. Hao, Positive solutions of singular differential equations on measure chains, *Comput. Math. Appl.*, **49** (2005), 651–663.

[18] K. L. Boey, Patricia J.Y.Wong, Existence of triple positive solutions of two-point right focal boundary value problems on time scales, *Comput. Math. Appl.*, **50** (2005), 1603-1620.

[19] R. P. Agarwal, D. O'Regan, Nonlinear boundary value Problems on time scales, *Nonlinear Analysis (TMA)*, **44** (2001), 527-535.

[20] R. I. Avery, D. R. Anderson; Existence of three positive solutions to a second-order boundary value problem on a measure chain, *J. Comput. Appl. Math.*, **141** (2002), 65-73.

[21] R. I. Avery, A. C. Peterson, Three positive fixed point of nonlinear operators on ordered Banach spaces, *Comput. Math. Appl.*, **42** (2001), 313-322.

[22] S. Hilger, Analysis on measure chains-a unified approach to continuous and discrete calculus, *Results Math.*, **18** (1990), 18-56.

[23] V. Lakshmikantham, S. Sivasundaram, B. Kaymakcalan, *Dynamic systems on measure chains,* Kluwer Academic Publishers, boston, 1996.

[24] Y. Y. Wang, W. X. Zhao and W. G. Ge, Multiple positive solutions for boundary value problems of second order delay differential equations with one-dimensional p-Laplacian, *J. Math. Anal. Appl.*, **326** (2007), 641-654.

[25] Y. H. Lee, I. Sim, Global bifurcation phenomena for singular one-dimensional p-Laplacian, *J. Diff. Equ.*, **229**, 1 (2006), 229-256.

[26] Z. M. He, Double positive solutions of three-point boundary value problems for p-Laplacian dynamic equations on time scales, *J. Comput. Appl. Math.*, **182** (2005), 304-315.

[27] Z. C. Hao, J. Liang, T. J. Xiao, Existence results for time scale boundary value problem. *J. Comput. Appl. Math.*, **1, 197** (2006), 156–168.

[28] Z. C. Hao, T. J. Xiao, J. Liang, Existence of positive solutions for singular boundary value problem on time scales, *J. Math. Anal. Appl.* **325** (2007), 517–528.

In: Progress in Evolution Equations
Editor: Gaston M. N'Guerekata, pp. 183-206
ISBN: 978-1-60456-328-3

Chapter 11

LEVINSON THEOREM FOR 2 X 2 SYSTEM AND APPLICATIONS TO THE ASYMPTOTIC STABILITY AND SCHRÖDINGER EQUATION

Gro R. Hovhannisyan*
Kent State University, Stark Campus
6000 Frank Ave. NW, Canton, OH 44720-7599, USA

Abstract

We prove new asymptotical stability and instability theorems for non autonomous 2×2 system of first-order differential equations by using a new version of the classical Levinson asymptotic theorem for 2×2 systems. The proof of this version is based on the construction of approximate fundamental solution of the original system in the special form with unknown phase function and the error estimates formulated in the terms of generalized characteristic functional. In the case of constant matrix A generalized characteristic functional turns to the usual characteristic polynomial and by choosing phase functions as eigenvalues of the matrix A the error could be eliminated. As another application we derive a transition probability formula for the two level atom in the external electromagnetic field described by Schrödinger system.

Key words: Asymptotic stability, asymptotic solutions, characteristic function, fundamental matrix, integral representation, stability estimates, first order system of differential equations, Schrödinger equation, two level atom, transition probability

1. Main Results

Consider the system of linear ordinary differential equations

$$u'(t) = A(t)u(t), \quad t > T, \tag{1.1}$$

where $u(t)$ is a 2-vector function, and

$$A(t) = \begin{pmatrix} a_{11}(t) & a_{12}(t) \\ a_{21}(t) & a_{22}(t) \end{pmatrix}$$

*E-mail address: ghovhann@kent.edu

is a 2×2 matrix-function differentiable by $t \in (T, \infty)$.

The rest state $u(t) = 0$ of (1.1) is called stable if for any $\varepsilon > 0$ there exists $\delta(T, \varepsilon) > 0$ such that if $|u(T)| < \delta(T, \varepsilon)$ then $|u(t)| < \varepsilon$ for all $t \geq T$. The rest state $u(t) = 0$ of (1.1) is called asymptotically stable if it is stable, and attractive:

$$\lim_{t \to \infty} u(t) = 0 \tag{1.2}$$

for every solution of (1.1).

The usual method of investigation of asymptotic stability of differential equations is the Lyapunov's method that uses energy functions and Lyapunov stability theorems [2, 3, 5, 13, 6, 8, 12, 14, 16, 17].

Here we continue the development of another approach started in [9, 10, 11]. This approach based on the usage of different asymptotic solutions [15] investigated in [7] (instead of construction of energy functions in Lyapunov's method), and the error estimates [4]. To prove stability inequalities for system (1.1) we use a new version of the Levinson theorem (see Theorem 2.1 below) for 2×2 systems about asymptotic solutions with explicit estimate of the error term, which may be used also for finding actual asymptotic solutions (see Remark 2.2). The classical Levinson theorem uses a decomposition of the right side matrix function $A = B + R$, where the leading matrix B is diagonal and the perturbation matrix R is integrable. In our version we prove the error estimate using the decomposition with the leading matrix B such that corresponding system is explicitly solvable. To prove the error estimates we use a construction of approximate fundamental matrix of system (1.1) in the special form with an unknown phase function $\varphi(t)$, which may be chosen by using known asymptotic solutions. In the paper we illustrate on examples some choices of the function $\varphi(t)$. For instance one of the choices of $\varphi(t)$ is based on the Green-Liouville asymptotic solutions (see (1.30))

Examples show that asymptotic solutions approach works better than Lyapunov's method for the systems with complex valued coefficients (see Example 1.3).

There is a bridge connecting asymptotic solutions approach with Lyapunov's method: when the asymptotic fundamental matrix solution Ψ of (1.1) is chosen the appropriate energy function of Lyapunov may be constructed by the formula

$$E(t, u(t)) = ||\Psi^{-1}(t) u(t)||^2.$$

Indeed $E(t) \geq 0$, and if Ψ is the exact fundamental matrix function of (1.1) then conservation law $E'(t) = 0$ is true.

Furthermore we deduce the transition probability formula for the Schrödinger system that describes the interaction of two-level atom with electromagnetic field, and we give the comparison of two approximate solutions.

Denote by $L^1(T, \infty)$ the class of Lebesgue integrable in (T, ∞) functions and by $C^1(T, \infty)$ the class of differentiable functions on (T, ∞).

Denote

$$TrA(t) = a_{11}(t) + a_{22}(t), \quad |A(t)| = det(A(t)).$$

Asymptotic behavior of solutions of autonomous systems is described by eigenvalues of corresponding matrix A. The key step of finding behavior of solutions of non autonomous system (1.1) is to find the phase functions θ_j that are minimizing (or eliminating) the generalized characteristic functional

$$Char(\theta) = -\theta^2 - \theta' + \theta\left(Tr(A) + \frac{a'_{12}}{a_{12}}\right) - |A| - \frac{W[a_{11}, a_{12}]}{a_{12}}, \tag{1.3}$$

where $W[\cdot,\cdot]$ is a Wronskian:

$$W[a,b] = a(t)b'(t) - a'(t)b(t).$$

Here and further in the text we often suppressed dependence on t for simplicity.

Note that in the case of constant matrix A in (1.1) the characteristic functional (1.3) turns to the usual characteristic polynomial:

$$Char(\theta) = -\theta^2 + \theta TrA - |A|,$$

so we can eliminate the characteristic functional by choosing phase functions as eigenvalues of the matrix A.

Using Liouville's formula that gives the connection between the functions θ_j we can start from a single unknown phase function $\xi(t) \in C^2[T,\infty)$, and the matrix function A(t) to construct the phase functions θ_j :

$$\theta_{1,2}(t) = \pm\xi(t) + \frac{Tr(A(t))}{2} + \frac{a'_{12}(t)}{2a_{12}(t)} - \frac{\xi'(t)}{2\xi(t)}. \tag{1.4}$$

Introducing the shifted phase function $\varphi(t) = TrA/2 - \theta_1$ we define the functional

$$H(\varphi(t)) = Char(TrA/2 - \varphi) = \left(\frac{Tr(A)}{2}\right)^2 - |A| + a_{12}\left(\frac{2\varphi + a_{11} - a_{22}}{2a_{12}}\right)' - \varphi^2. \tag{1.5}$$

Note that the function $\varphi(t)$ is connected with the function $\xi(t) = (\theta_1 - \theta_2)/2$ via transformation

$$\xi(t) = \frac{a_{12}(t)e^{\int_T^t 2\varphi(z)dz}}{2(C - \int_T^t a_{12}(s)e^{\int_T^s 2\varphi(z)dz}ds)}, \quad C = const. \tag{1.6}$$

Theorem 1.1. *Assume $A \in C^1(T,\infty), a_{12} \in C^2(T,\infty)$, $a_{12}(t) \neq 0$ on (T,∞), and there exists a function $\varphi \in C^1(T,\infty)$, such that*

$$\int_T^\infty \left|\frac{H(\varphi(s))}{\xi(s)}\right| e^{\pm 2\int_T^s \Re[\xi(y)]dy} ds < \infty. \tag{1.7}$$

Then the rest state of (1.1) is asymptotically stable if and only if

$$\int_T^\infty \Re[\theta_j(s)ds = -\infty, \quad j = 1,2, \tag{1.8}$$

$$\lim_{t\to\infty}\left|\frac{\theta_j(t,u) - a_{11}(t,u)}{a_{12}(t,u)}\right|(t)e^{\int_T^t \Re[\theta_j(y,u)]dy} = 0, \quad j = 1,2. \tag{1.9}$$

Remark 1.1. *The best choice of the function* φ *in Theorem 1.1 is such that* $H(\varphi(t)) \equiv 0$, *which means that the error of approximation is equal to zero and condition (1.7) disappears. It is well known that Riccati equation* $H(\varphi(t)) = 0$ *can not be solved analytically in general case, so we don't expect to find the best* φ *in general, but in many cases using theory of asymptotic solutions [7] one can find function* φ, *such that the function* $\frac{H(\varphi)}{\xi}$ *is so small that condition (1.7) is satisfied.*

Remark 1.2. *The main function* φ *could be constructed also by using specific asymptotic fundamental matrix solution* Ψ *of (1.1). Indeed (see formula (2.11)) from the given* $\Psi = \begin{bmatrix} \Psi_{11} & \Psi_{12} \\ \Psi_{21} & \Psi_{22} \end{bmatrix}$ *the phase function* ξ *may be found from the formula*

$$\xi(t) = \frac{d}{dt} ln\left(\frac{\Psi_{12}(t)}{\Psi_{11}(t)}\right).$$

and the function φ *from (1.6). For example, using Liouville-Green (or WKB) asymptotic solutions we deduce Corollary 1.6 below about asymptotic stability from Theorem 1.1.*

In the case $a_{12} \equiv 0$ Theorem 1.1 is not applicable, but system (1.1) can be solved explicitly, so the following theorem is trivial.

Theorem 1.2. *Assume* $a_{12}(t) \equiv 0, \quad A \in L_1(T,\infty)$. *Then the rest state of (1.1) is asymptotically stable if and only if*

$$\lim_{t\to\infty} \int_T^t a_{jj}(s)ds = -\infty, \quad j = 1,2, \tag{1.10}$$

$$\lim_{t\to\infty} \left(e^{\int_T^t a_{11}(y)dy} \int_T^t a_{21}(s) e^{\int_t^s (a_{11}-a_{22})(y)dy} ds \right) = 0, \quad j = 1,2. \tag{1.11}$$

Example 1.1. *Consider the system of linear equations*

$$u_1'(t) = f(t)u_2(t), \quad u_2'(t) = -g(t)u_2(t),$$

$$u_1(t_0) = u_{10}, \quad u_2(t_0) = u_{20},$$

with

$$f(t) = t^{-2a}, \quad g(t) = bt^{2a-2\gamma}, \quad 1 < \gamma < 2a, \quad b \neq 0.$$

For this example asymptotic stability follows from Theorem 1.1.

In the cases when one of the quantities $\pm\Re[\xi(t)]$ is unbounded condition (1.7) is very restrictive. In the next Theorem 1.3 under additional conditions (1.13), (1.14) below we prove the asymptotic stability of the rest state under condition (1.12) less restrictive than (1.7).

Theorem 1.3. *Assume* $a_{12}(t)$ *is not equal to zero for* $t > T$, *and there exists a function* $\varphi \in C^1(T,\infty)$, *such that (1.8) and*

$$\int_T^\infty \left| \frac{H(\varphi(s))}{\xi(s)} \right| ds < \infty. \tag{1.12}$$

$$\Re[\theta_j(t)] \le 0, \quad j = 1,2, \tag{1.13}$$

$$\left|\frac{\theta_j(t) - a_{11}(t)}{a_{12}(t)}\right| \le C, \quad j = 1,2, \tag{1.14}$$

are satisfied for all $t \ge T$.

Then the rest state of (1.1) is asymptotically stable.

Remark 1.3. *If for some positive number p we have*

$$\Re[\xi(t)] \le 0, \quad \Re[\theta_2(t)] \le -p < 0, \quad t > T, \tag{1.15}$$

then condition (1.8), (1.13) of theorem 1.3 may be removed, because they follow from condition (1.15).

Introduce the functions

$$\xi(t) = -\frac{a_{12}(t)}{2\int_T^t a_{12}(s) e^{\int_t^s (2S_{n+1} + a_{22} - a_{11})dy} ds}, \tag{1.16}$$

$$S_0 \equiv 0, \quad S_{n+1}(t) = a_{12}(t) \int_T^t \frac{(S_n^2 - a_{12}a_{21})(s)}{a_{12}(s)} e^{\int_t^s (a_{11} - a_{22})dy} ds, \quad n = 0,1,\dots. \tag{1.17}$$

Corollary 1.4. *Assume that the matrix-function $A(t)$ is real valued, and for some $T_1 > T$*

$$a_{12}(t) > 0, \quad t \ge T_1, \tag{1.18}$$

$$\int_{T_1}^{\infty} \left|\frac{S_{n+1}^2(t) - S_n^2(t)}{\xi(t)}\right| dt < \infty, \quad \textit{for some } n, \tag{1.19}$$

$$\int_{T_1}^{\infty} (2\xi + S_{n+1} - a_{11})(t)dt = \infty, \tag{1.20}$$

$$2\xi(t) + S_{n+1}(t) - a_{11}(t) \ge 0 \quad t > T_1, \tag{1.21}$$

$$\left|\frac{S_{n+1}}{a_{12}}\right|(t) \le const, \quad t \ge T_1. \tag{1.22}$$

Then (1.1) is asymptotically stable.

Example 1.2. *Consider the linear system (1.1) with*

$$A(t) = \begin{pmatrix} 0 & 1 \\ -g(t) - 2f'(t) & -2f(t) \end{pmatrix}.$$

Denote

$$S_1(t) = \int_T^t (g(s) + 2f'(s)) e^{\int_t^s 2f(y)dy} ds.$$

If $f \in C^1(T,\infty)$ and for some numbers g_0, f_0

$$0 < f_0 \le f(t), \quad 0 \le g(t) + 2f'(t) \le g_0 < 2f_0^2, \quad t \ge T \tag{1.23}$$

$$\int_{T_1}^{\infty} S_1(t)dt = \infty, \tag{1.24}$$

$$S_1(t) + 2\xi(t) \geq 0, \quad \int_{T_1}^{\infty} \left| \frac{S_1^2(s)}{\xi(s)} \right| ds < \infty, \tag{1.25}$$

then the problem (1.1) is asymptotically stable because all conditions of Corollary 1.4 are satisfied.

Note that it is well known [19] that in the large damping case (1.23) Wintner-Smith condition (1.24) is necessary and sufficient condition of asymptotic stability. So it is possible to get rid of extra conditions (1.25), but we don't know if it could be done in this approach.

Corollary 1.5. *Assume that the matrix-function $A(t)$ is real valued, $a_{12}(t) > 0$ on (T, ∞), for all $t \in (T, \infty)$, and*

$$\int_T^{\infty} \int_T^t \left| a_{21}(s) e^{\int_t^s (a_{22}(y) - a_{11}(y))dy} \right| ds dt < \infty, \tag{1.26}$$

$$\int_T^{\infty} \left(a_{11}(s) + \frac{a_{12}(s)}{\int_T^t a_{12}(s) e^{\int_t^s (a_{22} - a_{11})(y)dy} ds} \right) dt = -\infty, \tag{1.27}$$

$$a_{11}(t) + \frac{a_{12}(t)}{\int_T^t a_{12}(s) e^{\int_t^s (a_{22}(y) - a_{11}(y)dy} ds} \leq 0. \tag{1.28}$$

Then (1.1) is asymptotically stable.

Corollary 1.6. *Assume $a_{12}(t) \neq 0$ on (T, ∞), $A \in C^2(T, \infty)$, $a_{22} - a_{11}, a_{12} \in C^3(T, \infty)$, and*

$$\int_T^{\infty} |k'(t) + k^2(t)\xi(t)| e^{\pm 2 \int_T^t \Re[\xi(s)]ds} dt < \infty, \tag{1.29}$$

where

$$\xi(t) = \sqrt{\left(\frac{TrA}{2}\right)^2 - |A| + a_{12}\left(\frac{a_{11} - a_{22}}{2a_{12}}\right)'}, \quad k(t) = \frac{a_{12}}{2\xi^2}\left(\frac{\xi}{a_{12}}\right)'. \tag{1.30}$$

Then (1.1) is asymptotically stable if and only if (1.8),(1.9) are satisfied.

Corollary 1.7. *Assume $a_{22} - a_{11}, a_{12} \in C^3(T, \infty), A \in C^2(T, \infty)$, $a_{12}(t) \neq 0$ on (T, ∞), (1.8),(1.13), (1.14) and*

$$\int_T^{\infty} |k'(t) + k^2(t)\xi(t)| dt < \infty \tag{1.31}$$

are satisfied. Here functions ξ, θ_j, k are defined in (1.30), (1.4).

Then (1.1) is asymptotically stable.

Note that condition (1.31) is close to the main assumption of asymptotic stability theorems in Pucci and Serrin [16, 17], that $k(t)$ is the function of bounded variation $(\int_T^{\infty} |k'(t)| dt) < \infty$.

Example 1.3. *Consider system (1.1) with*

$$A(t)=\begin{pmatrix}0 & 1\\ -1 & -2f(t)\end{pmatrix},\tag{1.32}$$

where

$$f(t)=t^{\alpha}+it^{\beta}.\tag{1.33}$$

For the small damping case:

$$-1<\alpha<0,\quad \beta<0,\quad \alpha+\beta<-1\tag{1.34}$$

conditions of Corollary 1.6 are satisfied and this system is asymptotically stable.

From Corollary 1.7 it follows that this system is asymptotically stable in the more general case:

$$-1<\alpha<1,\quad \beta\le\frac{\alpha+1}{2}.\tag{1.35}$$

Here we consider another application of our approach in optical physics. We deduce the transition probability formula for dynamic system described by a general system (1.1) with antihermitian matrix function $A(t)$ and initial conditions:

$$u_1(0)=1,\quad u_2(0)=0.\tag{1.36}$$

The associated probability for an atom initially in state a to make a transition after excitation for a time t to state b is

$$P(t)=|u_1(t)|^2$$

Note that if the matrix A in equation (1.1) is antihermitian: $A^*=-A$, then the normalization of the wave function is constant at all times:

$$|u_1(t)|^2+|u_2(t)|^2=1.$$

Introducing the auxiliary functions

$$g(t)=g(0)-2\int_0^t a_{12}(t)e^{\int_0^t 2\varphi(y)dy},\tag{1.37}$$

$$\alpha(t)=\frac{1}{2}\Re\ln\left(\frac{g(0)(a_{11}(0)-\theta_2(0))}{g(t)(\theta_1(0)-a_{11}(0))}\right),\quad \beta(t)=\frac{1}{2}\Im\ln\left(\frac{g(0)(a_{11}(0)-\theta_2(0))}{g(t)(\theta_1(0)-a_{11}(0))}\right),\tag{1.38}$$

$$B(t)=\frac{|g(t)g(0)(a_{11}(0)-\theta_2(0))(a_{11}(0)-\theta_1(0))|}{|a_{12}(0)|^2e^{-2\int_0^t\Re(\theta_1)dy}}\tag{1.39}$$

we have general transition probability formulas

$$|u_1(t)|^2=B(t)\left(sinh^2\alpha(t)+cos^2\beta(t)\right),\quad |u_2(t)|^2=1-|u_1(t)|^2.\tag{1.40}$$

Note that in view of (1.4),(1.6), (1.37)-(1.40) to calculate transition probability we need to know only the function $\varphi(t)$.

Formula (1.40) allows quickly calculate transition probability for any approximation given via a function φ. Anyway the best choice of φ is such that minimizes $H(\varphi)/\xi$.

Example 1.4. *Consider the dynamic system (1.1) which describes an interaction of two-level atom in the external monochromatic electromagnetic field with frequency* ω*:*

$$u'(t) = \begin{pmatrix} 0 & iWe^{itE}\cos(t\omega) \\ iWe^{-itE}\cos(t\omega) & 0 \end{pmatrix} u(t). \tag{1.41}$$

where E is the difference of energy levels of the atom.

From (1.5) we have

$$H(\varphi) = -\varphi^2 - W^2\cos^2(t\omega) + e^{itE}\cos(t\omega)\left(\frac{\varphi e^{-itE}}{\cos(t\omega)}\right)'$$

$$= \varphi' + \varphi(-iE + \omega\tan(t\omega)) - W^2\cos^2(t\omega) - \varphi^2. \tag{1.42}$$

The function

$$\varphi = a_{12} = iWe^{itE}\cos(t\omega) \tag{1.43}$$

gives a good approximation from a mathematical point of view. From (1.40) we get

$$|u_1(\tau)|^2 = \sin^2\left(W\int_0^t \cos(s\omega)\cos(sE)ds\right) + \sinh^2\left(W\int_0^t \cos(s\omega)\sin(sE)ds\right). \tag{1.44}$$

Note that using expression (1.42) we get

$$\frac{H(\varphi(t))}{\xi(t)} = \frac{W^2\cos^2(t\omega)(e^{2itE}-1)}{-iWe^{itE}\cos(t\omega)} = -2W\sin(tE)\cos(t\omega),$$

or

$$\int_0^t \frac{|H(\varphi(s))|ds}{|\xi(s)|} \le 2W\int_0^t |\sin(sE)\cos(s\omega)|ds \le tW, \tag{1.45}$$

which is small for small tW.

Using rotating wave approximation from optical physics (see[1, 18]) we get another function φ :

$$\varphi_0 = \frac{i}{2}(\omega - E + \Delta), \quad \Delta = \sqrt{(\omega-E)^2 + W^2}. \tag{1.46}$$

From (1.40)

$$|u_1(\tau)|^2 = B\sin^2\left(\frac{t\Delta + t\omega - \eta(t)}{2}\right) + B\sinh^2\left(\frac{1}{2}\ln\frac{R}{(\Delta+\omega)(m-1)}\right). \tag{1.47}$$

where

$$\eta(t) = \tan^{-1}\left(\frac{\omega\tan(t\omega)}{\Delta+\omega}\right), \quad R = \sqrt{(\Delta+\omega)^2\cos^2(t\omega) + \omega^2\sin^2(t\omega)}.$$

$$B = \frac{(\Delta+\omega-E)^2(m-1)(\Delta+\omega)\sqrt{(\Delta+\omega)^2+\omega^2}}{\Delta^2(\Delta+2\omega)^2}, \quad m = \frac{2\Delta(\Delta+2\omega)}{(\Delta+\omega)(\Delta+\omega-E)}. \tag{1.48}$$

If $\omega = 0$*, then from (1.47) we get*

$$|u_1(\tau)|^2 = \frac{W^2}{E^2+W^2}\left[\sin^2\left(\frac{t\Delta}{2}\right)+\sinh^2\left(\frac{1}{2}ln\frac{\Delta-E}{\Delta+E}\right)\right].$$

If $E=\omega=0$, *then*

$$|u_1(\tau)|^2 = \sin^2(tW/2),$$

which is often referred as the Rabi formula [1]. Note that one can estimate the error function for each approximation by using Theorem 2.1 below.

2. Levinson Theorem for 2×2 System and Proofs of Main Results

Suppose we can find the exact solutions of the system

$$\psi'(t) = B(t)\psi(t), \quad t > T, \tag{2.1}$$

with the matrix-function

$$B(t) = \begin{pmatrix} b_{11}(t) & b_{12}(t) \\ b_{21}(t) & b_{22}(t) \end{pmatrix}$$

close to the matrix-function A, which means that the condition (2.7) below is satisfied. Let $\Psi(t)$ is the 2×2 fundamental matrix function of the auxiliary system (2.1). Then the solutions of (2.1) can be represented in the form

$$u(t) = \Psi(t)(C+\varepsilon(t)), \tag{2.2}$$

where $u(t),\varepsilon(t),C$ are the 2-vector columns: $u(t)=\text{colomn}(u_1(t),u_2(t)),\varepsilon(t)=\text{colomn}(\varepsilon_1(t),\varepsilon_2(t))$, $C=\text{colomn}(C_1,C_2),C_k$ are arbitrary constants. We can consider representation (2.2) as a definition of the error vector-function $\varepsilon(t)$.

Following theorem is a version of the Levinson Theorem [15, 7] about asymptotic solutions for 2×2 systems:

Theorem 2.1. *Assume there exist a function* $\xi\in C^1[T,\infty)$ *such that*

$$L(t)\equiv \max_{j=1,2}\left|\frac{Char(\theta_j(t))}{2\xi(t)}e^{(-1)^{j-1}\int_T^t 2\Re[\xi(y)]dy}\right| \in L^1(T,\infty). \tag{2.3}$$

Then every solution of (1.1) can be represented in form (2.2) and the error vector-function $\varepsilon(t)$ *can be estimated as*

$$\|\varepsilon(t)\| \le \|C\|\left(-1+\exp\int_t^\infty |L(s)|ds\right), \tag{2.4}$$

where C *is the constant vector and* $\|\cdot\|$ *is the Euclidean vector (or matrix) norm:* $\|\varepsilon(t)\| = \sqrt{\varepsilon_1^2(t)+\varepsilon_2^2(t)}$.

Remark 2.1. *From (2.3),(2.4) it follows that* $\varepsilon(t)=o(1), \quad t\to\infty$. *Also if* $Char(\theta_1) = Char(\theta_2)\equiv 0$, *then* $\varepsilon(t)\equiv 0$.

Remark 2.2. *Trying to find asymptotic solutions that are minimizing the error or corresponding function H given by formula (1.5), one can choose the function* φ *for example by the formula (see also (1.30))*

$$\varphi^2(t) = \left(\frac{TrA}{2}\right)^2 - |A| + a_{12}\left(\frac{a_{11}-a_{22}}{2a_{12}}\right)'.$$

Then asymptotic solutions obtained by this choice via formulas (1.4) will coincide with the well known Liouville-Green functions. Another choice of φ *is given in (2.30) below.*

Proof of Theorem 2.1. The substitution $u(t) = \Psi(t)v(t)$ transforms (1.1) into

$$v'(t) = M(t)v(t), \quad M(t) = \Psi^{-1}(A\Psi - \Psi')(t).$$

By integration we get

$$v(t) = C - \int_t^b M(s)v(s)ds, \quad T < t < b, \tag{2.5}$$

where the constant vector C is chosen as in (2.2).

Estimating $v(t)$

$$\|v(t)\| \le \|C\| + \int_t^b \|M(s)\|\|v(s)\|ds,$$

and using Gronwall's inequality we have

$$\|v(t)\| \le \|C\|e^{\int_t^b \|M(s)\|ds}.$$

From representation (2.2) we have

$$\varepsilon(t) = \Psi^{-1}u - C = v - C = -\int_t^b M(s)v(s)ds,$$

and using previous estimate we get

$$\begin{aligned}\|\varepsilon(t)\| &\le \int_t^b \|Mv\|ds \\ &\le \|C\|\int_t^b \|M(s)\|\exp\Big(\int_s^b \|M\|dy\Big)ds \\ &= \|C\|\Big(-1 + \exp\Big(\int_t^b \|M\|ds\Big)\Big),\end{aligned}$$

or

$$\|\varepsilon(t)\| \le \|C\|\Big(-1 + \exp\Big(\int_t^\infty \|\Psi^{-1}(A\Psi - \Psi')(s)\|ds\Big)\Big). \tag{2.6}$$

Note that error function $\varepsilon(t)$ is bounded if

$$\int_t^\infty \|\Psi^{-1}(A\Psi - \Psi')(s)\|ds < \infty. \tag{2.7}$$

To finish the proof we should calculate matrix function M in terms of characteristic functions $Char(\theta_j)$ using the construction of approximate fundamental matrix solution of (1.1).

To construct the approximate fundamental matrix function Ψ let us seek approximate solutions of (1.1)

$$u_1' = a_{11}u_1 + a_{12}u_2, \quad u_2' = a_{21}u_1 + a_{22}u_2,$$

as a linear combination of exponential functions

$$u_1 = C_1 e^{\int_T^t \theta_1(y)dy} + C_2 e^{\int_T^t \theta_2(y)dy}. \tag{2.8}$$

Substituting this representation for u_1 in the first equation

$$a_{12}u_2 = u_1' - a_{11}u_1 = (\theta_1 - a_{11})C_1 e^{\int_1^t \theta_1(y)dy} + C_2(\theta_2 - a_{11})e^{\int_T^t \theta_2(y)dy},$$

and solving for u_2 we have

$$u_2(t) = U_1(t)C_1 e^{\int_T^t \theta_1(y)dy} + U_2(t)C_2 e^{\int_T^t \theta_2(y)dy},$$

$$U_1(t) = \frac{\theta_1 - a_{11}}{a_{12}}, \quad U_2(t) = \frac{\theta_2 - a_{11}}{a_{12}}, \quad U_1(t) - U_2(t) = \frac{2\xi(t)}{a_{12}(t)}, \tag{2.9}$$

or

$$u(t) = \Psi(t)C, \tag{2.10}$$

where the fundamental matrix $\Psi(t)$ is defined by the formula

$$\Psi(t) = \begin{pmatrix} 1 & 1 \\ U_1(t) & U_2(t) \end{pmatrix} \begin{pmatrix} e^{\int_T^t \theta_1(y)dy} & 0 \\ 0 & e^{\int_T^t \theta_2(y)dy} \end{pmatrix}. \tag{2.11}$$

Define

$$\xi(t) = \frac{\theta_1(t) - \theta_2(t)}{2}, \quad Char_j(t) = Char(\theta_j(t)). \tag{2.12}$$

If $A \in C^1(T,\infty)$, $a_{12} \in C^2(T,\infty)$, $a_{12}(t)$ is not equal to zero on (T,∞), then following formulas are true

$$|\Psi(t)| = det[\Psi(t)] = -\frac{2\xi(t)}{a_{12}(t)} e^{\int_T^t (\theta_1 + \theta_2)dy}, \tag{2.13}$$

$$\frac{Char_2(t) - Char_1(t)}{2\xi} = \theta_1 + \theta_2 - Tr(A) + \frac{\xi'}{\xi} - \frac{a_{12}'}{a_{12}} \tag{2.14}$$

$$\Psi M \Psi^{-1} = A - \Psi'\Psi^{-1} = \frac{1}{2\xi} \begin{pmatrix} 0 & 0 \\ U_1 Char_2 - U_2 Char_1 & Char_1 - Char_2 \end{pmatrix}, \tag{2.15}$$

$$M(t) = \Psi^{-1}A\Psi - \Psi^{-1}\Psi'(t) = \frac{1}{2\xi(t)} \begin{pmatrix} Char_1(t) & e^{-2\int_T^t \xi dy} Char_2(t) \\ -e^{2\int_T^t \xi dy} Char_1(t) & -Char_2(t) \end{pmatrix}. \tag{2.16}$$

From Liouville's formula

$$\frac{|\Psi|'}{|\Psi|} = Tr(A) = a_{11} + a_{22} \tag{2.17}$$

in view of (2.11) we have

$$\frac{(\theta_1 - \theta_2)'}{\theta_1 - \theta_2} - \frac{a_{12}'}{a_{12}} + \theta_1 + \theta_2 - a_{11} - a_{22} = 0$$

or another version of Liouville's formula

$$\theta_1 + \theta_2 = Tr(A) - \frac{\xi'}{\xi} + \frac{a_{12}'}{a_{12}}. \tag{2.18}$$

It easy to check that the functions θ_j from (1.4) satisfy (2.18).

Remark 2.3. *From (1.3) and (2.18) it follows that*

$$Char_1(t) = Char_2(t) \equiv H(t), \tag{2.19}$$

and formula (2.16) turns to

$$M(t) = \Psi^{-1}A\Psi - \Psi^{-1}\Psi' = \frac{H(t)}{2\xi(t)} \begin{pmatrix} -1 & -e^{-2\int_T^t \xi(y,u(y))dy} \\ e^{2\int_T^t \xi(y,u(y)dy} & 1 \end{pmatrix}. \tag{2.20}$$

Formulas (2.13)-(2.16) can be checked by direct calculations. Indeed,

$$Char_2(t) - Char_1(t) = (\theta_1 - \theta_2)\left(\theta_1 + \theta_2 - Tr(A) + \frac{W[a_{12}, \theta_1 - \theta_2]}{a_{12}(\theta_1 - \theta_2)}\right).$$

From (2.11)

$$\Psi^{-1}(t) = \frac{1}{U_2 - U_1} \begin{pmatrix} e^{-\int_T^t \theta_1 dy} & 0 \\ 0 & e^{-\int_T^t \theta_2 dy} \end{pmatrix} \begin{pmatrix} U_2 & -1 \\ -U_1 & 1 \end{pmatrix}$$

$$\Psi'(t) = \begin{pmatrix} \theta_1 & \theta_2 \\ \Lambda_1 U_1 & \Lambda_2 U_2 \end{pmatrix} \begin{pmatrix} e^{\int_T^t \theta_1 dy} & 0 \\ 0 & e^{\int_T^t \theta_2 dy} \end{pmatrix}$$

where

$$\Lambda_j = \theta_j + \frac{(\theta_j - a_{11})'}{\theta_j - a_{11}} - \frac{a_{12}'}{a_{12}} = \theta_j + \frac{W[a_{12}, \theta_j - a_{11}]}{a_{12}(\theta_j - a_{11})}, \quad j = 1, 2.$$

So

$$\Psi'\Psi^{-1} = \begin{pmatrix} a_{11} & a_{12} \\ a_{21} + \frac{Char_1 U_2 - Char_2 U_1}{2\xi} & a_{22} + \frac{Char_2 - Char_1}{2\xi} \end{pmatrix}$$

Indeed,

$$\Psi'\Psi^{-1} = \frac{1}{U_2 - U_1} \begin{pmatrix} \theta_1 & \theta_2 \\ \Lambda_1 U_1 & \Lambda_2 U_2 \end{pmatrix} \begin{pmatrix} U_2 & -1 \\ -U_1 & 1 \end{pmatrix} =$$

$$\frac{1}{U_2-U_1}\begin{pmatrix} \theta_1 U_2-\theta_2 U_1 & \theta_2-\theta_1 \\ (\Lambda_1-\Lambda_2)U_1U_2 & \Lambda_2 U_2-\Lambda_1 U_1 \end{pmatrix}=$$

$$\begin{pmatrix} a_{11} & a_{12} \\ a_{21}+\frac{Char_2U_1-Char_1U_2}{2\xi} & a_{22}+\frac{Char_2-Char_1}{2\xi} \end{pmatrix},$$

in view of

$$\frac{\theta_1 U_2-\theta_2 U_1}{U_2-U_1}=\theta_1+\frac{(\theta_1-\theta_2)U_1}{U_2-U_1}=\theta_1+a_{11}-\theta_1=a_{11},$$

and

$$\frac{\Lambda_2 U_2-\Lambda_1 U_1}{U_2-U_1}=\frac{(\Lambda_2-\Lambda_1)U_1}{U_2-U_1}+\Lambda_2=\frac{(a_{11}-\theta_1)(\Lambda_2-\Lambda_1)}{2\xi}+\Lambda_2$$

$$=\frac{a_{11}-\theta_1}{2\xi}\left[-2\xi+\frac{(\theta_2-a_{11})'}{\theta_2-a_{11}}-\frac{(\theta_1-a_{11})'}{\theta_1-a_{11}}\right]+\theta_2+\frac{(\theta_2-a_{11})'}{\theta_2-a_{11}}-\frac{a_{12}'}{a_{12}}=$$

$$\theta_1-a_{11}+\frac{(\theta_2-a_{11})'}{\theta_2-a_{11}}\left(\frac{a_{11}-\theta_1}{2\xi}+1\right)+\frac{(\theta_1-a_{11})'}{2\xi}+\theta_2-\frac{a_{12}'}{a_{12}}=$$

$$=\theta_1+\theta_2-a_{11}+\frac{\xi'}{\xi}-\frac{a_{12}'}{a_{12}}=a_{22}+\frac{Char_2-Char_1}{2\xi},$$

and

$$\frac{U_1U_2(\Lambda_1-\Lambda_2)}{U_2-U_1}=\frac{(\theta_1-a_{11})(\theta_2-a_{11})}{-2\xi a_{12}}\left[2\xi+\frac{(\theta_1-a_{11})'}{\theta_1-a_{11}}-\frac{(\theta_2-a_{11})'}{\theta_2-a_{11}}\right]$$

$$=-\frac{(\theta_1-a_{11})(\theta_2-a_{11})}{a_{12}}+\frac{(\theta_2-a_{11})'(\theta_1-a_{11})-(\theta_1-a_{11})'(\theta_2-a_{11})}{2a_{12}\xi}=$$

$$=a_{21}+\frac{Char_1U_2-Char_2U_1}{2\xi}=a_{21}-\frac{Char_1}{a_{12}}+\frac{(Char_1-Char_2)(\theta_1-a_{11})}{2a_{12}\xi},$$

where we use the calculations

$$\frac{Char_2U_1-Char_1U_2}{2\xi}=$$

$$\frac{\theta_2-a_{11}}{2a_{12}\xi}\left[|A|+\theta_1^2-\theta_1(a_{11}+a_{22})+(\theta_1-a_{11})'-\frac{a_{12}'(\theta_1-a_{11})}{a_{12}}\right]+$$

$$-\frac{\theta_1-a_{11}}{2a_{12}\xi}\left[|A|+\theta_2^2-\theta_2(a_{11}+a_{22})+(\theta_2-a_{11})'-\frac{a_{12}'(\theta_2-a_{11})}{a_{12}}\right]=$$

$$=\frac{|A|}{-a_{12}}+\frac{\theta_1\theta_2-a_{11}(\theta_1+\theta_2)}{a_{12}}+\frac{a_{11}+a_{22}}{2a_{12}\xi}[\theta_2(\theta_1-a_{11})-\theta_1(\theta_2-a_{11})]+$$

$$+\frac{(\theta_2-a_{11})(\theta_1-a_{11})'-(\theta_1-a_{11})(\theta_2-a_{11})'}{2a_{12}\xi}=$$

$$a_{21}+\frac{(\theta_1-a_{11})(\theta_2-a_{11})}{a_{12}}+\frac{W[\theta_2-a_{11},\theta_1-a_{11}]}{2a_{12}\xi}.$$

The final estimate (2.4) follows from (2.6) and (2.16). □

Proof of Theorem 1.1. From condition (1.7) of Theorem 1.1 and formula (2.20) it follows that

$$\|M(t)\| \in L_1(T,\infty),$$

and condition (2.3) of Theorem 2.1 is satisfied. Applying Theorem 2.1 we obtain representation (2.2) for solutions of (1.1). From (2.2) and (2.4) we get stability inequality

$$\|u(t)\| \leq c\cdot\|\Psi(t)C\|. \tag{2.21}$$

Because of this estimate all solutions of (1.1) are stable and attractive if and only if

$$\lim_{t\to\infty}(\|\Psi(t)\|) = 0.$$

This condition is satisfied because of conditions (1.8),(1.9) of Theorem 1.1 and formula (2.11). □

Proof of Example 1.1. We have

$$A = \begin{pmatrix} 0 & f(t) \\ -g(t) & 0 \end{pmatrix},$$

and

$$a_{12} = f(t), \quad Tr(A) = 0, \quad |A| = f(t)g(t), \quad H(t) = -fg - F^2 + f(t)\left(\frac{F(t)}{f(t)}\right)'.$$

Choosing

$$\xi = \frac{ib}{t^\gamma}$$

we get

$$k = O(t^{\gamma-1}), \quad \frac{H(t)}{\xi} = -\frac{fg}{\xi} - \xi(1-k^2) + k' = O(t^{\gamma-2}) \in L_1(T,\infty), \quad \text{if} \quad \gamma > 1,$$

and condition (1.7) is satisfied. From

$$\theta_{1,2} = \xi + \frac{a'_{12}}{2a_{12}} - \frac{\xi'}{2\xi} = \pm\frac{ib}{t^\gamma} + \frac{\gamma - 2a}{2t}$$

$$\Re[\theta_j] = \frac{\gamma-2a}{2t} < 0, \quad \text{if} \quad \gamma < 2a,$$

it follows that condition (1.8) is satisfied.

If $\gamma > 1$ then condition (1.9) is satisfied as well:

$$\frac{\theta_j - a_{11}}{a_{12}} e^{\int_1^t \Re[\theta_j]dy} = t^{2a}\left(\frac{ib}{t^\gamma} + \frac{\gamma-2a}{2t}\right)\exp\left(\int^t \frac{(\gamma-2a)dy}{y}\right) = O(t^{\gamma-1}) \to 0,$$

when $t \to \infty$ □

Denote by $G(t,s) = \Psi(t)\Psi^{-1}(s)$ the Cauchy matrix function of (1.1).

Lemma 2.2. *Assume that conditions (1.13), (1.14) are satisfied. Then*

$$|G(t,s)| = |\Psi(t)\Psi^{-1}(s)| \leq C\left|\frac{a_{12}(s)}{\xi(s)}\right|, \quad T \leq s \leq t, \tag{2.22}$$

$$\|\Psi(t)M(t)\Psi^{-1}(t)\| \leq C\left|\frac{H(t)}{a_{12}}\right|, \quad t \geq T. \tag{2.23}$$

Proof of Lemma 2.2. From condition (1.14) it follows that

$$|U_j(t)| \leq C, \quad j = 1,2, \quad \text{for all} \quad t \geq T.$$

By direct calculations

$$G(t,s) = \frac{1}{U_2(s) - U_1(s)} \begin{pmatrix} e^{\int_s^t \theta_1 dy} & e^{\int_s^t \theta_2 dy} \\ U_1(t)e^{\int_s^t \theta_1 dy} & U_2(t)e^{\int_s^t \theta_2 dy} \end{pmatrix} \begin{pmatrix} U_2(s) & -1 \\ -U_1(s) & 1 \end{pmatrix}$$

So estimate (2.22) follows from

$$|G_{kj}(t,s)| \leq \frac{C}{|U_2(s) - U_1(s)|} = C\left|\frac{a_{12}(s)}{2\xi(s)}\right|, \quad k,j = 1,2.$$

The estimate (2.23) follows from the formula (2.15). □

Proof of Theorem 1.2. Proof follows directly from the explicit formula for fundamental matrix function in the case $a_{12} \equiv 0$:

$$\Psi(t) = \begin{pmatrix} e^{\int_T^t a_{11}(y)dy}, & 0 \\ e^{\int_T^t a_{22}(y)dy}[C + \int_T^t a_{21}(s)e^{\int_T^s (a_{11}-a_{22})(y)dy}ds], & e^{\int_T^t a_{22}(y)dy} \end{pmatrix}.$$

□

Proof of Theorem 1.3. Consider the system (1.1). By substitution

$$u(t) = \Psi(t)v(t),$$

we get

$$v'(t) = M(t)v(t), \quad v(t) = C + \int_T^t M(s)v(s)ds,$$

or

$$\Psi^{-1}(t)u(t) = C + \int_T^t M(s)\Psi^{-1}(s)u(s)ds, \quad T \leq s \leq t, \tag{2.24}$$

$$u(t) = \Psi(t)C + \int_T^t G(t,s)\Psi(s)M(s)\Psi^{-1}(s)u(s)ds. \tag{2.25}$$

From this representation and Lemma 2.2 we obtain the estimates

$$\|u(t)\| \leq \|\Psi(t)C\| + \int_T^t \|G(t,s)\| \cdot \|\Psi(s)M(s)\Psi^{-1}(s)u(s)\|ds$$

$$\leq \|\Psi(t)C\| + \int_T^t \left|\frac{H(s)}{\xi(s)}\right| \|u(s)\| ds.$$

Applying Gronwall's inequality (see for example [10]) we get

$$\|u(t)\| \leq \|\Psi(t)C\| + \int_T^t \|\Psi(s)C\| \left|\frac{H(s)}{\xi(s)}\right| \exp\left(\int_T^s \left|\frac{H(y)}{\xi(y)}\right| dy\right) ds,$$

$$\|u(t)\| \leq \|\Psi(t)C\|_\infty \left(1 + \int_T^t \left|\frac{H(s)}{\xi(s)}\right| \exp\left(\int_T^s \left|\frac{H(y)}{\xi(y)}\right| dy\right) ds\right),$$

where $\|u(t)\|_\infty = sup_{t\geq T}\|u(t)\|$.

So we obtain the stability estimate

$$\|u(t)\| \leq \|\Psi(t)C\|_\infty \exp\left(\int_T^t \left|\frac{H(s)}{\xi(s)}\right| ds\right). \tag{2.26}$$

Using this inequality we can estimate (2.25) again

$$\|u(t) - \Psi(t)C\| \leq \|\Psi(t)C\|_\infty \left(\exp \int_T^t \left|\frac{H(s)}{\xi(s)}\right| ds - 1\right). \tag{2.27}$$

From conditions (1.13),(1.14) of Theorem 1.3 and formula (2.11) we have

$$\|\Psi(t)C\| \leq const.$$

So from stability inequality (2.26) and condition (1.12) of Theorem 1.3 we get stability of (1.1).

From (1.8),(1.9) we have

$$\lim_{t\to\infty} \|\Psi(t,u)C\| = 0,$$

and asymptotic stability of (1.1) follows from the estimate (2.26).

□

Proof of Corollary 1.4. We deduce Corollary 1.4 from Theorem 1.3 by choosing ξ as in (1.16), and

$$\varphi = S_{n+1} + \frac{a_{22} - a_{11}}{2}. \tag{2.28}$$

From the condition $a_{12} > 0$ it follows that for all $t \geq T_1 > T$

$$\int_T^t a_{12} e^{\int_t^s (2S_{n+1} + a_{22} - a_{11}) dy} \geq \int_T^{T_1} a_{12} e^{\int_t^s (2S_{n+1} + a_{22} - a_{11}) dy} = \gamma_1(T_1) > 0,$$

and from (1.16)

$$-\frac{a_{12}}{2\gamma_1(T_1)} \leq \xi < 0, \quad t \geq T_1 > T. \tag{2.29}$$

By direct calculations we get from (1.5),(2.28)

$$H(t) = \left(\frac{a_{11} - a_{22}}{2}\right)^2 + a_{12}a_{21} + a_{12}\left(\frac{2\varphi + a_{11} - a_{22}}{2a_{12}}\right)' - \varphi^2 =$$

$$\left(\frac{a_{11}-a_{22}}{2}\right)^2 + a_{12}a_{21} + a_{12}\left(\frac{S_{n+1}}{a_{12}}\right)' - (S_{n+1} + \frac{a_{22}-a_{11}}{2})^2 =$$

$$a_{12}a_{21} + a_{12}\left(\frac{S_{n+1}}{a_{12}}\right)' + (a_{11}-a_{22})S_{n+1} - S_{n+1}^2 = S_n^2 - S_{n+1}^2,$$

if S_{n+1} are the solutions of first order equations:

$$S_0 \equiv 0, \quad \left(\frac{S_{n+1}}{a_{12}}\right)' + (a_{11}-a_{22})\frac{S_{n+1}}{a_{12}} = \frac{S_n^2}{a_{12}} - a_{21}, \quad n = 0, 1, 2, ...$$

and given by formulas (1.17). So condition (1.12) of Theorem 1.3 turns to (1.19).

In view of (1.6):

$$\frac{\xi'}{2\xi} - \frac{a_{12}'}{2a_{12}} = \varphi + \xi$$

we have from (1.21),(2.28)

$$\theta_1 = \xi + \frac{TrA}{2} + \frac{a_{12}'}{2a_{12}} - \frac{\xi'}{2\xi} = \frac{TrA}{2} - \varphi = a_{11} - S_{n+1} \le a_{11} - S_{n+1} - 2\xi \le 0$$

$$\theta_2 = \theta_1 - 2\xi = a_{11} - S_{n+1} - 2\xi \le 0.$$

From condition (1.20) it follows condition (1.8):

$$\int^t \theta_j(s)ds \to -\infty, \quad j = 1, 2, \quad t \to \infty.$$

Finally condition (1.14) of Theorem 1.3 follows from (2.29) and (1.22):

$$\left|\frac{2\xi}{a_{12}}\right| = \frac{1}{\int_T^t a_{12} e^{\int_t^s (2S_{n+1}+a_{22}-a_{11})dy}} \le \frac{1}{\gamma_1}, \quad \left|\frac{\theta_1 - a_{11}}{a_{12}}\right| = \left|-\frac{S_{n+1}}{a_{12}}\right| \le const.$$

□

Proof of Example 1.2. From (1.16),(1.17) with $n = 0$ we have

$$\xi(t) = -\frac{1}{2\int_T^t e^{\int_t^s 2(S_1 - f)dy}}, \quad S_1(t) = \int_T^t (g + 2f')e^{\int_t^s 2f dy} ds,$$

and conditions (1.19), (1.21) of Corollary 1.4 turns to condition (1.25). From the estimate

$$S_1(t) \le g_0 \int_T^t e^{\int_t^s 2f dy} ds \le g_0 \int_T^t e^{2f_0(s-t)} ds \le \frac{g_0}{2f_0}$$

condition (1.22) is fulfilled. The condition (1.20) follows from the estimates

$$\int_T^t e^{\int_T^s 2(S_1 - f)dy} ds \le \int_T^\infty e^{(s-T)(\frac{g_0}{2f_0} - f_0)} ds = \frac{1}{f_0 - \frac{g_0}{2f_0}} \equiv \frac{1}{f_1} < \infty$$

$$\int_{T_1}^t 2\xi(\tau)d\tau = -\int_{T_1}^t \left(\ln \int_T^\tau e^{\int_T^s 2(S_1 - f)dy} ds\right)'(\tau)d\tau =$$

$$\ln\left(\int_T^{T_1} e^{\int_T^s 2(S_1-f)dy}\right) - \ln\left(\int_T^{t} e^{\int_T^s 2(S_1-f)dy}\right) \geq \ln(f_2(T,T_1)) + \ln(f_1).$$

So (1.20) follows from (1.24):

$$\int_{T_1}^t (2\xi + S_1)dy \geq \ln(f_1 f_2) + \int_{T_1}^t S_1 dy \to \infty, \quad t \to \infty.$$

□

Proof of Corollary 1.5. Choosing

$$\varphi = \frac{a_{22} - a_{11}}{2} \tag{2.30}$$

we get

$$\xi = -\frac{a_{12}(t)}{2\int_T^t a_{12}(s) e^{\int_t^s (a_{22}-a_{11})dy} ds} \leq 0, \quad \theta_1 = a_{11},$$

$$\theta_2 = a_{11} - 2\xi = a_{11} + \frac{a_{12}}{\int_T^t a_{12}(s) e^{\int_t^s (a_{22}-a_{11})dy} ds},$$

$$H(t) = \left(\frac{Tr(A)}{2}\right)^2 - |A| + a_{12}\left(\frac{2\varphi + a_{11} - a_{22}}{2a_{12}}(t)\right)' - \varphi^2(t) =$$

$$\left(\frac{a_{11} - a_{22}}{2}\right)^2 + a_{12}a_{21} - \varphi^2 = a_{12}a_{21},$$

$$\frac{H(t)}{2\xi} = -a_{21}(t)\int_T^t a_{12}(s) e^{\int_t^s (a_{22}-a_{11})dy} ds.$$

So condition (1.12) of Theorem 1.3 turns to (1.26). The rest of the proof is similar to the proof of Corollary 1.4.

□

Proof of Corollary 1.6. From

$$\varphi = \frac{\xi'}{2\xi} - \xi - \frac{a_{12}'}{2a_{12}} = \frac{a_{12}}{2\xi}\left(\frac{\xi}{a_{12}}\right)' - \xi = (k-1)\xi,$$

we get from (1.5)

$$H(t) = \left(\frac{Tr(A)}{2}\right)^2 - |A| + a_{12}\left(\frac{a_{11} - a_{22}}{2a_{12}}\right)' - \xi^2 + k^2\xi^2 + k'\xi.$$

Choosing ξ as in (1.30) we get

$$\frac{H(t)}{\xi(t)} = k'(t) + k^2(t)\xi(t), \tag{2.31}$$

and Corollary 1.6 follows from Theorem 1.1.

□

Proof of Corollary 1.7. Corollary 1.7 follows from Theorem 1.3 by choosing ξ as in (1.30). □

Proof of Example 1.3. From (1.30)

$$\xi = \sqrt{f^2(t) - 1 + f'(t)}.$$

To check conditions of Corollary 1.6 denote

$$P = \Re[f^2(t) - 1 + f'(t)] = t^{2\alpha} - t^{2\beta} - 1 + \alpha t^{\alpha-1},$$

$$Q = \Im[f^2(t) - 1 + f'(t)] = 2t^{\alpha+\beta} + \beta t^{\beta-1}.$$

From $\alpha < 0, \beta < 0$ we get

$$\sqrt{P^2+Q^2} + P = \frac{Q^2}{\sqrt{P^2+Q^2} - P} = \frac{Q^2}{2}(1+o(1)), \quad t \to \infty$$

and

$$\Re[\xi] = \Re[P+iQ] = \frac{\sqrt{P+\sqrt{P^2+Q^2}}}{\sqrt{2}} = \frac{Q}{2}(1+o(1)).$$

So condition (1.29) follows from

$$|\xi| = O(1), \quad k' + k^2\xi = O(t^{-2}), \quad t \to \infty,$$

and

$$\pm\int_T^\infty \Re[\xi]dt = \frac{1}{2}\int_T^\infty [2t^{\alpha+\beta} + \beta t^{\beta-1} 2(1+o(1))]dt < \infty.$$

From $\alpha > -1$ it follows that conditions (1.8) (1.9) are satisfied as well:

$$\Re[\theta_j] = \Re\left[\pm\xi - f(t) - \frac{\xi'}{2\xi}\right] = -t^\alpha(1+o(1)) + O(1/t), \quad t \to \infty.$$

Further we will show that conditions of Corollary 1.7 are satisfied if conditions (1.35) are fulfilled. Denote

$$V(t) = \Re[f - \xi] = t^\alpha - \sqrt{(P+R)/2}, \quad R = \sqrt{P^2+Q^2}.$$

By calculations

$$V(t) = \frac{t^{2\alpha} - (P+R)/2}{t^\alpha + \sqrt{(P+R)/2}} = \frac{2t^{2\alpha} - P - R}{2t^\alpha + \sqrt{2P+2R}} = \frac{K(t)}{(2t^\alpha + \sqrt{2P+2R})(2t^{2\alpha} - P + R)},$$

$$K = (2t^{2\alpha} - P)^2 - R^2 = 4t^{2\alpha}[1 - \alpha t^{\alpha-1} - \beta t^{2\beta-1-\alpha} - \beta^2 t^{2\beta-2-\alpha}].$$

From the formulas for K and Q it follows that

$$K = t^{2\alpha}(1+o(1), \quad Q = 2t^{\alpha+\beta}(1+o(1)).$$

To prove

$$\int_T^{\infty} V(t)dt = \infty$$

we divide the plane (α,β) on 3 regions:

$$\{\alpha \geq \beta, \alpha \geq 0\}, \quad \{\beta \geq \alpha, \beta \geq 0\}, \quad \{\alpha \leq 0, \beta \leq 0\},$$

and prove it in each region separately.

Region 1: $\alpha \geq \beta, \quad \alpha \geq 0$.

From $P = t^{2\alpha}(1+o(1)), \quad R = t^{2\alpha}(1+o(1)), \quad R-P = \frac{Q^2}{P+R} = t^{2\beta}(1+o(1))$,

$$V(t) = \frac{t^{2\alpha}(1+o(1))}{t^{\alpha}+\sqrt{(P+R)/2}} = \frac{t^{2\alpha}(1+o(1))}{t^{3\alpha}} = t^{-\alpha}(1+o(1)),$$

and if $\alpha < 1$ then the formula is true.

Region 2: $\beta \geq \alpha, \quad \beta \geq 0$.

From $P = -t^{2\beta}(1+o(1)), \quad R = t^{2\beta}(1+o(1)), \quad R-P = t^{2\beta}, \quad R+P = \frac{Q^2}{R-P} = t^{2\alpha}(1+o(1))$,

$$V(t) = \frac{t^{2\alpha}(1+o(1))}{t^{\alpha}+\sqrt{(P+R)/2}} = \frac{t^{2\alpha}(1+o(1))}{t^{\alpha+2\beta}} = t^{\alpha-2\beta}(1+o(1)),$$

and if $\alpha+1-2\beta > 0$ then the formula is true.

Region 3: $\alpha \leq 0, \quad \beta \leq 0$.

From $P = -1+o(1), \quad R = 1+o(1), \quad R-P = R = 2+o(1), \quad R+P = \frac{Q^2}{R-P} = t^{2\alpha+2\beta}(1+o(1))$,

$$V(t) = \frac{t^{2\alpha}(1+o(1))}{t^{\alpha}+\sqrt{(P+R)/2}} = \frac{t^{2\alpha}(1+o(1))}{t^{\alpha}+t^{\alpha+\beta}} = t^{\alpha}(1+o(1)),$$

and if $\alpha+1 > 0$ then the formula is true.

Now we are ready to check conditions (1.8):

$$e^{\int_T^t \Re[\theta_1]ds} = e^{\int_T^t \Re[\xi - f - \frac{\xi'}{2\xi}]ds} =$$

$$\left|\frac{\xi(T)}{\xi(t)}\right|^{1/2} e^{-\int_T^t \Re[f-\xi]ds} \leq Ce^{-\int_T^t V(s)ds} \to 0, \quad t \to \infty.$$

From $\Re[\xi] \geq 0$ we get

$$e^{\int_T^t \Re[\theta_2]ds} = e^{-\int_T^t \Re[\xi+f-\frac{\xi'}{2\xi}]ds} = \left|\frac{\xi(T)}{\xi(t)}\right|^{1/2} e^{-\int_T^t \Re[f]ds} \leq Ce^{-\int_T^t V(s)ds} \to 0, \quad t \to \infty.$$

Conditions (1.13) are obviously true.

To check conditions (1.14) note that $\alpha > 0$ or $\beta > 0$ then

$$|f(t)| = \sqrt{t^{2\alpha}+t^{2\beta}} \to \infty, \quad t \to \infty, \quad \frac{|f'|}{|f|} \leq \frac{C}{t}$$

$$|\xi - f| = \sqrt{f^2 + f' - 1} - f = \frac{f' - 1}{f + \sqrt{f^2 + f' - 1}} = \frac{f' - 1}{f(1 + o(1))} \le C\left(\frac{f'}{f} + \frac{1}{f}\right) \le C$$

If $\alpha \le 0$ or $\beta \le 0$ then

$$|\xi - f| \le |\xi| + |f| \le C.$$

From these estimates and $\frac{\xi'}{\xi} = \frac{1+o(1)}{t}$ it follows that conditions (1.14) are satisfied. At last (1.31) is satisfied in view of:

$$k(t) = \frac{\xi'}{2\xi^2} = \frac{1 + o(1)}{t}, \quad k'(t) = \frac{1 + o(1)}{t^2}, \quad t \to \infty.$$

□

Proof of transition probability formula (1.40). From representation (2.10) we get

$$u_1(t) = C_1 e^{\int_0^t \theta_1 dy} + C_2 e^{\int_0^t \theta_2 dy} =$$

$$\sqrt{C_1 C_2} e^{\int_0^t \frac{\theta_1 + \theta_2}{2} dy} \left(\sqrt{\frac{C_1}{C_2}} e^{\int_0^t \frac{\theta_1 - \theta_2}{2} dy} + \sqrt{\frac{C_2}{C_1}} e^{\int_0^t \frac{\theta_2 - \theta_1}{2} dy} \right).$$

In view of $x + \frac{1}{x} = e^{\ln x} + e^{-\ln x} = 2\cosh(\ln x)$, and

$$\theta_1 - \theta_2 = -\frac{g'(t)}{g(t)}, \quad g(t) = g(0) - \int_0^t b e^{\int_0^s 2\varphi dy} ds, \quad \frac{\theta'_{12}}{\theta_{12}} = 2\varphi + \frac{b'}{b} + \theta_{12}$$

we have

$$u_1(t) = 2\sqrt{C_1 C_2} e^{\int_0^t \frac{\theta_1 + \theta_2}{2} dy} \cosh\left(\frac{1}{2} \ln \frac{C_1 g(0)}{C_2 g(t)}\right),$$

and

$$|u_1(t)|^2 = 4|C_1 C_2| e^{\int_0^t \Re(\theta_1 + \theta_2) dy} \left|\cosh\left(\frac{1}{2} \ln \frac{C_1 g(0)}{C_2 g(t)}\right)\right|^2.$$

From initial conditions (1.36) we get

$$C_1 + C_2 = 1, \quad C_1 U_1(0) + C_2 U_2(0) = 0,$$

or

$$C_1 = \frac{-U_2(0)}{U_1(0) - U_2(0)} = \frac{a_{11}(0) - \theta_2(0)}{2\xi(0)}, \quad C_2 = \frac{U_1(0)}{U_1(0) - U_2(0)} = \frac{\theta_1(0) - a_{11}(0)}{2\xi(0)}.$$

Further we have

$$|u_1(t)|^2 = 4|C_1 C_2| e^{\int_0^t \Re(\theta_1 + \theta_2) dy} \left|\cosh\left(\frac{1}{2} \ln \frac{C_1 g(0)}{C_2 g(t)}\right)\right|^2 = B|\cosh(\alpha + i\beta)|^2,$$

where α, β are defined in (1.38) and

$$B = 4|C_1 C_2| e^{\int_0^t \Re(\theta_1 + \theta_2) dy} = \frac{4|a_{11}(0) - \theta_1(0)||a_{11}(0) - \theta_2(0)|}{|\theta_1(0) - \theta_2(0)|^2} e^{\int_0^t \Re(\theta_1 + \theta_2) dy}.$$

Formula (1.40) follows from this formula in view of

$$|\cosh(\alpha + i\beta)|^2 = \sinh^2(\alpha) + \cos^2(\beta).$$

□

Proof of (1.44). By direct calculations

$$g(t)=g(0)-2\int_0^t a_{12}(t)e^{\int_0^t 2\varphi(y)dy}=g(0)-2\int_0^t \varphi e^{\int_0^s 2\varphi(z)dz}=-e^{\int_0^s 2\varphi(z)dz},$$

$$\xi=-\frac{g'(t)}{2g(t)}=-\varphi(t),\quad \theta_1=\xi+\frac{a'_{12}}{2a_{12}}-\frac{\xi'}{2\xi}=-\varphi+\frac{\varphi'}{2\varphi}-\frac{\varphi'}{2\varphi}=-\varphi,\quad \theta_2=\varphi,$$

$$B(t)=\frac{4|\theta_1(0)\theta_2(0)|}{|\theta_1(0)-\theta_2(0)|^2}e^{\int_0^t \Re(\theta_1+\theta_2)dy}=1,\quad \varphi=iW\cos(t\omega)[\cos(tE)+i\sin(tE)],$$

$$\alpha(t)=\frac{1}{2}\Re\ln\left(\frac{-g(0)\theta_2(0)}{g(t)\theta_1(0)}\right)=-\frac{1}{2}\Re\ln\left(\frac{g(t)}{g(0)}\right)=-\frac{1}{2}\Re\ln\left(e^{i\pi+\int_0^t 2\varphi}\right),$$

$$\alpha(t)=-\frac{1}{2}\Re(i\pi+\int_0^t 2\varphi ds)=-\int_0^t \Re[\varphi]ds=W\int_0^t \cos(s\omega)\sin(sE)ds,$$

$$\beta(t)=-\frac{1}{2}\Im(i\pi+\int_0^t 2\varphi ds)=-\frac{\pi}{2}-W\int_0^t \cos(s\omega)\cos(sE)ds.$$

□

Proof of (1.50). From (1.37) we have

$$g_0(t)=g(0)-2iW\int_0^t e^{is(\omega+\Delta)}\cos(s\omega)ds=\frac{-2W[(\Delta+\omega)\cos(t\omega)-i\omega\sin(t\omega)]e^{it(\Delta+\omega)}}{\Delta(\Delta+2\omega)}$$

$$\xi=-\frac{g'(t)}{2g(t)}=-\frac{i\Delta(\Delta+2\omega)\cos(t\omega)}{2(\Delta+\omega)\cos(t\omega)-2i\omega\sin(t\omega)},\quad m=\frac{2\Delta(\Delta+2\omega)}{(\Delta+\omega)(\Delta+\omega-E)}\geq 1,$$

$$\theta_1=\xi+\frac{a'_{12}}{a_{12}}-\frac{\xi'}{2\xi}=\frac{i(E-\Delta-\omega)}{2},\quad \theta_2=\frac{i}{2}\left(E-\Delta-\omega+\frac{2\Delta(\Delta+2\omega)\cos(t\omega)}{(\Delta+\omega)\cos(t\omega)-i\omega\sin(t\omega)}\right)$$

$$B=B(t)=\frac{4|\theta_1(0)\theta_2(0)|}{|\theta_1(0)-\theta_2(0)|^2}e^{\int_0^t \Re(\theta_1+\theta_2)dy}=\frac{(\Delta+\omega-E)^2(m-1)(\Delta+\omega)\sqrt{(\Delta+\omega)^2+\omega^2}}{\Delta^2(\Delta+2\omega)^2} \tag{2.32}$$

$$\alpha=\frac{1}{2}\Re\ln\left(\frac{e^{it(\Delta+\omega)}[i\omega\sin(t\omega)-(\Delta+\omega)\cos(t\omega)]}{(\Delta+\omega)(m-1)}\right)=\frac{1}{2}\ln\frac{R}{(\Delta+\omega)(m-1)} \tag{2.33}$$

$$\beta=\frac{1}{2}\Im\ln\left(\frac{e^{it(\Delta+\omega)}[i\omega\sin(t\omega)-(\Delta+\omega)\cos(t\omega)]}{(\Delta+\omega)(m-1)}\right)=\frac{t(\Delta+\omega)-\eta(t)}{2} \tag{2.34}$$

where

$$i\omega\sin(t\omega)-(\Delta+\omega)\cos(t\omega)=Re^{-i\eta(t)}.$$

□

References

[1] Allen, J.H. Eberly, *Optical Resonance and Two-level Atoms.* Dover, New York, 1987.

[2] Z. Arstein and E. F. Infante; *On asymptotic stability of oscillators with unbounded damping, Quart. Appl. Mech.* **34** (1976), 195-198.

[3] R. J. Ballieu and K. Peiffer; *Asymptotic stability of the origin for the equation,* $x''(t)+f(t,x,x'(t))|x'(t)|^{\alpha}+g(x)=0$ *J.Math Anal. Appl* **34** (1978) 321-332

[4] J. D. Birkhoff *Quantum mechanics and asymptotic series. Bull. Amer. Math., Soc.* **32** 1933, 681-700

[5] L. Cesary *Asymptotic behavior and stability problems in ordinary differential* , 3rd ed., Springer Verlag, Berlin, 1970.

[6] L. H. Duc, A. Ilchmann, S. Siegmund, P. Taraba *On stability of linear time-varying second-order differential equations Quart. Appl. Math.* **64** (2006), 137-151.

[7] M. S. P. Eastham *The asymptotic solution of linear differential systems. Applications of the Levinson theorem*, Oxford Science Publications, 1989.

[8] L. Hatvani Integral conditions on asymptotic stability for the damped linear oscillator with small damping, *Proceedings of the American Mathematical Society.* 1996, Vol. 124 No.2, p.415-422.

[9] G. R. Hovhannisyan Asymptotic stability for second-order differential equations with complex coefficients, *Electronic Journal of Differential Equations* . Vol. 2004(2004), No. 85, pp. 1-20.

[10] G. R. Hovhannisyan Asymptotic Stability and Asymptotic Solutions of Second-order Differential Equations, *Journal of Mathematical Analysis and Applications,* **327** (2007) 47-62

[11] G. R. Hovhannisyan Asymptotic Stability for Dynamic Equations on Time Scales, *Advances in Difference Equations*, vol. 2006, Article ID 18157 (2006), 17 pages

[12] A. O. Ignatyev Stability of a linear oscillator with variable parameters, *Electronic Journal of Differential Equations* Vol.1997(1997), No. 17, p.1-6.

[13] V. Lakshmikantham, S.Leela, and A. A Martynuk *Stability analysis of nonlinear systems*, Marcel Dekker Inc., New York, 1989

[14] J. J. Levin and J. A. Nobel Global asymptotic stability of nonlinear systems of differential equations to reactor dynamics, *Arch. Rational Mech. Anal.* **5**(1960), 104-211.

[15] N. Levinson The asymptotic nature of solutions of linear systems of differential equations, *Duke Math. J.* **15**,111-126, (1948).

[16] P. Pucci and J. Serrin Precise damping conditions for global asymptotic stability for nonlinear second order systems, *Acta Math.* **170**(1993), 275-307.

[17] P. Pucci and J. Serrin Asymptotic stability for ordinary differential systems with time dependent restoring potentials, *Archive Rat. Math. Anal.* **113**(1995), 1-32.

[18] J.H. Sherley, W.D.Lee and R. E. Drullinger, Accuracy evaluation of the primary frequency standard *Metrologia* 2001, 38, 427-458

[19] R. A. Smith Asymptotic stability of $x'' + a(t)x' + x = 0$, *Quart. J.Math Oxford Ser.*(**2**),12(1961),123-126.

In: Progress in Evolution Equations
Editor: Gaston M. N'Guerekata, pp. 207-218
ISBN: 978-1-60456-328-3

Chapter 12

EXACT SOLUTIONS AND CONSERVATION LAWS FOR IBRAGIMOV-SHABAT EQUATION WHICH DESCRIBE PSEUDO-SPHERICAL SURFACE

***S. M. Sayed*[1,2*], *A. M. Elkholy*[1,2] *and G.M. Gharib*[2]**
[1]Mathematics Department, Faculty of Science,
Beni-Suef University, Beni-Suef, Egypt.
[2]Mathematics Department, P. O. Box 1144, Tabouk Teacher College,
Deputy for Teacher College, Ministry of Higher Education,
Tabouk, Kingdom of Saudi Arabia.

Abstract

Travelling wave solution for Ibragimov-Shabat equation, is obtained by using an improved sine-cosine method and the Wu's elimination method. An infinite number of conserved quantities for the above equation are also obtained by solving a set of coupled Riccati equations.

MSC: 35; 53C; 58J; 58Z05

Keywords: Nonlinear evolution equations; Conservation laws; Pseudo-spherical surfaces

1. Introduction

Sine-cosine method and the Wu's elimination method have been useful in the calculation of soliton solutions of certain nonlinear evolution equations (NLEEs) of physical significance [1-6] restricted to one space variable x and a time coordinate t.

Khater, et al. [7,8] used the notion of a differential equation (DE) for a function $u(x,t)$ that describes a pseudo-spherical surface (pss), and they derived some Bäcklund transformations and conservation laws for NLEEs which are integrability condition of $sl(2,R)$-valued linear problems [9-18].

*E-mail address: S_M_Sayed71@yahoo.com. (Corresponding author)

It is well-known [19-23] that a DE for a real-valued function $u(x,t)$ or a differential system for a two-vector valued function $u(x,t)$, is said to describe pss if it is the necessary and sufficient condition for the existence of smooth real functions f_{ij}, $1 \leq i \leq 3$, $1 \leq j \leq 2$, depending only on u and a finite number of its derivatives, such that the one-forms

$$\omega_i = f_{i1}dx + f_{i2}dt, \quad 1 \leq i \leq 3, \tag{1}$$

satisfying the structure equations of a surface of constant Gaussian curvature $K = -1$,

$$d\omega_1 = \omega_3 \wedge \omega_2, \quad d\omega_2 = \omega_1 \wedge \omega_3, \quad d\omega_3 = \omega_1 \wedge \omega_2. \tag{2}$$

The inverse scattering method (ISM) was introduced first for the Korteweg-de Vries equation (KdVE) [24]. Later it was extended by Zakharov and Shabat [25] to a 2×2 scattering problem for the nonlinear Schrödinger equation (NLSE) and that was subsequently generalized by Ablowitz, Kaup, Newell and Segur (AKNS) [26] to include a variety of NLEEs. Khater, et al. [7] generalized the results of Konno and Wadati [27] by considering ν as a three component vector and Ω as a traceless 3×3 matrix one-form. The above definition of a DE is equivalent to saying that the DE for u is the integrability condition for the problem

$$d\nu = \Omega\nu, \quad \nu = \begin{pmatrix} \nu_1 \\ \nu_2 \\ \nu_3 \end{pmatrix}, \tag{3}$$

where ν is a vector and the 3×3 matrix Ω $(\Omega_{ij}, i, j = 1,2,3)$ is traceless

$$tr\Omega = 0, \tag{4}$$

and consists of a one-paramter (η), family of one-forms in the independent variables (x,t), the dependent variable u and its derivatives.

Khater et al. [7] introduced the inverse scattering problem (ISP):

$$\nu_{1x} = f_{31}\nu_2 - f_{11}\nu_3, \quad \nu_{2x} = -f_{31}\nu_1 - \eta\nu_3, \quad \nu_{3x} = -f_{11}\nu_1 - \eta\nu_2, \tag{5}$$

$$\nu_{1t} = f_{32}\nu_2 - f_{12}\nu_3, \quad \nu_{2t} = -f_{32}\nu_1 - f_{22}\nu_3, \quad \nu_{3t} = -f_{12}\nu_1 - f_{22}\nu_2. \tag{6}$$

The integrability condition for Eq. (3) is given by

$$d\Omega - \Omega \wedge \Omega = 0, \tag{7}$$

or in component form

$$f_{12,x} - f_{11,t} = f_{31}f_{22} - \eta f_{32}, \tag{8}$$

$$f_{22,x} = f_{11}f_{32} - f_{12}f_{31}, \tag{9}$$

$$f_{32,x} - f_{31,t} = f_{11}f_{22} - \eta f_{12}. \tag{10}$$

By various choices of the coefficients f_{ij}, it can be shown that the conditions (8)-(10) are equivalent to a large class of NLEEs. The procedure is clarified in the following example:

(a) The Ibragimov-Shabat equation (ISE)

$$u_t = u_{xxx} + 3u^2u_{xx} + 9uu_x^2 + 3u^4u_x, \tag{11}$$

$$\Omega = \begin{pmatrix} 0 & C_1 & -A_1 \\ -C_1 & 0 & -B_1 \\ -A_1 & -B_1 & 0 \end{pmatrix}, \tag{12}$$

where

$$A_1 = (\frac{u_x}{u} + u^2)dx + (\frac{u_{xxx}}{u} + u^6 + 8u_x^2 + 5uu_{xx} + 9u^3u_x)dt,$$

$$B_1 = \eta dx + \eta(\frac{u_{xx}}{u} + u^4 + 4uu_x)dt, \tag{13}$$

$$C_1 = -\eta dx - \eta(\frac{u_{xx}}{u} + u^4 + 4uu_x)dt.$$

The essence of the first step of the ISM is summarized as follows [7]. Find nine one - forms ω_i^j, $i = 1,2,3$, $j = 1,2,3$ consisting of independent and dependent variables and their derivatives, such that the NLEE is given by

$$\Theta \equiv d\Omega - \Omega \wedge \Omega = 0, \quad \Omega = \begin{pmatrix} \omega_1^1 & \omega_1^2 & \omega_1^3 \\ \omega_2^1 & \omega_2^2 & \omega_2^3 \\ \omega_3^1 & \omega_3^2 & \omega_3^3 \end{pmatrix}, \quad Tr\Omega = 0. \tag{14}$$

It should be noted that the solution of these equations are of very special kind. In general, Eq. (14) gives three different equations, which cannot be satisfied simultaneously by one dependent variable u. It has been pointed out [7,28] that Ω can be interpreted as a connection one-form for the principle $SL(3,R)$ bundle on R^3 and Θ as its curvature two form. The geometrical explanation of the $SL(3,R)$ structure is given in section 2.

The main aim of this paper is to extend the fundamental equations of pseudo-spherical surfaces in reference [28] by considering ν as a three component vector and Ω as a traceless 3×3 matrix one-form. In the present paper we use sine-cosine method and the Wu's elimination method derived in [1-6] in the construction of exact soliton solution for ISE describing pss. We also obtain an infinite number of conserved charges by solving a set of coupled Riccati equations and apply the geometrical method to ISE which describe pss.

The paper is organized as follows. In section 2 the geometry of pss is described and the correspondence between the soliton equations and their families of pss are established. Section 3 The sine-cosine method and the Wu's elimination method is used to obtain travelling wave solutions for the ISE. Section 4 contains the derivation of an infinite number of conserved charges from the Riccati equations. Finally, we give some conclusions in section 5.

2. On Equations Describing pss

In this section we shall show that the fundamental equations of pss, can be written in the form of Eq. (14). Let us start with the general description of a three-dimensional Riemannian manifold S following reference [28]. An orthonormal basis is

$$\{e_i\}, \quad i = 1,2,3, \quad e_i.e_j = \delta_{ij}, \tag{15}$$

with respect to the Riemannian metric introduced on the tangent plane T_p at each point $p \in S$. Then the structure equations for S read

$$dp = \omega_i e_i, \quad i = 1,2,3 \tag{16}$$

$$de_i = \sum_{j=1}^{3} \omega_i^j e_j, \tag{17}$$

where ω_i are one forms dual to $\{e_i\}$ and ω_i^j, $(i,j = 1,2,3)$ are called the connection one form. The integrability conditions are obtained by differentiating Eq. (16) and using (17). These conditions are

$$d\omega_i = \sum_{j=1}^{3} \omega_i \wedge \omega_i^j, \tag{18}$$

$$d\omega_i^j = \sum_{k=1}^{3} \omega_i^k \wedge \omega_k^j. \tag{19}$$

To sum up, a set of one forms $(\omega_i,\ \omega_i^j)$ satisfying Eqs. (18) and (19) describes a pss locally through the structure Eqs. (16) and (17). It is easy to show that Eqs. (18) and (19) can be written in the form of Eq. (14) by choosing

$$\Omega = \begin{pmatrix} 0 & \omega_3 & -\omega_1 \\ -\omega_3 & 0 & -\omega_2 \\ -\omega_1 & -\omega_2 & 0 \end{pmatrix}, \tag{20}$$

from Eqs. (14) and (20) we obtain this relations

$$\omega_1^1 = \omega_2^2 = \omega_3^3 = 0, \ \omega_1 = -\omega_1^3 = -\omega_3^1, \ \omega_2 = -\omega_2^3 = -\omega_3^2, \ \omega_3 = \omega_1^2 = -\omega_2^1. \tag{21}$$

Let M^2 be a two - dimensional differentiable manifold parametrized by coordinates x,t. We consider a metric on M^2 defined by ω_1, ω_2. The first two equations in (2) are the structure equations which determine the connection form ω_3, and the last equation in (2), the Gauss equation, determines that the Gaussian curvature of M^2 is -1, i.e. M^2 is a pss. Moreover, an evolution equation must be satisfied for the existence of forms (1) satisfying (2). This justifies the definition of a DE which describes a pss that we considered in the introduction.

It has been known, for a long time, that the sine-Gordon equation describes a pss. KdVE and mKdVE, were also shown to describe such surfaces [26]. Here we show that ISE equation also describe pss as well. The latter equation proved to be of great importance in many physical applications [7-14,29,30]. The procedure is clarified in the following example: Let M^2 be a differentiable surface, parametrized by coordinates x,t.

(a) The ISE

Consider

$$\omega_1^3 = \omega_3^1 = -(\frac{u_x}{u} + u^2)dx - (\frac{u_{xxx}}{u} + u^6 + 8u_x^2 + 5uu_{xx} + 9u^3 u_x)dt,$$

$$\omega_2^3 = \omega_3^2 = -\eta dx - \eta(\frac{u_{xx}}{u} + u^4 + 4uu_x)dt, \tag{22}$$

$$\omega_1^2 = -\omega_2^1 = -\eta dx - \eta(\frac{u_{xx}}{u} + u^4 + 4uu_x)dt.$$

Then M^2 is a pss iff u satisfies ISE (11).

3. Exact Solution for ISE

Now we shall find a travelling wave solution $u(x,t)$ for ISE

$$u_t = u_{xxx} + 3u^2u_{xx} + 9uu_x^2 + 3u^4u_x. \tag{23}$$

Wang [1] has found some exact solutions for compound KdV-Burgers equations by using the homogenous balance method. In this section we obtain travelling wave solution class for IES by using an improved sine - cosine method [4,5] and Wu's elimination method [2]. The main idea of the algorithm is as follows. Given a partial differential equation (PDE) of the form

$$f(u, u_x, u_t, u_{xx}, u_{xt}, u_{tt}, \cdots) = 0, \tag{24}$$

where f is a polynomial. By assuming travelling wave solutions of the form

$$u(x,t) = \phi(\rho), \quad \rho = \lambda(x - kt + c), \tag{25}$$

where k, η are constant parameters to be determined, and c is an arbitrary constant, from the two Eqs. (24) and (25) we obtain an ordinary differential equation (ODE)

$$f(\phi', \phi'', \phi''', \cdots) = 0, \tag{26}$$

where $\phi' = \frac{d\phi}{d\rho}$. According to the sine-cosine method [1-6], we suppose that Eq. (26) has the following formal travelling wave solution

$$\phi(\rho) = \sum_{i=1}^{n} \sin^{i-1}\psi(B_i \sin\psi + A_i \cos\psi) + A_0, \tag{27}$$

and

$$\frac{d\psi}{d\rho} = \sin\psi \quad or \quad \frac{d\psi}{d\rho} = \cos\psi, \tag{28}$$

where $A_0, \cdots, A_n$ and $B_1, \cdots, B_n$ are constants to be determined. Then we proceed as follows:

(i) Equating the highest order nonlinear term and highest order linear partial derivative in (26), yield the value of n.

(ii) Substituting Eqs. (27), (28) into (26), we obtain a polynomial equation involving $\cos\psi\sin^j\psi$, $\sin^j\psi$ for $j = 0, 1, 2, \cdots, n$, (with n being positive integer)

(iii) Setting the constant term and coefficients of $\sin\psi, \cos\psi, \sin\psi\cos\psi, \sin^2\psi, \cdots,$ in the equation obtained in (ii) to zero, we obtain a system of algebraic equations about the unknown numbers $k, \lambda, A_0, A_i, B_i$ for $i = 1, 2, \cdots, n$.

(iv) Using the Mathematica and the Wu's elimination methods, the algebraic equations in (iii) can be solved.

These yield the solitary wave solutions for the system (26). We remark that the above method yield solutions that includes terms sech ρ or tanh ρ, as well as their combinations. There are different forms of those obtained by other methods, such as the homogenous balance method [1-6]. We assume formal solutions of the form

$$u(x,t)=\phi(\rho), \quad \rho=\lambda(x-kt+c), \tag{29}$$

where k, η are constant parameters to be determined later, and c is an arbitrary constant. Substituting from (29) and (23), we obtain an ODE

$$k\phi'+3\lambda\phi^2\phi''+9\lambda\phi\phi'^2+3\phi^4\phi'+\lambda^2\phi'''=0. \tag{30}$$

(i) We suppose that equation (30) has the following formal solutions

$$\phi(\rho)=A_0+B_1\sin\psi+A_1\cos\psi, \tag{31}$$

and

$$\frac{d\psi}{d\rho}=\sin\psi. \tag{32}$$

(ii) From two Eqs. (31) and (32), we get

$$\begin{aligned}
&k\phi'+3\lambda\phi^2\phi''+9\lambda\phi\phi'^2+3\phi^4\phi'+\lambda^2\phi'''=\\
&[12A_1B_1A_0^3+12A_1^3B_1A_0+3\lambda B_1A_0^2+3\lambda A_1^2B_1]\sin\psi+\\
&[3B_1A_0^4+18A_1^2B_1A_0^2+3A_1^4B_1+(k+\lambda^2)B_1+6\lambda A_1B_1A_0]\sin\psi\cos\psi+\\
&[-3A_1A_0^4+36A_1B_1^2A_0^2-18A_1^3A_0^2+12A_1^3B_1^2-3A_1^5-(k+4\lambda^2)A_1+\\
&6\lambda B_1^2A_0-12A_0A_1^2+9\lambda B_1^2A_0]\sin^2\psi+[12B_1^2A_0^3-12A_1^2A_0^3+36A_1^2B_1^2A_0+\\
&6\lambda A_1B_1^2-12A_0A_1^4-6\lambda A_1A_0^2-6\lambda A_1^3+9\lambda A_1B_1^2]\cos\psi\sin^2\psi+[-24A_1B_1A_0^3+36A_1B_1^3A_0+\\
&3\lambda B_1^3-60A_0B_1A_1^3-6\lambda B_1A_0^2-21\lambda A_1^2B_1+9\lambda B_1^3-18\lambda A_1^2B_1]\sin^3\psi+[18B_1^3A_0^2+\\
&18A_1^2B_1^3-54A_1^2B_1A_0^2-18A_1^4B_1-6\lambda^2B_1-24\lambda B_1A_1A_0-18\lambda A_1B_1A_0]\cos\psi\sin^3\psi+\\
&[12A_1B_1^4-54A_1B_1^2A_0^2+18A_1^3A_0^2-42A_1^3B_1^2+6A_1^5+6\lambda^2A_1-12\lambda B_1^2A_0+12A_1^2A_0-9\lambda A_0B_1^2+\\
&9\lambda A_1^2A_0]\sin^4\psi+[12B_1^4A_0-72A_1^2B_1^2A_0+12A_0A_1^4-18\lambda A_1B_1^2+6\lambda A_1^3-27\lambda A_1B_1^2+\\
&9\lambda A_1^3]\cos\psi\sin^4\psi+[-48A_1B_1^3A_0+48A_1^3B_1A_0-6\lambda B_1^3+18\lambda A_1^2B_1-9\lambda B_1^3+\\
&27\lambda A_1^2B_1]\sin^5\psi+[3B_1^5-30A_1^2B_1^3+15B_1A_1^4]\cos\psi\sin^5\psi+\\
&[30A_1^3B_1^2-3A_1^5-15A_1B_1^4]\sin^6\psi=0.
\end{aligned} \tag{33}$$

(iii) Setting the coefficients of $\sin^j\psi\cos^i\psi$ for $i=0,1$ and $j=1$ to 6, we have the following set of over determined equations in the unknowns A_0, B_1, A_1, λ and k:

$$12A_1A_0^3+12A_1^3A_0+3\lambda A_0^2+3\lambda A_1^2=0,$$

$$3A_0^4+18A_1^2A_0^2+3A_1^4+(k+\lambda^2)+6\lambda A_1A_0=0,$$

$$-3A_1A_0^4+36A_1B_1^2A_0^2-18A_1^3A_0^2+12A_1^3B_1^2-3A_1^5-(k+4\lambda^2)A_1+$$

$$6\lambda B_1^2A_0-12A_0A_1^2+9\lambda B_1^2A_0=0,$$

$$12B_1^2A_0^3-12A_1^2A_0^3+36A_1^2B_1^2A_0+6\lambda A_1B_1^2-12A_0A_1^4-6\lambda A_1A_0^2-6\lambda A_1^3+9\lambda A_1B_1^2=0,$$

$$-24A_1A_0^3+36A_1B_1^2A_0+3\lambda B_1^2-60A_0A_1^3-6\lambda A_0^2-21\lambda A_1^2+9\lambda B_1^2-18\lambda A_1^2=0,$$

$$18B_1^2A_0^2+18A_1^2B_1^2-54A_1^2A_0^2-18A_1^4-6\lambda^2-24\lambda A_1A_0-18\lambda A_1A_0=0,$$

$$12A_1B_1^4-54A_1B_1^2A_0^2+18A_1^3A_0^2-42A_1^3B_1^2+6A_1^5+6\lambda^2A_1-12\lambda B_1^2A_0+$$

$$12A_1^2A_0-9\lambda A_0B_1^2+9\lambda A_1^2A_0=0,$$

$$12B_1^4A_0-72A_1^2B_1^2A_0+12A_0A_1^4-18\lambda A_1B_1^2+6\lambda A_1^3-27\lambda A_1B_1^2+9\lambda A_1^3=0,$$

$$-48A_1B_1^3A_0+48A_1^3B_1A_0-6\lambda B_1^3+18\lambda A_1^2B_1-9\lambda B_1^3+27\lambda A_1^2B_1=0,$$

$$3B_1^4-30A_1^2B_1^2+15A_1^4=0,$$

$$30A_1^2B_1^2-3A_1^4-15B_1^4=0. \tag{34}$$

(iv) We now solve the above set of equations by using Mathematica and the Wu's elimination method, and obtain the following solution:

$$A_1=\frac{-\lambda}{4A_0},\quad B_1=\frac{-\lambda}{12.3106A_0},\quad A_0=-\frac{[-(5\lambda^2+8k)\pm4(\lambda^2+4k)^{1/2}(\lambda^2+k)^{1/2}]^{1/4}}{2\sqrt[4]{3}}, \tag{35}$$

by integrating (32) and taking the integration constant equal zero, we obtain

$$\sin\psi=\operatorname{sech}\rho,\qquad \cos\psi=\pm\tanh\rho. \tag{36}$$

Substituting (35) and (36) into (31), we obtain

$$u(x,t)=A_0+\frac{-\lambda}{12.3106A_0}\operatorname{sech}\rho\pm\frac{-\lambda}{4A_0}\tanh\rho, \tag{37}$$

where

$$A_0=-\frac{[-(5\lambda^2+8k)\pm4(\lambda^2+4k)^{1/2}(\lambda^2+k)^{1/2}]^{1/4}}{2\sqrt[4]{3}},\qquad \rho=\lambda(x-kt+c).$$

4. Infinite Number of Conserved Charges for ISE

It will be found that the Riccati form of ISM is sometimes useful. Introducing the variables [31-33]

$$\Gamma_1 = \frac{\nu_1}{\nu_3}, \quad \Gamma_2 = \frac{\nu_2}{\nu_3}, \tag{38}$$

which are related to the conserved charges $\alpha_{33}(x,t,\eta)$ in the following way, from Eq. (5)

$$\ln\alpha_{33}(\eta) = \ln\nu_3 = \int_{-\infty}^{\infty}(\eta\Gamma_2 - f_{11}\Gamma_1)dx. \tag{39}$$

Eqs. (5) and (6) can be rewritten as

$$\Gamma_{1x} = f_{31}\Gamma_2 + \eta\Gamma_1\Gamma_2 + f_{11}\Gamma_1^2 - f_{11}, \tag{40}$$

$$\Gamma_{2x} = -f_{31}\Gamma_1 + f_{11}\Gamma_1\Gamma_2 + \eta\Gamma_2^2 - \eta, \tag{41}$$

and

$$\Gamma_{1t} = f_{32}\Gamma_2 + f_{22}\Gamma_1\Gamma_2 + f_{12}\Gamma_1^2 - f_{12}, \tag{42}$$

$$\Gamma_{2t} = -f_{32}\Gamma_1 + f_{12}\Gamma_1\Gamma_2 + f_{22}\Gamma_2^2 - f_{22}. \tag{43}$$

The set of coupled NLPDEs for Γ_1 and Γ_2 in (40)-(43) are known as Riccati equations. It is obvious from (39) that the solutions of Riccati equations eventually determine the conserved quantities. Now in order to solve (40)-(43), we assume a series solutions for Γ_1 and Γ_2 as

$$\Gamma_1(x,t,\eta) = \sum_{n=0}^{\infty}\phi_n^1(x,t)\eta^{-n}, \quad \Gamma_2(x,t,\eta) = \sum_{n=0}^{\infty}\phi_n^2(x,t)\eta^{-n}. \tag{44}$$

Substituting (44) into (40) and (41), the following recursion relations are obtained:

$$\phi_0^1 = \phi_1^2 = 0, \ \phi_0^2 = 1, \ \phi_1^1 = f_{11} - f_{31}, \ \phi_2^1 = (f_{11} - f_{31})_x, \ \phi_2^2 = \frac{1}{2}(f_{11} - f_{31})^2, \tag{45}$$

$$\phi_{kx}^1 = f_{31}\phi_k^2 + \sum_{m=0}^{k+1}\phi_m^1\phi_{k-m+1}^2 + \sum_{m=0}^{k} f_{11}\phi_m^1\phi_{k-m}^1, \ k = 1,2,\ldots, \tag{46}$$

$$\phi_{kx}^2 = -f_{31}\phi_k^1 + \sum_{m=0}^{k+1}\phi_m^2\phi_{k-m+1}^2 + \sum_{m=0}^{k} f_{11}\phi_m^1\phi_{k-m}^2, \ k = 1,2,\ldots. \tag{47}$$

The infinite number of Hamiltonians (conserved quantities) may explicitly be determined in terms of smooth real functions f_{ij} and their derivatives by expanding $\alpha_{33}(x,t,\eta)$ in the form

$$\ln\alpha_{33}(\eta) = \sum_{l=0}^{\infty} H_l\eta^{-l}, \tag{48}$$

and thus comparing (48) with (39) and (44), H_l becomes

$$H_l = \int(\eta\phi_l^2 - f_{11}\phi_l^1)dx, \quad l = 0,1,2,\ldots. \tag{49}$$

The explicit expressions of the first few order Hamiltonians are

$$H_0 = \eta x,\ H_1 = -\int 2(qr+q^2)dx,\ H_2 = \int [-2(r+q)q_x + \frac{\eta}{2}q^2]dx,$$

$$H_3 = 2\int (r+q)[q^2(q+r+2) - q_{xx}]dx, \tag{50}$$

$$H_4 = \int [\eta q^3(2-r-3q) - \eta q q_{xx} - (r+q)((2q_{3x} - q^2(9q_x+r_x) - 4qq_x(r+4))]dx,$$

where

$$f_{11} - f_{31} = 2q,\ f_{11} + f_{31} = 2r,$$

$$2A = f_{22},\ 2B = f_{12} - f_{32},\ 2C = f_{12} + f_{32},$$

such that the functions r, q, A, B and C satisfy the equations [7]

$$A_x = qC - rB,$$

$$q_t - 2Aq - B_x + \eta B = 0, \tag{51}$$

$$C_x = r_t + 2Ar - \eta C.$$

This section ends with the following example:

(a) The ISE

For any solution u of the ISE (11), we consider the functions

$$r = \frac{1}{2}(\frac{u_x}{u} + u^2 - \eta), \qquad q = \frac{1}{2}(\frac{u_x}{u} + u^2 + \eta),$$

$$A = \frac{\eta}{2}(\frac{u_{xx}}{u} + u^4 + 4uu_x),$$

$$\tag{52}$$

$$B = \frac{1}{2}[(\frac{u_{xxx}}{u} + u^6 + 8u_x^2 + 5uu_{xx} + 9u^3u_x) + \eta(\frac{u_{xx}}{u} + u^4 + 4uu_x)],$$

$$C = \frac{1}{2}[(\frac{u_{xxx}}{u} + u^6 + 8u_x^2 + 5uu_{xx} + 9u^3u_x) - \eta(\frac{u_{xx}}{u} + u^4 + 4uu_x)].$$

Eq. (49) implies that the first few order Hamiltonians are determined by the relation

$$H_1 = -\frac{1}{2}\int [(\frac{u_x}{u} + u^2)^2 + (\frac{u_x}{u} + u^2 + \eta)^2 - \eta^2]dx,$$

$$H_2 = \int [\frac{\eta}{8}(\frac{u_x}{u} + u^2 + \eta)^2 - (\frac{u_x}{u} + u^2)(\frac{u_x}{u} + u^2)_x]dx, \tag{53}$$

$$etc.$$

Whenever $u(x,t)$ is a solution (37) of the ISE. This Hamiltonians relations yields a sequence of conserved charges given by the coefficients of the series in η

$$\eta x + \sum_{l=1}^{\infty} H_l \eta^{-l}. \tag{54}$$

5. Conclusions

In this paper, the fundamental equations of pseudo-spherical surfaces in reference [28] may be rewritten by considering v as a three component vector and Ω as a traceless 3×3 matrix one-form [7]. The latter yields directly the curvature condition (Gaussian curvature equal to -1, corresponding to pseudo-spherical surfaces). This geometrical method is considered for the ISE.

We obtain travelling wave solution for ISE by using an improved sine-cosine method and Wu's elimination method. This geometrical method allows some further generalization of the work on conserved charges given by Wadati, Sanuki and Konno [34]. An infinite number of conserved charges for ISE mentioned above are derived in this way.

References

[1] Wang, M.L., Exact solution for a compound KdV-Burgers equation, *Phys. Lett.* **A213**(1996), 279-287.

[2] Wu, W.T., Polynomial equations-solving and its applications, Algorithms and Computation, (Beijing 1994), 1-9, *Lecture Notes in Comput. Sci.* **834**, Springer-Verlag, Berlin, (1994).

[3] Xia, T.C., Zhang, H.Q., Yan, Z.Y., New explicit exact travelling wave solution for a compound KdV-Burgers equation, *Chinese Phys.* **8**(2001), 694-699.

[4] Yan, C.T., A simple transformation for nonlinear waves, Phys. Lett. A224(1996), 77-82.

[5] Yan, T.Z., Zhang, H.Q., New explicit and exact travelling wave for a system variant Boussinesq equation in mathematical physics, *Phys. Lett.* **A252**(1999), 291-296.

[6] Zhang, X.D., Xia, T.C., Zhang, H.Q., New explicit exact travelling wave solution for compound KdV-Burgers equation in mathematical physics, *Applied Mathematics E-Notes.* **2**(2002), 45-50.

[7] Khater, A.H., Callebaut, D.K., Sayed, S.M., Conservation laws for some nonlinear evolution equations which describe pseudospherical surfaces, *J. of Geometry and Phys.* **51**(2004), 332-352.

[8] Khater, A.H., Callebaut, D.K., Sayed, S.M., Exact solutions for some nonlinear evolution equations which describe pseudo-spherical surfaces, *J. of Comput. and Appl. Math.*, **189**(2006), 387-411.

[9] Khater, A.H., Callebaut, D.K., Abdalla, A.A., Sayed, S.M., Exact solutions for self-dual Yang-Mills Equations, *Chaos Solitons & Fractals* **10**(1999), 1309-1320.

[10] Khater, A.H., Callebaut, D.K., Abdalla, A.A., Shehata, A.M., Sayed, S.M., Bäcklund Transformations and Exact solutions for self-dual $SU(3)$ Yang-Mills Equations *Il Nuovo Cimento* **B114**(1999), 1-10.

[11] Khater, A.H., Callebaut, D.K., El-Kalaawy, O.H., Bäcklund transformations and exact solutions for a nonlinear elliptic equation modelling isothermal magentostatic atmosphere, *IMA J. of Appl. Math.* **65**(2000) 97-108.

[12] Khater, A.H., Callebaut, D.K., El-Kalaawy, O.H., Bäcklund transformations and exact soliton solutions for nonlinear Schrödinger-type equations, *Il Nuovo Cimento* **B113**(1998), 1121-1136.

[13] Khater, A.H., Callebaut, D.K., Ibrahim, R.S., Bäcklund transformations and Painlevé analysis: exact solutions for the unstable nonlinear Schrödinger equation modelling electron-beam plasma, *Phys. Plasmas* **5** (1998), 395-400.

[14] Khater, A.H., El-Kalaawy, O.H., Callebaut, D.K., Bäcklund transformations for Alfvén solitons in a relativistic electron-positron plasma, *Physica Scripta* **58**(1998), 545-548.

[15] Khater, A.H., Helal, M.A., El-Kalaawy, O.H., Two new classes of exact solutions for the KdV equation via Bäcklund transformations, *Choas, Solitons & Fractals* **8**(1997), 1901-1909.

[16] Khater, A.H., Sayed, S.M., Exact Solutions for Self-Dual SU(2) and SU(3)YangMills Fields, *International J. of Theoretical Physics* **41**(2002) 409-419.

[17] Khater, A.H., Shehata, A.M., Callebaut, D.K., Sayed, S.M., Self-Dual Solutions for SU(2) and SU(3) Gauge Fields on Euclidean Space, *International J. of Theoretical Physics* **43**(2004) 151-159.

[18] Rogers, C., Shadwick, W., Bäcklund transformations and their applications (Academic Press, New York, 1982).

[19] Bäcklund, A.V., Einiges über Curven-und Flächentransformationen, Lunds. Univ. Arsskr. Avd. X, For Ar, (1873); Afdelningen för Mathematik Och Naturetens kap, ibid, (1873-1874), 1-12.

[20] Beals, R., Coifman, R.R., Inverse scattering and evolution equations, *Commun. Pure Appl. Math.* **38**(1985), 29-42.

[21] Dodd, R.K., Bullough, R.K., Bäcklund transformations for the AKNS inverse method, *Phys. Lett.* **A62**(1977), 70-74.

[22] Lie, S., Zür Theorie der Flächen Konstanter Krmmung III, *Arch. Math. Naturvidensk. V Heft* **3**, 282-306(1880), 282-306; 3(1880), 328-358.

[23] Terng, C.L., Soliton equations and differential geometry, *J. of Diff. Geometry* **45**(1997), 407-445.

[24] Gardner, C. S., Greene, J. M., Kruskal, M. D., Miura, R. M., Method for solving the korteweg-de Vries equation *Phys. Lett.* **19** (1967), 1095-1097.

[25] Zakharov, V.E., Shabat, A.B., Exact theory of two-dimensional self-focusing and one-dimensional self-modulation of waves in nonlinear Media, *Soviet Phys. JETP* **34**(1972), 62-69.

[26] Ablowitz, M.J., Kaup, D.J., Newell, A.C., Segur, H., The inverse scattering transform-Fourier analysis for nonlinear problems, *Stud. Appl. Math.* **53**(1974), 249-315.

[27] Konno, K., Wadati, M., Simple derivation of Bäcklund transformation from Riccati Form of inverse method, *Progr. Theor. Phys.* **53**(1975), 1652-1656.

[28] Sasaki, R., Soliton equation and pseudospherical surfaces, *Nucl. Phys.* **B154**(1979), 343-357.

[29] Foursov, M. V., Olver, P. J., Reyes, E. G., On formal integrability of evolution equations and local geometry of surfaces *Diff. Geometry and its Applications* **15**(2001), 183-99.

[30] Chern, S.S., Tenenblat, K., Pseudospherical surfaces and evolution equations, *Stud. Appl. Math.* **74**(1986), 55-83.

[31] Ghosh, S., Kundu, A., Nandy, S., Soliton Solutions, Liouville Integrability and Gauge Equivalence of Sasa Satsume Equation *J. Math. Phys.* **40** (1999), 1993-2000.

[32] Ghosh, S., Nandy, S., Inverse Scattering Method and Vector Higher Order Nonlinear Schrödinger Equation *Nucl. Phys.* **B561** (1999), 451-66.

[33] Nandy, S., Inverse scattering approach to coupled higher-order nonlinear Schrödinger equation and N-soliton solutions *Nucl. Phys.* **B679** (2004), 647-58.

[34] Wadati, M., Sanuki, H., Konno, K., Relationships among inverse method, Bäcklund transformation and an infinite number of conservation laws *Progr. Theor. Phys.* **53** (1975), 419-36

INDEX

A

B

C

D

E

F

G

S

T

U

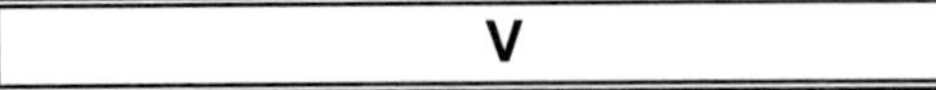

V

W

Y